FOOD INDUSTRY

FOOD INDUSTRY

DESIGN, TECHNOLOGY AND INNOVATION

Volume 2

Ethan Campbell
Lucy Wheeler
Eleanor Barnes

Kruger Brentt
Publishers

2026

Kruger Brentt Publishers UK. LTD.
Company Number 9728962

Regd. Office: 68 St Margarets Road, Edgware, Middlesex HA8 9UU

ISBN: 978-1-78715-671-5 (Vol. 2)

For information on all our publications visit our website at http://krugerbrentt.com/

PREFACE

The food industry is one of the most dynamic and rapidly evolving sectors worldwide, driven by changing consumer preferences, technological advancements, and the growing demand for safe, nutritious, and innovative food products. The design, technology, and innovation in food production play a pivotal role in ensuring quality, efficiency, and sustainability across the supply chain. From processing and packaging to quality assurance and product development, a deep understanding of modern food industry practices is essential for professionals, researchers, and students in food science and technology.

This book, Food Industry: Design, Technology and Innovation, provides a comprehensive exploration of the principles, technologies, and innovative approaches shaping the modern food industry. It is intended as a reference for students, researchers, industry professionals, and entrepreneurs seeking insights into food processing technologies, industrial design, and innovation management within the sector.

The initial chapters focus on the fundamental concepts of food industry design, including facility layout, equipment selection, workflow optimization, and regulatory considerations. Emphasis is placed on designing food production systems that ensure operational efficiency, safety, and compliance with hygiene standards.

Subsequent chapters explore technological advancements in food processing, preservation, packaging, and automation. The book covers traditional and emerging methods, such as thermal processing, high-pressure processing, extrusion, nanotechnology, and smart packaging, highlighting their applications, benefits, and limitations. Special attention is given to integrating sustainability and energy efficiency into food industry operations.

Innovation is a central theme throughout the book, with discussions on product development, sensory evaluation, consumer trends, and market-driven strategies. The role of research and development (R&D) in creating novel food products, improving nutritional quality, and enhancing shelf life is thoroughly examined.

Case studies from leading food companies illustrate how technology and innovation are successfully applied to meet market demands and regulatory standards.

The book also addresses critical challenges in the food industry, such as food safety, quality control, traceability, and the impact of globalization on supply chains. Strategies for managing these challenges using modern technologies, data analytics, and process optimization are highlighted to provide a practical perspective.

Food Industry: Design, Technology and Innovation serves as a vital resource for anyone engaged in food science, food engineering, or food business management. By combining theoretical principles with practical applications and contemporary case studies, this book equips readers with the knowledge and tools needed to innovate, optimize, and lead in the modern food industry.

Ethan Campbell

Lucy Wheeler

Eleanor Barnes

CONTENTS

SECTION 3: FOOD SAFETY

VOLUME 2

17 Chapter

MOISTURE-DEPENDENT ENGINEERING CHARACTERISTICS OF CHIA (SALVIA HISPANICA L.) SEEDS

1. INTRODUCTION: AN OVERVIEW

Salvia hispanica L., often known as chia, is an annual herbaceous plant that belongs to the Lamiaceae family. This plant species, indigenous to southern Mexico and northern Guatemala, was a significant crop in pre-Columbian Mesoamerica alongside corn, beans, and amaranth. Chia seeds were esteemed not only for culinary purposes but also for medicinal applications and pigments [1]. The Spanish invaders prohibited its cultivation and substituted it with exotic crops such as wheat and barley. Chia seeds are currently being reintegrated into Western diets to enhance human health[2].These seeds have been investigated and recommended due to their oil content with the highest proportion of α-linolenic acid (omega-3) compared to other natural source known to date [3, 4], and also because of their high levels of protein, antioxidant, dietary fiber, vitamins and minerals [5, 6]. Chia seeds from Argentina exhibited 30.0 - 38.6 g oil/100 g, with 60.7 - 67.8 g/100 g of α-linolenic acid [7, 8]. Figure 1 shows the chemical composition of chia seed [9].

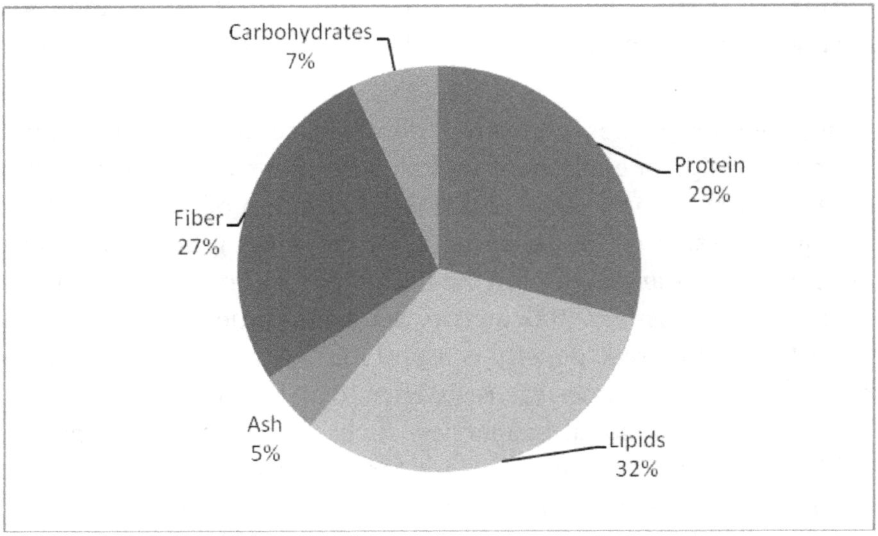

Figure 1. Chemical composition of chia seed (% d.b.)

Chia seeds are traditionally consumed throughout Mexico, the southwestern United States, and South America, however they remain relatively unknown in Europe. In 2009, the European Union authorized chia seeds as a new food, permitting them to constitute up to 5% of a bread product's total composition. Currently, chia is predominantly cultivated in Mexico, Bolivia, Argentina, Ecuador, Australia, and Guatemala, and evidence suggests that the species possesses significant promise as a future agricultural crop plant [7, 11].

Salvia hispanica fruit consist in four nutlets, similar to an indehiscent achene, which contains a single seed. These nutlets, are commercially named as seeds, and in the text, we will use this last term. The plant produces small white and dark seeds. Most of chia population that is commercially grown today contains a low percentage of white seeds. Their shapes are oval and in general, the white seeds are somewhat larger than the black ones. Ixtaina et al. [12], reported length, width and thickness value of 2.11, 1.32 and 0.81 mm for dark seeds and 2.15, 1.40 and 0.83 mm for white seeds, respectively. Chia seeds are shown in Figure 2.

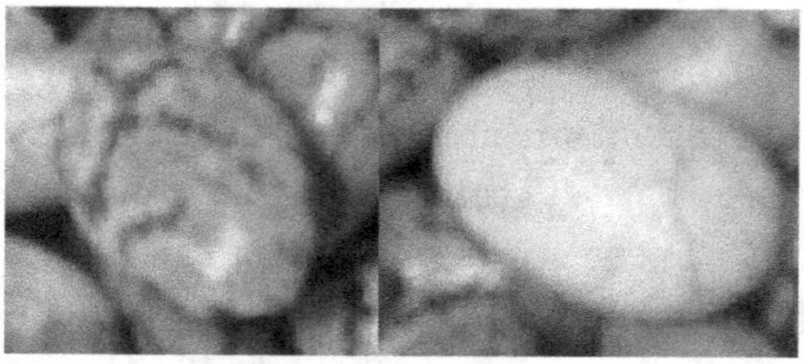

Figure 2. Photographs of dark and white chia seeds (13x)

The knowledge of engineering properties constitutes important and essential data for the design of machines, storage structures, and processes. The value of this basic information is not only important to engineers but also to food scientists, processors, and other scientists who may exploit these properties and find new uses.

Engineering seed properties and their dependence on moisture content are necessary in the design of equipment for transporting, storage and/or processing. The knowledge of the morphology and size distribution of chia seeds is essential for the adequate design of the equipment for cleaning, grading and separation. Gravimetric properties are useful for the design of equipment related to aeration, drying, storage and transport. Bulk density determines the capacity of storage and transport systems while true density is useful for separation equipment; porosity of the mass of seeds determines the resistance to airflow during the aeration and drying of seeds. The frictional properties, such as the angle of repose and the static coefficient of friction, are important for the design of grain bins and other storage structures whose operation is influenced by the compressibility and flow behaviour of materials. Several researchers investigated the moisture dependence of engineering properties of seed or grain and reported different behaviour of these properties as a function moisture content.

This study aimed to assess the engineering features of dark chia seeds in relation to moisture content and to compare their characteristics with those of other grains..

2. MATERIALS AND METHODS

2.1 Seeds

Chia seeds (20 kg) were obtained from commercial sources in Salta, Argentina (25° S, 65.5° W). Seeds were packed in hermetic plastic vessels and stored at 5 ± 1°C until use. The seeds were manually cleaned to remove all foreign matter, such as stones, dirt, and broken seeds. In this way, a randomized sample of chia seeds (about 2 kg) was picked by a sample splitter (CPASA, Centro Proveedor Agropecuario, Buenos Aires, Argentina). The seeds were manually separated according to their white or dark pericarp surface.

2.2 Sample preparation

The dark chia seeds were further divided into seven lots and they were conditioned to obtain a moisture content range of 4.6 - 17.7% (d.b.).To obtained a less moisture content, a predetermined quantity of seeds was dried down to the desired moisture content (convection air oven, 40 - 45°C). Higher moisture contents were reached indirectly through the saturation of the atmosphere in contact with the seeds. For this purpose, small, clean and dry vessels containing the seeds were placed into a container with 100 cm3 of water, which was then hermetically sealed for 36 - 240 h. These moisture levels were selected according to the conditions usually applied in harvesting and most processing operations of grains [13].

2.3 Moisture content

The moisture content of the samples was determined by the ASAE standard method used for rapeseed [14], using a convection air oven at 130 ±1°C for 4 h.

2.4 Engineering properties

The engineering properties of seeds were assessed at all moisture levels, as described below.

In order to determine the average size of the seeds, a sample of 100 seeds was randomly selected. For each individual seed, the three principal dimensions, namely length (L), width (W) and thickness (T), were measured using a digital micrometer (least count 0.01 mm). The three principal dimensions were used to calculate the geometric mean diameter (Dg) and surface area (S) of individual grains by assuming that the seeds were ellipsoid.

The geometric mean diameter (Dg) of chia seeds was calculated using the following relationship [15]:

$$D_g = (LWT)^{1/3} \tag{1}$$

The specific surface area (S) of chia seeds was calculated it according to equation 2 [16]:

$$S = \pi D_g^2 \tag{2}$$

To determine the mass of a thousand seeds (W1000), a set of 100 randomly selected seeds was weighed with an analytical balance (0.0001 g accuracy) and then extrapolated to 1000 seeds. This process was replicated 30 times [13].

The bulk density (ϱb), considered as the ratio of the mass sample of the seeds to its total volume, was determinate using a weight per hectoliter tester [16] (equipment of 90 cm3 of total volume). The true density (ϱt) defined as the ratio of the mass of the sample to its true volume, was determined using an electronic balance reading 0.001 g and a pycnometer (50±0.1 cm3 capacity, liquid displacement method) [17]. The toluene (C_7H_8; ϱ = 0.867 g cm-3) was used as a solvent to prevent water absorption in the seeds during the experiment [16]. Both measurements were performed in triplicate.

The porosity value (ε) defined as the fraction of space in the bulk seed which is not occupied by the seed [18], was determined from bulk and true densities using the relationship given by [19] and [20] as follows:

$$\varepsilon = \left(\frac{(\rho_t - \rho_b)}{\rho_t} \right) \times 100 \tag{3}$$

The seed volume (V) was determined from the following relationship [21]

$$V = \left(\frac{m}{\rho_t} \right) 10^3 \tag{4}$$

where m is the unit mass of the seed (g) determined in samples used to calculate the true density.

The equivalent diameter (De) and sphericity (φ), defined as the ratio between the surface area of the sphere having the same volume as that of the seed and the surface area of the seed, were calculated using equation 5 and equation 6 respectively [18]:

$$D_e = \left(\frac{6V}{\pi} \right)^{1/3} \tag{5}$$

$$\phi = \left(\frac{D_e}{L} \right) \times 100 \quad (6)$$

Table 1. References studying the engineering properties of several seeds

Seed	Moisture Content (%d.b.)*	Reference
Cumin	7.0 - 22.0	Singh and Goswami, 1996. [13]
Sunflower	4.0 - 20.0	Gupta and Das, 1997. [15]
Fenugreek	8.9 - 20.1	Altuntaş et al., 2005. [16]
Safflower	3.7 - 15.6	Baümler et al., 2006. [17]
Sunflower	2.0 – 18.8	de Figueiredo et al., 2006 [18]
Sunflower	2.9 – 20.1	de Figueiredo et al., 2006 [18]
Quinoa	4.6 - 25.8	Vilche et al., 2003. [22]
Soybean	8.7 - 25.0	Deshpande et al., 1993. [23]
Amaranth	7.7 - 43.9	Abalone et al., 2004. [24]
Chia	4.6 - 17.7	Guiotto et al., 2011 [25]
Cotton	8.3 - 13.8	Özarslan, 2002. [26]
Flaxseed	6.1 - 16.2	Coşkuner and Karababa, 2007 [27]
Rapeseed	4.7 - 24.0	Çalişir et al., 2005. [28]
* Moisture content range considered in each study		

The static coefficient of friction was measured using two structural materials, namely galvanized iron and aluminum. These materials are commonly used for transport, storage, and handling operations of grains, pulses, and seed and for building storage and drying bins. A PVC cylinder (50 mm diameter, 50 mm high, open at both ends) was placed on an adjustable tilting table, faced with the test surface, and filled with the sample. The structural surface with the cylinder resting on it was tilted gradually with a screw device until the cylinder just started to slide down [22]. The angle (α) was read on a graduated scale and the friction coefficient was calculated using the following relationship:

$$\mu = \tan \alpha \quad (7)$$

where μ is the static coefficient of friction, and α is the angle of tilt (degrees). The static coefficient of friction was replicated reading ten times for each moisture content.

Different authors evaluated the engineering properties of several seeds as a function of seed moisture content (Table 1). These data were considered to compare the behavior of different grain properties with regard to moisture content.

3. RESULTS AND DISCUSSION

Dark seeds represented 89% by mass of the samples studied. The initial moisture content was 10.0% and 10.9% d.b., for dark and white seeds, respectively. Averages of the three principal dimensions were L = 1.97 mm, W = 1.29 mm, and T = 0.84 mm for dark chia seeds. The statistical analysis (ANOVA) detected significant differences ($p \leq 0.05$) between different moisture levels for each principal dimension. Figure 3, shows a linearly increasing tendency between the dimensions (L, W, and T) and moisture content ($R2 > 0.95, p \leq 0.005$). Similar trends of increase have been reported in fenugreek [16], quinoa [22], soybean [23] and amaranth [24] seeds. Knowledge of the seed axial dimensions is important, for instance, in determining aperture sizes in the design of grain handling machinery.

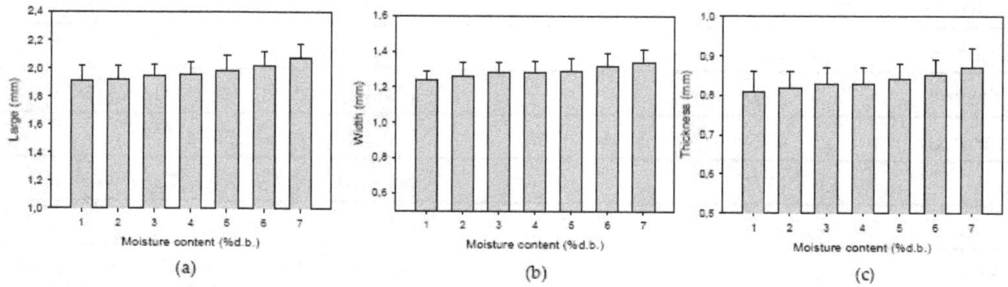

Figure 3. Principal dimensions (a) length (L); (b) width (W); (c) thickness (T) of dark chia seeds with different moisture content (1) 4.6%, (2) 6.5%, (3) 8.7%, (4) 10.0%, (5) 12.5%, (6) 15.3%, (7) 17.7%).

The geometric mean of the axial dimensions is useful in estimating the projected area of a particle moving in the turbulent or near-turbulent region of an air stream. It is therefore generally indicative of its pattern of behavior in air streams, especially with respect to the ease of separating extraneous materials from the particle during cleaning by pneumatic means [26]. Significant differences ($p < 0.05$) were found between different moisture contents for geometric diameter (Dg, mean 1.28 mm) and for specific area (S, mean 5.20 mm2). Both properties, Dg and S, increased linearly 8.14% and 16.9%, respectively, in function of increasing moisture content ($R2 > 0.97, p < 0.005$) (Table 2 and Figure 4). This Dg trend was similar to that observed for amaranth, flaxseed, quinoa and soybean. Similar behavior of S was reported for fenugreek and soybean seeds. The increase of S may be due to seed dilatation during moisture sorption, resulting in contact area enhancement.

Table 2. Regression equation as a function of moisture content (x, %d.b.) with their respective coefficient of determination (R2) for geometric diameter (Dg, mm).

Seed [Reference]	Regression equation	R2
Amaranth [24]	1.071 + 0.00385 x	0.99
Chia [25]	1.2075 + 0.0057	0.97
Flaxseed [27]	1.941 + 0.011 x	0.97
Quinoa [22]	1.359 + 0.009819 x	0.99
Soybean [23]	4.882 + 0.0253 x	0.99

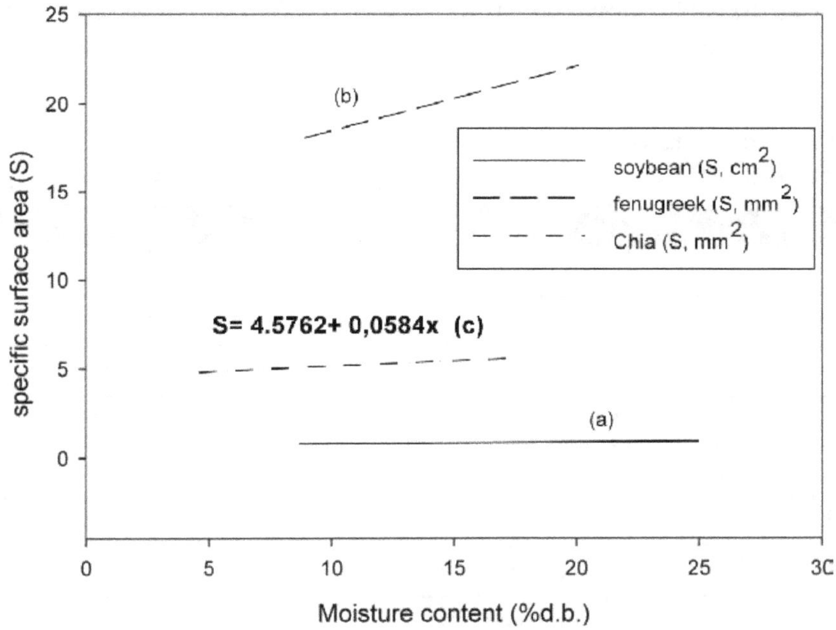

Figure 4. Effect of moisture content (x, % d.b.) on specific area (S). Date correspond to the adjusted function reported by (a) [23]; (b) [16] and (c) [25]

Bulk density, true density, and porosity (the ratio of inter granular space to the total space occupied by the grain) can be useful in sizing grain hoppers and storage facilities; they can affect the rate of heat and mass transfer of moisture during aeration and drying processes. The theories used to predict the structural loads for storage structures have bulk density as basic parameter. Bulk density and porosity are major considerations in designing near-ambient drying and aeration systems, as these properties affect the resistance to airflow of the stored mass. Grain bed with low porosity will have greater resistance to water vapor scape during the drying process, which may lead to higher power to drive the aeration fans.

True density average of dark chia seeds with different moisture content was 1.069 g cm-3. Statistical analysis showed significant differences ($p \leq 0.05$) between different moisture levels. The true density of seed was found to increase linearly at a decreasing in moisture content from 4.6 to 17.7% d.b. The relationship between true density (ϱt, g cm-3) and the moisture content (x, % d.b.) of the seed can be represented for ϱt = 1.1457 - 0.008x (R2= 0.9912, $p < 0.0001$) [25]. This trend can be attributed to minor volumetric product contraction during drying with respect to the decrease of seed mass due to water loss. The negative linear relationship of true density with moisture content was also observed by other authors for fenugreek and soybean, but it was different for cumin, flaxseed, quinoa and sunflower seeds, which increased linearly. However, a non-linear decreasing relationship of true density with moisture content was reported for amaranth and safflower seed (Table 3).

Table 3. Regression equations as a function of moisture content (x, %d.b.) with their respective coefficient of determination (R2) for true density (ϱt, kg m-3).

Seed [Reference]	Regression equation	R2
Cumin [13]	**1010 + 6.05 x**	**0.98**
Flaxseed [27]	**826.92 + 28.91 x**	**0.99**
Fenugreek [16]	**1275.8 – 37.555 x**	**0.99**
Quinoa [22]	**853.19 + 13.16 x**	**0.98**
Sunflower [15]	**694.6 + 3.72 x**	**0.92**
Sunflower* [18]	**706.12 + 4.06 x**	**0.94**
Soybean [23]	**1254.8 – 5.258 x**	**0.99**
Amaranth [24]	**(1411(100 + x))/**	**0.99**
	(100 + 1,25 x)	
Safflower [17]	**0,7887+(0,000298 x)- (0,0004551 x2)**	**0.99**
*** black-hull oilseed**		

Chia seed bulk density decreased from 0.713 to 0.644 g cm-3 as a function of the increase of moisture content. Table 4 shows the relationship of bulk density with moisture content reported by different authors for several seeds. Regarding bulk density, chia seeds presented a behavior similar (non-linear) to those of cumin and safflower seeds. The negative linear relationship of bulk density with moisture content was observed by other authors for amaranth, flaxseed, fenugreek, quinoa, rapeseed, sunflower and soybean (Table 4). The decrease in bulk density of seed with increase in moisture content indicates that the increase in volumetric expansion is greater than weight.

Table 4. Regression equations as a function of moisture content (x, %d.b.) with their respective coefficient of determination (R2) for bulk density (ϱb, kg m-3).

Seed [Reference]	Regression equation	R2
Amaranth [24]	869 – 3.50 x	0.87
Chia [25]	697.7+ 5.0 x – 0.4 x2	0.99
Cumin [13]	407 + 15.67 x – 0.70 x2	0.97
Flaxseed [27]	822.43 – 15.308 x	0.99
Fenugreek [16]	726.76 – 27.675 x	0.98
Quinoa [22]	771.5 – 3.94 x	0.98
Rapeseed [28]	616.74 – 1.4518 x	0.93
Safflower [17]	452.2 – 0.12 x2	0.96
Sunflower [15]	472.4 – 1.69 x	0.88
Sunflower* [18]	446.35 – 201 x	0.87
Sunflower** [18]	528,75 – 2.24 x	0.80
Soybean [23]	748.9 – 1.6626 x	0.99
* black-hull oilseed, **striped-hull oilseed		

These discrepancies observed for as much true density as for bulk density could be due to cell structure and the volume and mass increase characteristics of the seeds as moisture content increases [29].

The average porosity observed for all dark samples was 35.3%, with significant differences ($p \leq 0.05$) among moisture contents. The polynomial relationship between the porosity (ε, %) and moisture content (x, % d.b.) for chia seeds can be represented by ε = 40.0745 – 1.0991 x + 0.0049 x2) (R2 = 0.9701, p < 0.0010) [25]. Nevertheless, different authors reported linear increase in porosity with increase in moisture content for amaranth, cumin, flaxseed, fenugreek, quinoa, rapeseed, safflower, sunflower seeds, and trend linear decrease in porosity for soybean (Table 5).

Table 5. Regression equations as a function of moisture content (x, %d.b.) with their respective coefficient of determination (R2) for porosity (ε, %).

Seed [Reference]	Regression equation	R2
Amaranth [24]	28 + 0.16 x	0.75
Cumin [13]	48 + 0.643 x	0.93
Flaxseed [27]	11.453 + 2.7621 x	0.99
Fenugreek [16]	42.987 + 0.555 x	0.95
Quinoa [22]	13.1 + 1.22 x	0.98

Rapeseed [28]	44.659 + 0.6656 x	0.99
Safflower [17]	39.53 + 0.342 x	0.93
Sunflower [15]	32.27 + 0.54	0.95
Sunflower* [18]	36.99 + 0.58 x	0.92
Sunflower** [18]	30.10 + 0.37 x	0.91
Soybean [23]	40.5 – 0.1365 x	0.98
* black-hull oilseed, **striped-hull oilseed		

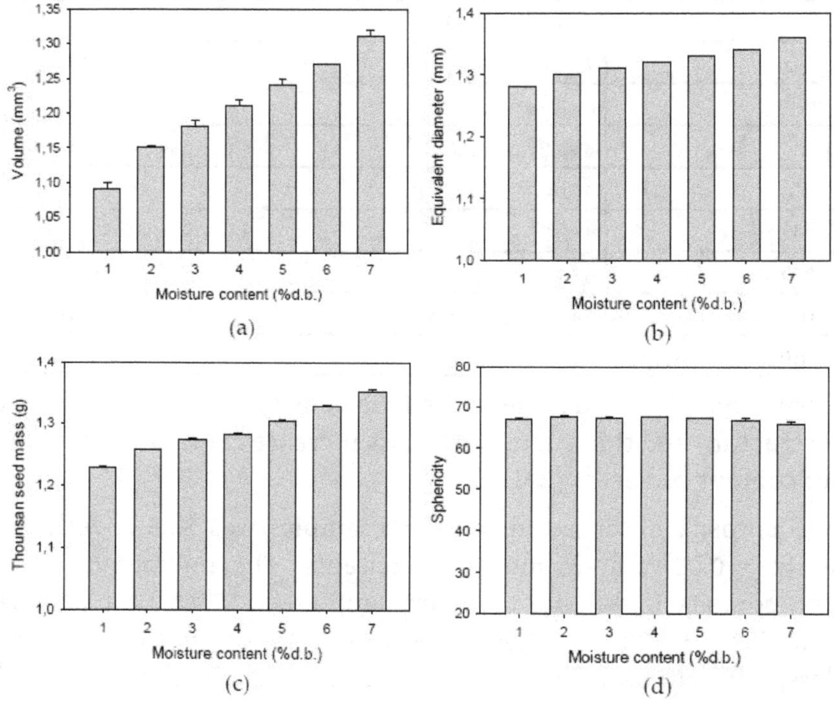

Figure 5. (a) volume (V), (b) equivalent diameter (De), (c) thousand seed mass (W1000), and (d) sphericity ( ) of dark chia seeds with different moisture content (1) 4.6%, (2) 6.5%, (3) 8.7%, (4) 10.0%, (5) 12.5%, (6) 15.3%, (7) 17.7%).

Since porosity depends on bulk and true densities, the magnitude of its variation depends mainly on these properties. Therefore, the porosity of each type of seed or grain could respond differently with increasing moisture content. This fact could be attributed to the seeds' morphological characteristics; the relative changes in their length, width, and thickness; and the associated bulk and true densities. Taking into account the high level of polyunsaturated fatty acids, chia seeds can be easily affected by temperature. For this reason, aeration is an important process to maintain a low uniform temperature and prevent the moisture migration. The resistance to airflow or pressure drop is affected by different factors, such as the bulk density, porosity, and moisture content. Due to the low bulk density and size of chia seeds, the grain bed will have an important pressure drop, requiring a high level of power for driving the aeration fans [25].

The variation of volume (V), equivalent diameter (De), thousand seed mass (W1000), and sphericity (φ) of chia seeds with moisture content (4.6 - 17.7 % d.b.) is shown in Figure 5.; the average values were 1.21 mm3, 1.32 mm, 0.129 g and 66.7 % respectively. Statistical analysis revealed significant differences ($p < 0.05$) between seeds with different moisture content for V, De and W1000. Nevertheless, sphericity did not present significant differences ($p > 0.05$).

The sphericity varied between 65.8% and 67.6%, values higher than the data reported for sunflower and safflower seed, but lower than those of amaranth, quinoa, rapeseed and soybean seed (Table 6).

Table 6. Sphericity (φ) of different seeds

Seed [Reference]	Sphericity (%)
Amaranth [24]	82 #
Fenugreek[16]	60.79 – 64.06
Quinoa [22]	77 – 80
Rapeseed [28]	93 – 92
Safflower [17]	58 – 62
Sunflower* [18]	49 – 52
Sunflower** [18]	47 – 50
Soybean [23]	80.6 – 81.6
# mean value, * black-hull oilseed, **striped-hull oilseed	

The high φ value thus suggests that the seeds tend towards a spherical shape. Thus, the value of the φ generally indicates a likely difficulty in getting the seed to roll. This tendency to either roll or slide should be necessary in the design of hoppers and dehulling equipment for the seed.

As can be seen in Figure 6, the equivalent diameter (De, mm) increased linearly from 1.28 to 1.36 mm when the moisture content was increased from 4.6% d.b. to 17.7% d.b., representing a variation of 6.3%.

The V and W1000 of dark chia seeds linearly increased with moisture content (Figure 6). The similar trend was reported for fenugreek [16], safflower [17] sunflower [18] soybean [23] and rapeseed [28] seeds.

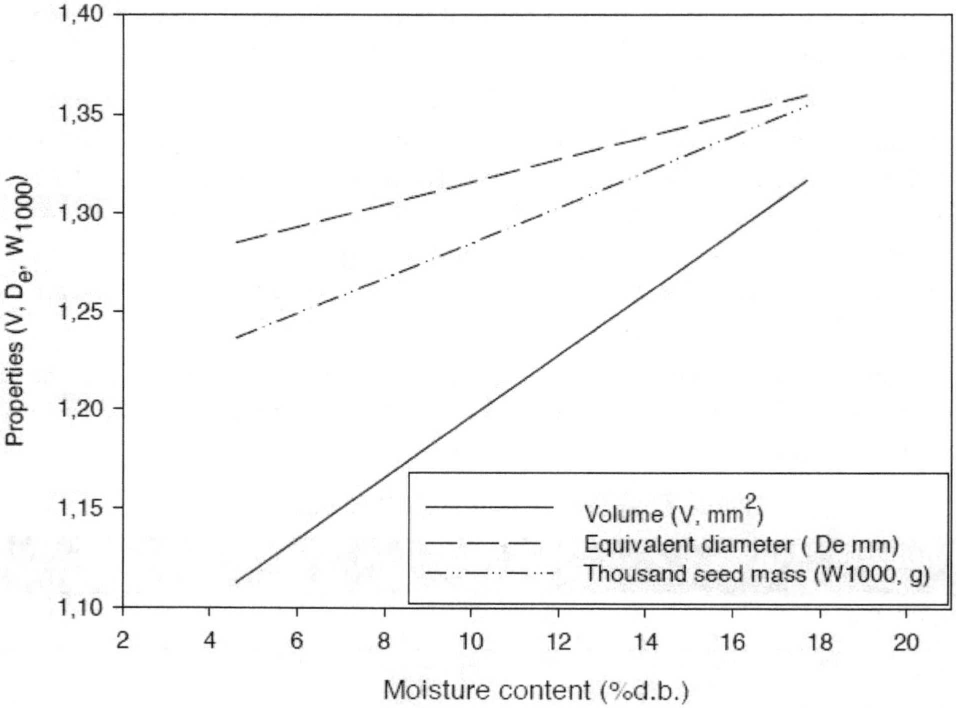

Figure 6. Effect of moisture content (x, % d.b.) on volume seed (V), equivalent diameter (De) and thousand seed mass (W1000). Date corresponds to the adjusted function reported by [25].

The frictional characteristics are important for the proper design of agricultural product handling equipment. Friction between a seed and a surface has an influence on the movement of particles on oscillating conveyors, separation on oscillating sieves and unloading and loading operations. The static coefficient of friction of dark chia seeds was determined on two structural surfaces: galvanized iron and aluminum. The values obtained were higher for aluminum (mean: 0.30, minimum: 0.26, maximum: 0.37) than for galvanized iron (mean: 0.28, minimum: 0.25, maximum: 0.34). Increments of 28.4% and 29.5% were recorded for the galvanized iron and aluminum surfaces, respectively, as the moisture content increased from 4.6% to 17.7% d.b. The reason for the increased friction coefficient at higher moisture content may be that the water present in the seed offered a cohesive force on the contact surface and the seed became rougher and sliding characteristics are diminished [22, 26]. For both structural surfaces, the coefficient of static friction increased linearly with an increase in moisture content. Similar trends were reported for cumin, flaxseed, quinoa and sunflower (Figure 7).

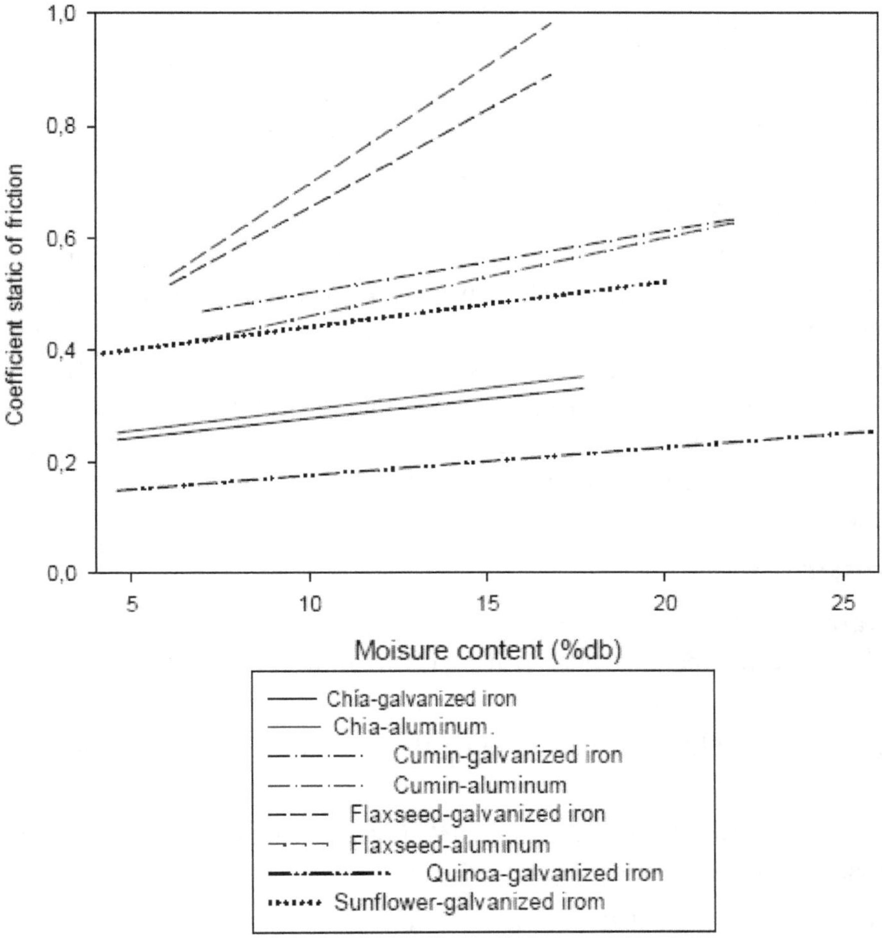

Figure 7. Effect of moisture content (x, % d.b.) on coefficient of static friction. Dates correspond to the adjusted function reported by () [25], () [13], () [27], () [22] and () [15].

4. CONCISE SUMMARY

The engineering characteristics of dark chia seeds were assessed based on moisture content, ranging from 4.6% to 17.7% d.b., and their performance was compared with that of amaranth, cumin, flaxseed, fenugreek, quinoa, rapeseed, safflower, soybean, and sunflower. The primary dimensions of dark chia seeds (length, width, and thickness), geometric diameter, specific surface area, volume, equivalent diameter, thousand seed mass, and static coefficient of friction on galvanized sheet and aluminum exhibited a linear increase with rising seed moisture content. Chia seed is among the tiniest, comparable to amaranth and quinoa, and possesses a relatively light weight.

The sphericity exhibited no significant variations across the range of moisture content examined for dark chia seeds. The seeds most comparable in sphericity to chia seeds were rapeseed, amaranth, soybean, and quinoa. An elevation in moisture content results in a reduction of both bulk and actual density. The bulk density and porosity of chia seeds exhibited a nonlinear relationship, demonstrating a quadratic concave pattern in relation to moisture content.

The friction generated by the aluminum surface was marginally more than that of the galvanized iron surface. The fluctuation of the engineering properties of chia seeds in relation to moisture content exhibited a tendency analogous to that documented for other seeds, albeit with certain variations. Nonetheless, they exhibited distinct regions of variation. This may be ascribed to the morphological and physiological properties of the seeds. The analysis of data from various seeds is crucial for the design and adaptation of equipment for transportation, storage, and processing.

REFERENCES

1. Álvarez-Chávez, L. M., Valdivia-López, M. A., Alberto-Juárez, M. L., & Tecante, A. (2008). Chemical characterization of the lipid fraction of mexican chia seed (Salvia hispanica L.). Int. J. Food Prop., 11, 687-697.
2. Cahill, J. 2003. Ethnobotany of chia, Salvia hispanica L. (Lamiaceae). Econ. Bot. 57(4): 604-618.
3. Palma, F., Donde, M., Lloyd, W.R., 1947. Fixed oils of Mexico. Part 1. Oil of chia-Salvia hispanica. J. Am. Oil Chem. Soc. 24, 27.
4. Ayerza Jr., R., 1995. Oil Content and fatty acid composition of Chia (Salvia hispanica L.) from five northwestern locations in Argentina. J. Am. Oil Chem. Soc. 72, 1079– 1081.
5. Bushway, A.A., Belyea, P.R., Bushway, R.J., 1981. Chia seed as asource of oil, polysaccharide, and protein. J. Food Sci. 46, 1349–1350.
6. Capitani, M.I.; Spotorno, V.; Nolasco S.M.; Tomás, M.C. (2012). Physicochemical and functional characterization of by-products from chia (Salvia hispanica L.) seeds of Argentina LWT - Food Sci. Technol. 45, 94-102.
7. Coates, W., & Ayerza, R. (1996). Production potencial of chia in northwestern Argentina. Ind. Crops and Prod., 5, 229-233.
8. Ixtaina, V.Y., Martínez, M.L., Spotorno, V., Mateo, C.M., Maestri, D.M., Diehl, B.W., (2011). Characterization of chia seed oils obtained by pressing and solvent extraction. J. Food Comp. Anal., 24, 166-174.
9. Ixtaina, V. Y., M. I. Capitani, S. M. Nolasco, and M. C. Tomás. 2010. Caracterización microestructural de la semilla y el mucílago de chia (Salvia

hispanica L.). In Proc. XXVIII Congreso Argentino de Química. Buenos Aires, Argentina: Asociación Química Argentina.

10. Commission of the European Communities. 2009. Commission Regulation (EC) 827/2009. Official J. European Union 52: 12-13.

11. Brann, M. 2008. Chia: The Ord Valley's new super crop. ABC Rural. Sydney, Australia: Australian Broadcasting Corporation. Available at: www.abc.net.au/rural/content/ 2008/s2367335.htm. Accessed 12 July 2010.

12. Ixtaina, V. Y., S. M. Nolasco, and M. C. Tomas. 2008. Physical properties of chia (Salvia hispanica L.) seeds. Ind. Crops Prod. 28(3): 286-293.

13. Singh, K. K., and T. K. Goswami. 1996. Physical properties of cumin seed. J. Agric. Eng. Res. 64(2): 93-98.

14. ASAE Standards. 1999. S352.2: Moisture measurement: Unground grain and seeds. St. Joseph, Mich.: ASAE.

15. Gupta, R. K., and S. K. Das. 1997. Physical properties of sunflower seeds. J. Agric. Eng. Res. 66(1): 1-8.

16. Altuntaş, E., E. Özgöz, and Ö. F. Taşer. 2005. Some physical properties of fenugreek

17. (Trigonella foenum-graceum L.) seeds. J. Food Eng. 71(1): 37-43.

18. Baümler, E., A. Cuniberti, S. M. Nolasco, and I. C. Riccobene. 2006. Moisture-dependent physical and compression properties of safflower seed. J. Food Eng. 72(2): 134-140.

19. A.K. de Figueiredo, E. Baümler, I.C. Riccobene, S.M. Nolasco Moisture-dependent engineering properties of sunflower seeds with different structural characteristics J. Food Eng. 102 (2011) 58–65.

20. Thompson, R. A., and G. W. Isaacs. 1967. Porosity determination of grains and seeds with air comparison pycnometer. Trans. ASAE 10(5): 693-696.

21. Mohsenin, N. N. 1970. Physical Properties of Plant and Animal Materials. New York, N.Y.: Gordon and Breach Science

22. Özarslan, C. 2002. Physical properties of cotton seed. Biosyst. Eng. 83(2): 169-174.

23. Vilche, C., M. Gely, and E. Santalla. 2003. Physical properties of quinoa seeds. Bio syst. Eng. 86(1): 59-65.

24. Deshpande, S. D., S. Bal, and T. P. Ojha. 1993. Physical properties of soybean. J. Agric. Eng. Res. 56(2): 89-98.

25. Abalone, R., A. Cassinera, A. Gastón, and M. A. Lara. 2004. Some physical properties of amaranth seeds. Biosyst. Eng. 89(1): 109-117.

26. Guiotto, E., Ixtaina, V.Y., Tomás, M. C. and Nolasco, S.M. 2011. Influence of moisture content on physical properties of chia (Salvia hispanica L.) seeds, Trans.ASABE 54 (2): 527-533.

27. Omobuwajo, T.O., Sanni, L.A., Balami, Y.A. 2000. Physical properties of sorrel (Hibiscus sabdariffa) seeds. J. Food Eng. 45, 37-41.

28. Coşkuner, Y., and E. Karababa. 2007. Some physical properties of flaxseed (Linum usitatissimum L.). J. Food Eng. 78: 1067-1073.

29. Çalişir, S., T. Marakoğlu, H. Öğüt, and Ö. Öztürk. 2005. Physical properties of rape seed (Brassica napus oleifera L.). J. Food Eng. 69(1): 61-66.

30. Baryeh, E. A. 2002. Physical properties of millet. Baryeh, E. A. 2002. Physical properties of millet. J. Food Eng., 51, 39–46.

Chapter 18 ENHANCEMENT OF POLYGALACTURONASE PRODUCTION BY SOLID STATE FERMENTATION

1. INTRODUCTION: AN OVERVIEW

Polygalacturonases are pectinolytic enzymes that facilitate the hydrolytic breakdown of polygalacturonic acid chains by incorporating water via hydrogen bonds. These enzymes possess technological, functional, and organic applications in food processing and plant-fungus interactions.

They are generated by plants, fungus, bacteria, and yeasts. Fungi are favored in industrial applications because approximately 90% of the enzymes they generate can be secreted into the media, resulting in substantial enzyme yields [1,2].

Polygalacturonases that facilitate the hydrolysis of pectic compounds include endo-polygalacturonase and exo-polygalacturonase. Exo-polygalacturonase acts on the terminal monomers of polygalacturonic acid, yielding monogalacturonic acids. Endo-polygalacturonase acts randomly on polygalacturonic acid, yielding oligogalacturonic acid [6].

Pectic enzymes account for one-quarter of global food production enzymes. These are extensively utilized in the fruit juice sector to diminish viscosity, enhance filtration and clarity efficiency, serve as preliminary treatment in the grape wine industry, facilitate tomato pulp extraction, and for many other purposes.

The production of fruit juices from pomaceous fruits (e.g., citrus) and red fruits (e.g., grape) necessitates the incorporation of enzymes to transform the viscous macerated or triturated fruit semi-gel, which is a result of the partial solubility of pectins and the high water retention capacity of solids, in order to optimize juice extraction during the pressing phase of the process. Pectinases that can degrade highly methylated pectins are more appropriate and include endopolygalacturonase and pectin methylesterase. [7].

Principal application of enzymes that pectins hydrolyze is clarification or extraction of juices. The turbidity may be desirable in some juices (e.g. orange), but not for apple and grape juice, in which are translucent over sold. Turbidity is conferred by colloidal particles consisting of proteins coated with pectin. Pectinases depolymerize pectins, promoting flocculation and facilitating clarification [7].

Polygalacturonases produced by fungi are more active in pH range of 3.5 to 6.0 and temperature of 40-55°C. The pratical result of activity of these enzymes is that middle lamella is disrupted and the viscosity of pectin solutions is decreased as the action of enzyme is maintained [7]. In fact, the characteristics of enzyme related with temperature and pH effects will depend on factors of production process, e.g. microorganism used, available nutrients, fermentation temperature, among others.

Industrially, pectinases are produced either by submerged fermentation and solid state fermentation with Aspergillus niger strains, however, the solid state fermentation technique is generally considered more susceptible to higher yields of pectin esterase and polygalacturonase. Some authors state that this preference occurs because the solid state fermentation allows the production of crude enzymes more concentrated and therefore a lower cost of extraction and purification [5,8].

Solid state fermentation process is defined as a process that occurs over a non-soluble material, acting as support and nutrients source, with small quantity of water, under the action of fermenting agent [9].

One of the important factors in pectinases production by solid state fermentation is the medium composition. Appropriate balance between sources of nitrogen and carbon is so important to the nutritional requirement of microorganism that the effects of environmental conditions which affect mycelial growth [10].

Water in the system is a limiting factor. This amount of water is related to the medium through of moisture, as regards the percentage of water in total mass. Determination of its value in process is closely related to the substrate nature, requirements of microorganism used and type of end product desired. If moisture level is high, it will result in a decrease of substrate porosity and it will result in lower oxygen diffusion within the medium, consequently, decrease in gas exchange and increases the risk of contamination, especially bacterial. For lower levels of moisture needed by the microorganism, there will be greater difficulty in diffusion of nutrients, resulting in a lower growth and, consequently, lower production of desired product [11].

Temperature is also considered a critical factor, as well as moisture, due to the accumulation of metabolic heat generated during fermentation, which directly affects the microorganism germination and product formation. In composting process, this effect is desirable, however, for biotechnology processes, such as enzyme production, the heat must be dissipated immediately, so that temperature increase does not adversely affect the desired fermentation [11].

There are several types of reactors used in solid state fermentation process. Although there are many projects to industrial bioreactors, these have a limited extent for this type of process [12].

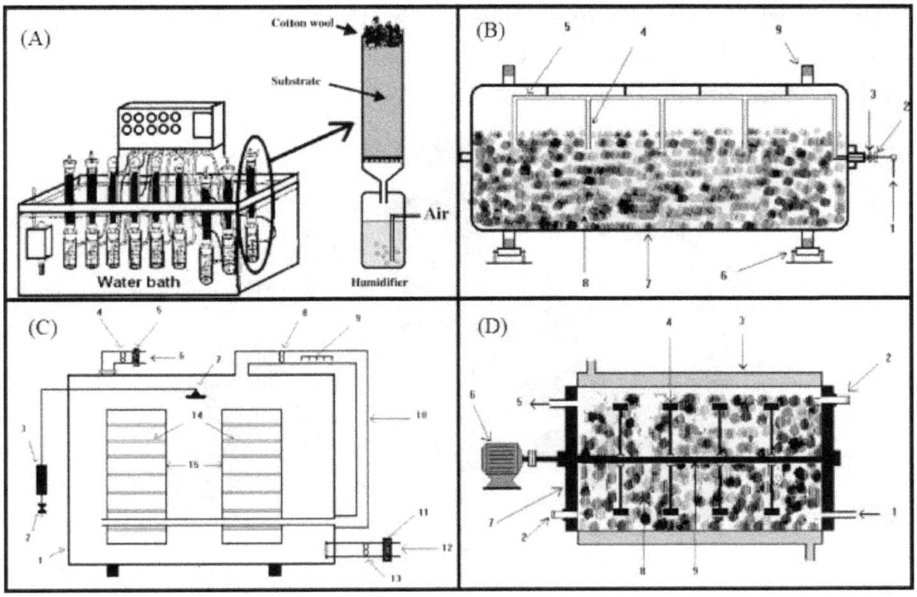

Figure 1. Type of bioreactors: (A) Column; (B) Rotating drum; (C) Koji-type; (D) Stirred horizontal [12].

Tray reactor consists of a chamber in which temperature and humidity air controlled circulates around a series of trays. Each tray contains a thin layer of depth. It is noteworthy that intermittent mixing of medium can be performed, but this generally occurs only once daily [13].

Scale-up using tray reactor cannot be simply by increasing the medium thickness, as this leads to problems of overheating. For this reactor type, scaling-up must be done by increasing the area of trays, which can be achieved using larger trays [14].

As described above, tray reactors are limited by heat transfer, but also by mass transfers, and may develop high temperatures and internal gas concentration gradients in height above 40 mm of substrate [15]. Still about the medium temperature, it is almost impossible to maintain this variable in the optimum value for production. However, fermentative process will vary in temperature that can be of 10°C above the optimum value [16].

Although there are numerous projects for industrial bioreactors, it is observed that these have limited development to processes using solid state fermentation. This occurs because there are some limitations of this process, such as the difficulty to remove heat generated by microbial metabolism, heterogeneity of mixture during fermentation, which makes the control of cell growth and various parameters such as temperature, pH, agitation, aeration, concentration of nutrients and products, making difficult the automation [17].

Despite these difficulties, the use of semi-solid medium may be advantageous, it allows the use of industrial residues (flours, bark and cake) as substrate, which is abundant raw material and low cost at Brazil.

Figure 2. Cashew fruit [22].

Industrial wastes have been used in bioproducts production through fermentation processes. Among these residues may be mentioned the cashew apple (Anacardium occidentale L.), which is rich in sugars, organic acids and fiber, that is why it has been used in the production of phenols [18], bioethanol, cashew wine [19], protein enrichment [20], pectinases [21], cellulases, among others.

This work seeks to delineate the scaling-up of the solid-state fermentation method for the synthesis of polygalacturonases, utilizing cashew apple dry bagasse as the substrate, Aspergillus niger CCT 0916 as the microbe, and a tray reactor as the operational equipment. This will be conducted to characterize the substrate in relation to its physicochemical qualities. The adsorption isotherms of the substrate will be developed and altered to illustrate their relationship with the fermentation process. Initially, the parameters that most significantly influence the fermentation process at a laboratory scale will be characterized. These features will facilitate the scaling-up of the solid-state fermentation process utilizing a tray reactor. The crude enzyme extract will be characterized, along with its stability concerning temperature and pH..

2. MATERIAL AND METHODS

2.1 Substrate

Cashew apple bagasse was obtained from fresh cashew fruit acquired at Empresa de Abastecimento e Serviços Agrícolas (EMPASA) at Campina Grande City, Brazil. First, cashew nut was removed. Next, apple was triturated and pressed to separate the juice. Humid bagasse was dried with air renewal and circulation at 55°C. After drying process, bagasse was ground in TECNAL knife mill.

2.2 Substrate characterization

Measurements of pH, moisture content and mineral waste (MW) followed the standards Brazil [23]. The pectin amount (PC) was determined by gravimetric precipitation method using calcium pectate [24]. Reducing sugars (RS) and saccharose were determined by HPLC (High performance liquid chromatography). Concentration of soluble solids (SS) was obtained by direct reading in refractometer after adding 9 mL of distilled water to 1 g of dry bagasse. It was used 100 g of material to determine the density. This mass was placed in a graduate to determine volume occupied without compression. Size distribution was performed using 100 g of residue in a Cotengo-Pavitest sieve shaker for 1o minutes in 14, 20, 24, 35, 48 and 60 mesh trays. Result was expressed as weight percentage. Protein was determined by semi-micro Kjeldhal method for nitrogen adjusted by spectrophotometry [25]. All characterizations were performed in triplicate. Standard deviation (SD) was based on means values.

2.3 Substrate adsorption isotherms

Triplicate samples were weighed approximately 1 g of product in aluminium crucibles and stored in airtight containers containing saturated salt solutions until reached the equilibrium moisture content for a certain relative humidity range (Table 1). Temperatures of 25, 30, 35 e 40°C were supplied by environmental chamber. Samples were weighed every 24 hours, reached constant weight. After that, samples were transferred to stove at 100°C for determination of dry weight. Equilibrium moisture content (xeq) was calculated by Equation 1.

$$xeq = \left(\frac{mi - ms}{ms}\right) \times 100 \quad (1)$$

In which: xeq = equilibrium moisture, % dry basis; mi = initial weight of sample, g; ms = dry weight of sample, g.

Table 1. Equilibrium relative humidity of saturated salt solutions [26]

Saturated salt solutions	Temperature (°C)			
	Relative humidity (%)			
25	30	35	40	
$K(C_2H_3O_2)$	23	23	23	23
$MgCl_2$	33	32	32	31
K_2CO_3	43	42	41	40
NaBr	57	57	57	57
NaCl	75	75	75	75
KCl	86	84	84	83

It was used the BET model [27] to fit experimental data (Equation 2):

$$\frac{xeq}{xm} = \frac{C.aw}{1 - aw}\left[\frac{1 - (n+1)(aw)^n + n(aw)^{n+1}}{1 - (1 - C)aw - C(aw)^{n+1}}\right] \quad (2)$$

In which: xeq = equilibrium moisture, % dry basis; aw = water activity, adimensional; C = BET constant; xm = moisture in the molecular monolayer; n = number of molecular layers.

Criteria used to observe the fit were the coefficient of determination (R2) between observed responses and predicted by the fitted model and average percentage deviation (P) (Equation 3). The best fits were those with the highest R2 and lowest value of P [28].

$$P = \frac{100}{n}\sum_{i=1}^{n}\frac{|Xexp - Xteo|}{Xexp} \quad (3)$$

In which: n = observation numbers; Xexp = humidity of experimental material; Xteo = humidity calculated by adjusted models.

2.4 Fermentation process in laboratory scale

Microorganism used was Aspergillus niger CCT0916, donated by Empresa Brasileira de Pesquisa Agropecuária (EMBRAPA, Fortaleza State – Brazil). Spore concentration was adjusted according to experimental design.

Substrate was hydrated with distilled water to obtain moisture content and ammonium sulfate was added to this volume. In a 250 mL Erlenmeyer flask, it was weighed 10 g of sterilized humidified medium. After spore inoculation, this medium was incubated at fermentation temperature by experimental design for 78 hours.

Enzyme extraction for fermented complex was performed by adding 2.5 mL/g of fermented medium using 200 mM acetate buffer pH 4.5. Samples were then left in water bath for 1 hour at 30°C and filtered through Wattman 1 filter paper.

A 24 factorial experimental design was conducted with 7 experiments at central point to determine the influence of spore concentration (E), initial moisture (U), ammonium sulfate concentration (N) and fermentation temperature (Tf) on polygalacturonase activity response (Table 2).

One unit of polygalacturonase activity was defined as the amount of enzyme that releases 1 µmol of galacturonic acid per minute of reaction at 35°C for 30 minutes.

2.5 Fermentation process using tray reactor

Microorganism used in this stage was also used for optimization in the laboratory scale. Spores concentration, moisture content and ammonium sulphate concentration were determined based on experiments performed on laboratory scale.

Substrate was hydrated with distilled water to obtain the initial moisture content and it was diluted ammonium sulphate in this water. On polypropylene trays (Figure 3), it was weighed 500 g of sterilized medium. Substrate thickness was equal to 40 mm. After spore inoculation, the medium was incubated at 23°C for 77 hours. Substrate thickness and fermentation temperature had been selected taking into account studies reported in literature [15,16].

Table 2. Concentrations and tests from factorial design

	Variables			
Test	U %(w.b.)	E (mL/g)	N %(w/w)	Tf (°C)
1	30 (-1)	106 (-1)	0.5 (-1)	25 (-1)
2	50 (+1)	106 (-1)	0.5 (-1)	25 (-1)
3	30 (-1)	108 (+1)	0.5 (-1)	25 (-1)
4	50 (+1)	108 (+1)	0.5 (-1)	25 (-1)
5	30 (-1)	106 (-1)	1.5 (+1)	25 (-1)
6	50 (+1)	106 (-1)	1.5 (+1)	25 (-1)
7	30 (-1)	108 (+1)	1.5 (+1)	25 (-1)
8	50 (+1)	108 (+1)	1.5 (+1)	25 (-1)
9	30 (-1)	106 (-1)	0.5 (-1)	35 (+1)
10	50 (+1)	106 (-1)	0.5 (-1)	35 (+1)
11	30 (-1)	108 (+1)	0.5 (-1)	35 (+1)
12	50 (+1)	108 (+1)	0.5 (-1)	35 (+1)
13	30 (-1)	106 (-1)	1.5 (+1)	35 (+1)
14	50 (+1)	106 (-1)	1.5 (+1)	35 (+1)
15	30 (-1)	108 (+1)	1.5 (+1)	35 (+1)
16	50 (+1)	108 (+1)	1.5 (+1)	35 (+1)
17	40 (0)	107 (0)	1.0 (0)	30 (0)
18	40 (0)	107 (0)	1.0 (0)	30 (0)
19	40 (0)	107 (0)	1.0 (0)	30 (0)
20	40 (0)	107 (0)	1.0 (0)	30 (0)
21	40 (0)	107 (0)	1.0 (0)	30 (0)
22	40 (0)	107 (0)	1.0 (0)	30 (0)
23	40 (0)	107 (0)	1.0 (0)	30 (0)

Figure 3. Tray reactor

Enzyme extraction for fermented complex was performed by adding 5.0 mL/g of fermented medium, using 200 mM acetate buffer pH 4.5. Samples were left in water bath for 1 hour at 30°C and filtered on Wattman 1 paper filter.

2.6 Enzymatic stability

To check the stability concerning temperature, crude extract samples were taken to a water bath for 20 min at temperatures between 10 and 90 ° C. For pH, crude extract was diluted in the buffers listed below, thus verifying the pH influence on polygalacturonase activity. After reaching the corresponding pH, samples were incubated for 24 hours at 2°C [46]: 0.1 M glycine-HCl pH 2.5; acetate buffer 200 mM pH 3.5-6.5; 0.1 M tris-HCl pH 7.5-8.5; 0.1 M glycine- NaOH pH 9.5. Results of thermostability and stability as to pH were expressed as relative activity (%).

3. RESULTS AND DISCUSSION

3.1 Physicochemical characterization of cashew apple dry bagasse

In a solid state fermentation process for enzymes production, it is important to know about the substrate composition, because the microorganism uses such as nutrient, for growth and reproduction so as to produce.

In case of pectinases, the inducing substance is pectin. Microorganism will adapt to the environment and for its maintenance, it will produce the enzyme needed to break these substances. Other substances also welcome in process are reducing sugars. These sugars (glucose, fructose, etc.) are sources of quick energy, and are also consumed by microorganism during fermentation process [21].

However, there must be a balance between nutrient sources. The literature indicates that high concentrations of sugars in fermentation medium, supplying the microorganism needs for its growth, and pectin little used, hence there is little enzymes production [29].

Cashew has on average 30% of reducing sugars (glucose and fructose) and about 10% of pectin in your composition. This implies the need for addition of inducer in a bagasse for use in a pectinases production.

Some authors observed that for pectinases production, using Aspergillus niger T005007-2 and wheat bran as substrate through a solid state fermentation process, the citrus pectin addition, up to 16% (w/w) led to increased enzyme production [30]. Based on these data, and in the proximity of same amount with sugar in cashew bagasse, it was decided not to add inductors in fermentation process.

In general, pH is an important variable in any biological process, with optimum values for microorganism growth. Generally, fungi prefer low pH between 4.5 and 5.0 [31,32]. Average value observed for cashew dry bagasse is 4.0, confirming proximity to other values found in the literature for cashew cake dry [17,33]. Thus, bagasse pH characterized can promote the pectinases production without requiring adjustment using buffer solution. Moreover, acid pH favors the storage at room temperature without contamination problems.

Bagasse was dried to below 15% (w.b), it became necessary to store a reasonable amount, noting that below 15%, the organic materials retain their properties over time and makes it difficult to contamination by bacterias.

Water amount in fermentation medium is a limiting factor and it directly affects the microorganism needs, and the final product type. To use this residue, it will be necessary to adjust the moisture content, since some microorganisms that produce pectinases require higher levels. This quantity of water is related with the environment via two variables: moisture and water activity.

Moisture regards percentage of water in the total mass of medium. And determination of this value in process is closely related to substrate nature, the requirements of microorganism used and the type of end product desired [11]. Water activity indicates that the organism may grow by fermentation, ensuring product quality.

Microorganism growth depends on the water activity, due to influence of osmotic pressure by on exchange membranes. And it can be related to moisture in the substrate used in fermentation using sorption isotherms for a given temperature [32].

Some authors [8,31] cite the use of various substrates for pectinase production by Aspergillus niger with water activity above 0.93. Importantly, low levels of water activity means low availability of water molecules near the cell, making exchange of solutes in solid phase, reducing metabolism and generating lower rates of growth or synthesis of metabolites. In contrast, high levels of water activity hinder the diffusion of air through solid particles, leading to a reduction in microbial growth [35]

Crude protein value is approximately 8% and it is close to the value observed by some authors [36,37]. This value is important for characterization as serve as a nitrogen source for microorganism. And from that data, it can be observed the necessity of supplementation with alternative sources of nitrogen such as urea or ammonium sulphate.

Physical characteristics with respect to substrate morphology are essential, particularly, in size and porosity, as these properties governing the accessible surface area of microorganism [17].

For granulometric distribution, 80% of bagasse was retained in sieves 20, 24 and 35 mesh, corresponding to particle size of 0.85, 0.70 and 0.42 mm, respectively. This particle size can be used in a solid state fermentation process by Aspergillus niger, which was already described in literature, particle sizes for pectinases production from 0.5 to 0.7 mm [39,40].

Average particle size of residual fermentation media must be obtained so that there have been no particles large or small. Particles of small size promote greater surface area and consequently a higher degree of processing. However, the process itself needs to have a particle size allowing the circulation of air through the mass and waste gases and heat produced, which could harm the efficiency of process [11]. Particles larger interparticles promote more space, reducing the efficiency of nutrients absorption for microorganisms. Furthermore, particle size analysis is important in enzyme complex extraction, since finely divided solid carriers facilitate access by the solvent [29].

It is important to remember that, in general, the crops characterization crops can vary dramatically depending on time of harvest, agricultural practices and phenomena related to planting. Is then very important to characterize the substrate for solid state fermentation process and therefore the adjustment of certain parameters, which is also an additional challenge is to be considered in using this process.

3.2 Adjust of adsorption isotherms from cashew apple dry bagasse

Data water activity (aw) and average equilibrium moisture (xeq) of material at temperatures studied (25, 30, 35 and 40°C) was adjusted BET model.

Table 3 are the values of equation BET parameters, average deviation percentage (P) and correlation coefficient (R2) for each temperature.

From Table 3, there is the BET model appropriately fit to experimental data because P value indicates a good fit when it is less than 10% and R2 must be as close to unity [38].

When comparing the monolayer humidity values of humidity (Xm) by BET equation, it is noted that the range of 25-30°C to 35-40°C, Xm increased by approximately 0.4%. This is not a common behaviour, but can be explained by two mechanisms: (1) increase of temperature may cause changes in product physical structure, providing a larger number of active sites with affinity for water molecules, (2) or may cause an increase in intrinsic solubility of solute to the product, causing a greater number of water molecules is retained on monolayer [39].

Observed that the curves 25 and 30°C overlap, indicating a similar behaviour as regards the adsorption of water to these their temperatures, in which case this difference (5°C) showed no influence of temperature on these isotherms, unlike curves 35 and 40°C, in which there is small, but clearly influence of this variable. There is also a greater effect of temperature interval 25-30°C to 35-40°C. An analysis of the equilibrium moisture behaviour in relation to water activity shows similarity to temperature behaviour.

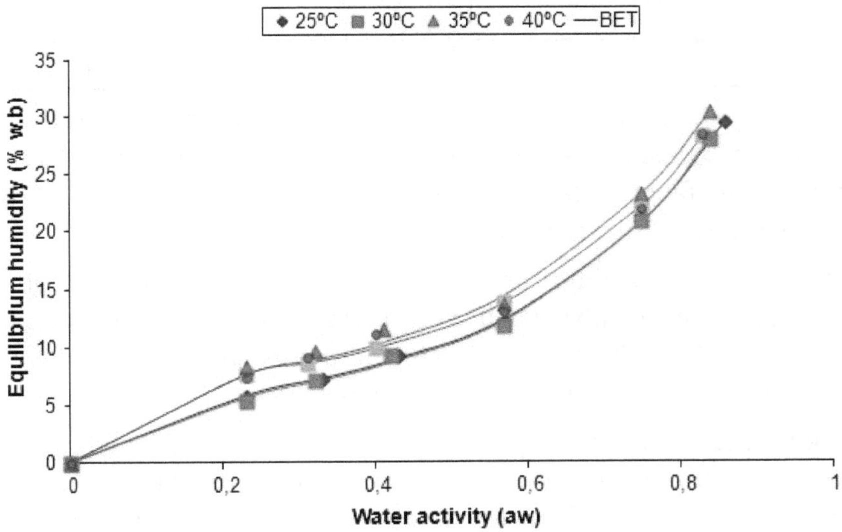

Figure 4. BET model to adsorption isotherms of dry cashew apple bagasse

Can be found in the literature several studies that relate initial content of water activity with development of various microorganisms responsible for synthesis products obtained in solid state fermentation process.

In particular, Aspergillus niger is described as the organism that best fits to the fermentation process, being around 0.7 the value of minimum water activity for development of their metabolic activities [11].

Thus, minimum value being 0.7 from water activity to grow Aspergillus niger, when using cashew apple dry bagasse as substrate in solid state fermentation process and adsorption isotherms obtained (Figure 4), has an indicative that moisture in substrate at intervals of 25-30°C and 35-40°C should not be less than 17.5 and 19.0%(d.b), respectively.

Table 3. Adsorption isotherms fitting parameters of cashew peduncle dry bagasse to the BET model

Parameters	Temperature (°C)			
25	30	35	40	
Xm	5.64	5.61	6.33	6.07
C	12.59	11.28	39.72	49.85
n	14.39	15.02	13.35	13.18
R2	0.9998	0.9986	0.9948	0.9941
P (%)	0.86	2.70	4.79	4.11

For pectinases production by Aspergillus niger in solid state fermentation process, several authors describe fermentation processes in which it can be seen that water activity which best favourable to synthesis of product is above 0.90. This implies that substrate moisture must be greater than 35%(d.b).

It is therefore extremely important to understand the hygroscopic behaviour of semisolid product used as substrate in a fermentation process, since the quantity of available water in through the microorganism to grow and synthesizing reactions during the production process is a limiting factor.

Most influential factor in solid state fermentation process on laboratory scale as previously noted, there are many factors that affect a solid state fermentation process: amount of water available to microorganism can be quantified by moisture of medium (U), amount of inoculum necessary to overcome the adaptation phase and microorganism to produce the desired products (E), addition of nitrogen source such as ammonium sulphate (N) and fermentation temperature (Tf). These are variables that will be studied as previously described. It was conducted an experimental design 24, taking an answer the polygalacturonase activity. Objective was to determine which variable most affects the process and maximize the amount of enzyme produced.

The highest activities found for each assay and the fermentation time it was noted, are available in the Table 4. The greatest polygalacturonase activity (33.27 U/g) found, during the execution of experimental design, was obtained under the initial conditions: 50% (w.b) initial moisture, 106 spores/g, 1.5% (w/w) ammonium sulfate and 35°C at 29 hours of fermentation.

Table 4. Highest polygacturonase activities (PG) observed for which solid state fermentation assay.

Test	PPG (U/g)	t (h)
1	1.06	21
2	14.93	54
3	0.42	70
4	15.91	46
5	0.59	5
6	15.38	54
7	5.00	46
8	10.78	21
9	0	---
10	24.23	46
11	0	---
12	7.69	29
13	1.66	54
14	33.27	29
15	0	---
16	11.31	70
17	2.01	70
18	5.07	29
19	10.28	46
20	14.29	70
21	5.53	78
22	11.37	21
23	8.30	46

From regression of polygalacturonase activity data and values of factors studied, it was constructed a first order model with 95% confidence for peaks of enzymatic activities.

$$PPG = \mathbf{8.66} + \mathbf{7.80U} - \mathbf{2.50E} + 0.86N + 0.88T_f - \mathbf{2.76UE} + 0.14UN + 1.56UT_f - 0.48EN - \mathbf{2.52ET_f} + 0.93NT_f \quad (4)$$

Model validation was done using Test F. This test shows the ratio between calculated value of F and F tabulated, knowing that the latter was equal to 2.75. When this ratio is greater than 1, regression is significant. Not only to a statistically significant regression, but also predictive value of ratio between this two F's must be greater than four [40].

Regarding the determination coefficient (R2), it is known that the maximum value is 1, meaning that between the experimental data and curve, there is no waste and any variation about the mean is explained by regression.

In Equation 4, calculated F was equal to 7.99. Thus, the ratio between calculated F and tabulated is equal to 2.91, meaning that this equation is statistically significant. The R2 was equal to 0.8695, showing a good fit of experimental data to this equation.

Figure 5 indicates the profile of curve representing synergistic effect of studied factors on peaks of polygalacturonase activity response. These figures show the influence of initial moisture content (U), concentration of spores in inoculation medium (E), ammonium sulfate concentration (N) and fermentation temperature (Tf) on polygalacturonase activity response (PPG).

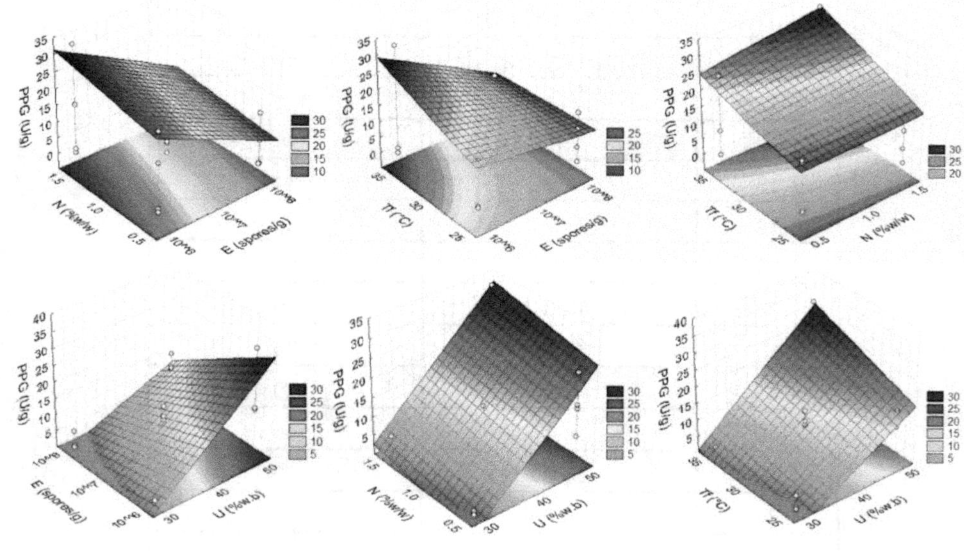

Figure 5. Response surface for polygalacturonase activity.

The highest peak of polygalacturonase activity calculated by the model (30 U/g) was obtained at higher levels for initial moisture content, for the lower concentration of inoculum and higher for fermentation temperature, observing that ammonium sulfate concentration did not influence significant on the response.

According to Pareto graphic (Figure 6), variable that most influenced fermentation process was initial moisture content of medium (U), and its interaction with variable concentration of inoculums (E), confirming claims reported in literature about the amount of water is thus a limiting factor [11,12].

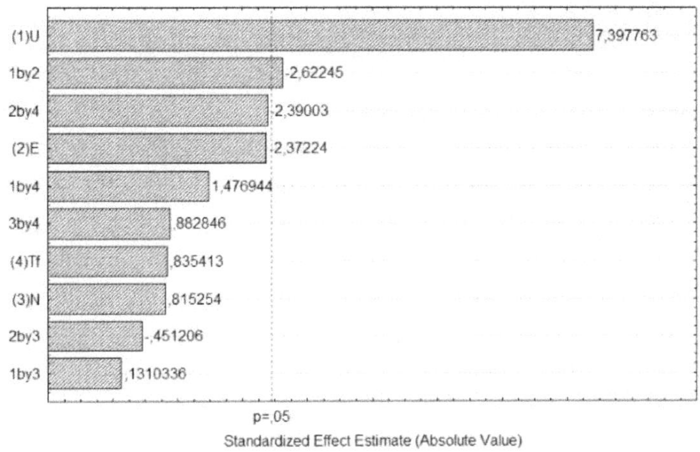

Figure 6. Pareto graphic

Authors [41] examined the effect of temperature on a solid state fermentation process using Aspergillus niger 163 and apple pomace as a substrate in a bioreactor rotary drum with 15L of solid medium. Temperature was studied within the range 22-60°C, observing their influence on polygalacturonase activity. Inoculated spore concentration was equal to 5x108 spores/ml. Temperature of 35°C was found to be more susceptible to production of polygalacturonases enzymes.

Similarly to what was described in this paper, authors [42] studied the influence of ammonium sulfate concentration (from 0.25 to 0.45%), pH (4.82 to 6.12) and fermentation time (50-90h) on endopectinase enzyme production in a solid state fermentation process, using as substrate apple pomace and Aspergillus niger PC5. It was observed that ammonium sulfate concentration have positive effect on enzymatic activity. However, the effect was insignificant compared to fermentation time.

3.4 Scaling-up of solid state fermentation process using tray reactor

It was evaluated the scale-up process using a tray reactor, setting the mass of humid medium in 500 g. Spores concentration, moisture content and ammonium sulphate concentration were determined based on experiments performed on laboratory scale. Thus the conditions of fermentation process was 50%(w.b) of initial moisture content, 106 spores/g of inoculum concentration, 1.5%(w/w) of ammonium sulphate concentration, 40 mm of substrate thickness [15] and 23°C of fermentation temperature [16].

In Figure 7, there are the behavior of polygalacturonase activity (PG) and process productivity (Prod) as function of fermentation time.

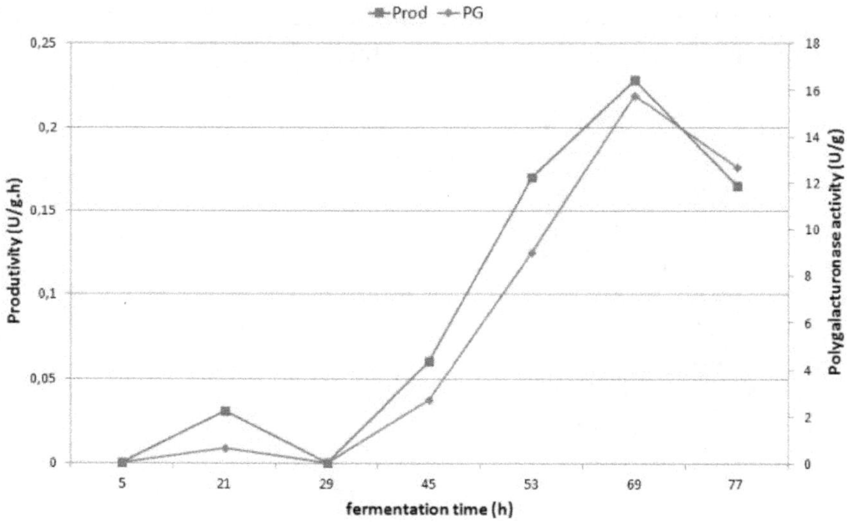

Figure 7. Polygalacturonase activity and productivity versus fermentation time.

Under conditions described previously, it was observed a peak of polygalacturonase activity of 15.76 U/g at 69 hours of fermentation, corresponding to the highest productivity.

Comparing the maximum activity obtained with reactor tray, and maximum activity obtained at laboratory scale, it is clear that production was 45 times lower. In this regard, the first fact to note is the internal temperature of medium. When using tray, there is an increase of at least 10°C above of initial value. Thus, there may be denatured enzyme.

Other factors can change during fermentation. One is the medium moisture, because with increase of temperature, there is an increased in evaporation rate of water, reducing the amount available to microorganism, which complicates the production process.

It can be observed that the time for adaptation of microorganism is greatly increased. Remembering, the greatest polygalacturonase activity in laboratorial scale was obtained at 29 h of fermentation. In the fermentation with tray, the highest activity obtained was reached at 69h of fermentation.

When the subject is solid state fermentation process, a large number of studies were conducted in a laboratory scale. It is relatively easy to control certain parameters for enzyme production. On this scale, many productions show great promise. However, beyond this process to pilot-scale bioreactors, which contain greater amounts of substrate, difficulties in carrying out the fermentation are revealed. Thus, it is characterized one of difficulties in scaling-up solid state fermentation processes [28].

3.5 Enzymatic stability as temperature and pH

In summary, the enzyme is maintained by a delicate balance of noncovalent forces such as hydrogen bonds, ion pairing, hydrophobic interactions and van der Waals force [45]. Thus, variations in temperature and pH are important in the analysis of enzyme activity.

Temperature has the activation and deactivation effect on enzyme activity. Continued increases of temperature beyond the maximum or optimum for enzyme activity leads to protein denaturation, which involves the deployment of large segments of polypeptide chain [45]. At the other extreme, enzymes inactivation by cooling can occur when nonpolar forces are involved and association of polypeptides. Low temperature reduces the strength of such interactions can promote the dissociation of subunits and compromise enzyme activity [7].

With respect to pH, all of ionisable groups of proteins undergo transitions pH dependent on the basis of intrinsic pKa values of amino acid residues. Many of these transitions will cause impacts on the stability of the enzyme and, in a narrow range of pH, can act together to destabilize it completely.

Knowing the pH and temperature stability is important for selection of enzymes compatible with the prevailing conditions for application potential so that the enzyme persist long enough to fulfil the expected function [7].

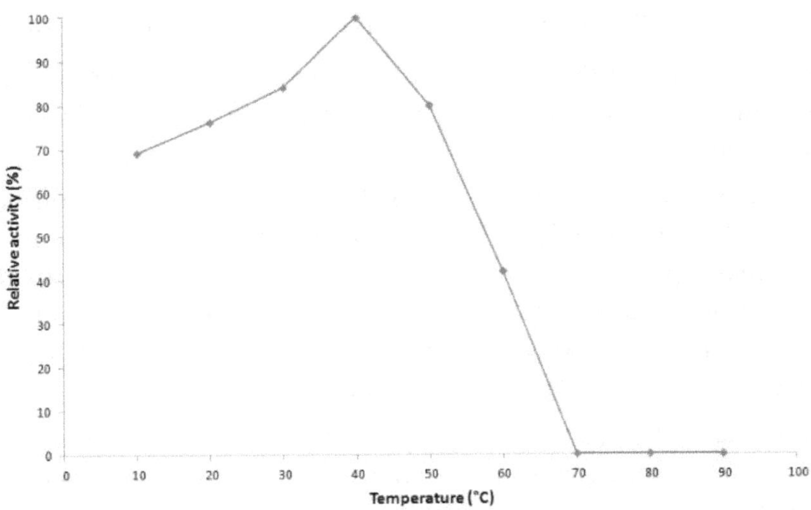

Figure 8. Thermostability of polygalacturonase enzyme

It can be observed (Figure 8) that temperatures of between 30 and 50°C, polygalacturonase activity remained with up to 80% of its maximum activity, and its optimum temperature at 40°C. After 50°C, relative enzymatic activity falls abruptly, reaching zero at 70°C.

Considering the greater use of pectic enzymes is in fruit juices industry, the processing of juices is normally done at 30-50°C. Thus, produced enzymes are been active during the process. For inactivation, the binomial time versus temperature must be considered, as well as chemical characteristics of juices. Usually, enzymatic inactivation is made between 70-90°C [7].

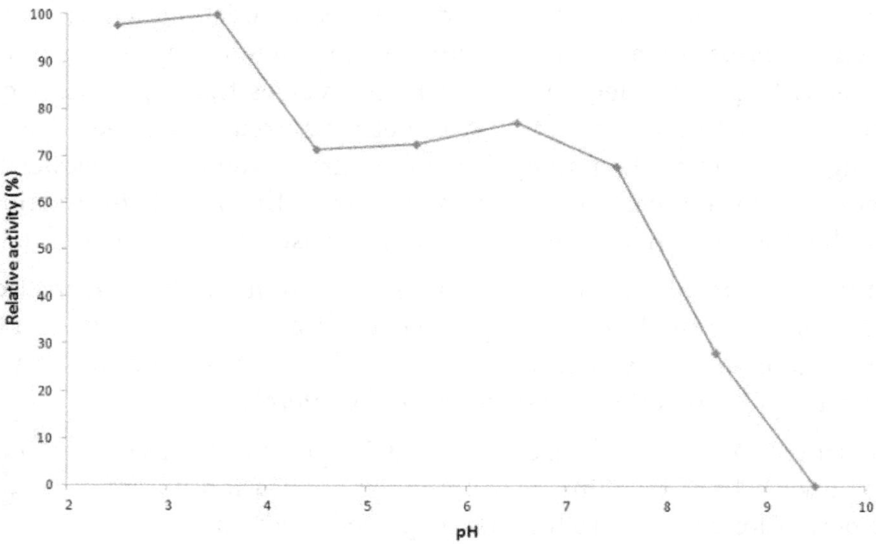

Figure 9. Stability of polygalacturonase enzyme in relation to pH

It is observed that the pH's of 2.5 and 3.5 (Figure 9), polygalacturonase activity was highest, almost 100%, characterizing the enzyme as acidic. Relative activity is equal to zero only at pH 9.5.

In literature some authors have reported the viability of various types of waste from processed fruits (apple, cranberry and strawberry), as substrates for polygalacturonases production, with Lentinus edodes as microorganism, using solid state fermentation process. These authors also observed the effects of temperature and pH on the enzymatic extract. Poligalactorunase produced has good thermal stability up to 50°C and high tolerance between pH 3.0 and 6.5 [46].

Other authors have studied the polygalacturonases production, using submerged fermentation with orange peel and passion fruit as substrate and Aspergillus niveus as microorganism. In terms of stability, polygalacturonase produced showed the highest activity at 40°C and pH between 3.0 and 4.5 [48].

In general, polygalacturonase enzyme activity has maximum pH ranges between 3.5 - 6.0 and temperature between 40 - 55°C [7].

For most industrial uses, fungal polygalacturonase is useful for high activity and optimal activity at low pH range, serving for most applications in the food industry

[16]. Thus, enzyme extract studied can be applied in fruit juice processes, such as Barbados cherry (pH 3.3), orange (pH 3.0), apple (pH 3.6), passion (pH 3.4), peach (pH 3.3) and grapes (pH 3.1) [47].

REFERENCES

1. Blandino A., Dravillas K., Cantero D., Pandiella S S., Webb C. Utilisation of whole wheat flour for the production of extracellular pectinases by some fungal strains. Process Biochemistry 2001; 37(5) 497–503.
2. Lara-Márquez A., Zaval-Pánamo M G., López-Romero E., Camacho H C. Biotechnological potential of pectinolytic complexes of fungi. Biotechnological Letters 2011; 33 859-868.
3. Gomes J., Zeni J., Cence K., Toniazzo G., Treichel H. Evaluation of production and characterization of polygalacturonase by Aspergillus niger ATCC 9642. Food and Bioproducts Processing 2011; 89(4) 281-287.
4. Alkorta I., Garbisu C., Llama M J., Serra J L. Industrial applications of pectic enzymes: a review. Process Biochemistry 1998; 33(1) 21-28.
5. Uenojo M., Pastore G M. Pectinases: aplicações industriais e perspectivas. Química Nova 2007; 30 388-394.
6. Abbasi H., Shafighzadeh H., Rahimi A. Continuos production of polygalacturonases (PGases) by Aspergillus awamori using wheat flour in surface culture fermentation. Iranian Journal of Biotechnology 2011; 9 50-55.
7. Damodaran S., Parkin K L., Fennema O R. Química de alimentos de Fennema. Porto Alegre: Artmed; 2010.
8. Antier P., Minjares A., Roussos S., Raimbault M., Vinergra-Gonzalez G. Pectinase-hyperproducing mutants of Aspergillus niger C28B25 for solid-state fermentation of coffee pulp. Enzyme Microbial Technology 1993; 15(3) 254-260.
9. Couto S R., Sanromán M A. Application of solid-state fermentation to food industry – A review. Journal of Food Engineering 2006; 76(3) 291-302.
10. Raimbault M., Deschamps A., Meyer F., Senez J C. Direct protein enrichment of starchy products by fungal solid fermentation. In: Proc. Giam-V, 1977, Marseilles, France.
11. Pinto, G A S., Brito E S., Silva F L H., Santos S F M., Macedo G R. Fermentação em estado sólido: uma alternativa para o aproveitamento e valorização de resíduos agroindustriais. Revista de Química Industrial 2006; 74 17-20.
12. Durand A. Bioreactor designs for solid state fermentation. Biochemical Engineering Journal 2003; 13(2-3) 113-125.

13. Mitchell D A., von Meien O F., Krieger N. Recent developments in modeling of solid- state fermentation: heat and mass transfer in bioreactors. Biochemical Engineering Journal 2003; 13(2-3) 137-147.

14. Mitchell D A., Krieger N., Stuart D M., Pandey A. New development in solid-state fermentation II. Rational approaches to the design, operation and scale-up of bioreactors. Process Biochemistry 2000; 35(10) 1211-1225.

15. Pandey A. Concise Enclyclopedia of Bioresource Technology. New York: Food Products Press; 2004.

16. Dalsenter F D H., Viccini G., Barga M C., Mitchell D A., Krieger N. A mathematical model describing the effect of temperature variations on the kinetics of microbial growth in solid state culture. Process Biochemistry 2005; 40(2) 801-807.

17. Santos S F M., Macedo G R., Silva F L H., Souza R L A., Pinto G A S. Aplicação da metodologia de superfície de resposta no estudo da produção e extração da poligalacturonase. Química Nova 2008; 31(8) 1973-1978.

18. Micjodjehoun-Mestres L., Souquet J., Fulcrand H., Bouchut C., Reynes M., Brillouet J. Monomeric phenols of cashew apple (Anacardium occidentale L.). Food Chemistry 2009; 112(4) 851-857.

19. Torres Neto A B., Silva M E., Silva W B., Swarnakar R., Silva F L H. Cinética e caracterização físico-química do fermentado do pseudofruto do caju (Anacardium occidentale L.). Química Nova 2006; 29 489-492.

20. Campos A R N., Santana R A C., Dantas J P., Oliveira L S C., Silva F L H. Enriquecimento proteico do bagaço do pedúnculo de caju por cultivo semi-sólido. Revista de Biologia e Ciências da Terra 2005; 5(2).

21. Alcântara S R., Almeida F A C., Silva F L H. Pectinases production by solid state fermentation with cashew apple bagasse: water activity and influence of nitrogen souce. Chemical Engineering Transactions 2010; 20 121-126.

22. Brasil Escola: Caju. http://www.brasilescola.com/frutas/caju.htm (acessed 12 July 2012).

23. Brasil. Instituto Adolfo Lutz (IAL). Métodos físico-químicos para análise de alimentos. Brasília: Ministério da Saúde; 2005.

24. Rangana S., Manual of analysis of fruit and vegetable products. New Delhi: Tata McGraw-Hill; 1979.

25. Le Poidevin N., Robinson L A. Métodos ou diagnosticos foliar utilizados nas plantações do grupo booken na Guiana Inglesa: amostra geral e técnica de analises. Fertilité 1964; 21 3-11.

26. Rockland L B. Satured salt solutions for static control of relative humidity between 5° and 40°C. Analytical Chemistry 1960; 32 1375-1376.

27. Brunauer S., Emmett P., Teller E. Adsorption of gases in multimolecular layers. Journal of American Chemical Society 1938; 60 309-319.

28. Santos M M., Rosa A S., Dal'Boit S., Mitchell D A., Krieger N. Thermal denaturation: is solid-state fermentation a really good technology for the production of enzymes? Bioresource Technology 2004; 93(3) 261-268.

29. Souza R L A., Oliveira L S C., Silva F L H., Amorim B C. Caracterização da poligalacturonase produzida por fermentação semi-sólida utilizando-se resíduo do maracujá como substrato. Revista Brasileira de Engenharia Agrícola e Ambiental 2010; 14(9) 987-992.

30. Fontana R C., Salvador S., Silveira M M. Efeito das concentrações de pectina e glicose sobre a formação de poligalacturonase por Aspergillus niger em meio sólido. In: SINAFERM2005, 2005, Recife, Brazil.

31. Taragano V M., Pilosof A M R. Application of Doehlert designs for water activity, pH, and fermentation time optimization for Aspergillus niger pectinolytic activities production in solid-state and submerged fermentation. Enzyme and Microbial Technology 1999; 25(3-5) 411-419.

32. Fawole O B., Odunfa S A. Some factors affecting production of pectic enzymes by Aspergillus niger. International Biodeterioration & Biodegradation 2003; 52(4) 223-227.

33. Brandão M C C., Maia G A., Lima D P., Parente E J S., Campello C C., Nassu R T., Feitosa T., Sousa P H M. Análises físico-químicas e sensoriais e pedúnculo de caju submetido à desidratação osmótico-solar. Revista de Ciências Agronômica 2003; 34 139-145.

34. Santin A P. Estudo da secagem da inativação de leveduras (Sacchatomyces cerevisiae). Masters dissertation. Universidade Federal de Santa Catarina; 1996.

35. Abud A K S., Silva G F., Narain N. Caracterização de resíduos de indústria de processaento de frutas visando à produção de pectinases por fermentação semi-sólida. In: SINAFERM2007, 2007, Curitiba, Brazil.

36. Matias M F., Oliveira E L., Gertrudes E., Magalhães M M A., Use of fibres obtained from the cashew (Anacardium ocidentale L.) and guava (Psidium guayava) fruits for enrichment of food products. Brazilian Archives of Biology and Technology 2005; 48 143-150.

37. Holanda J S., Oliveira J O., Ferreira A C. Enriquecimento proteico do pedúnculo de caju com emprego de leveduras, para alimentação animal. Pesquisa Agropecuária Brasileira 1998; 33 787-792.

38. Hennies P T. Produção de pectinases de Penicillium italicum através de fermentação em meio sólido. Masters dissertation. Universidade Estadual de Campinas; 1996.

39. Martins N., Silva D., Martins E C., Silva R., Gomes E. Seleção de fungos termofílicos produtores de pectinase em fermentação em estado sólido de resíduos agroindustriais. In: SINAFERM2005, 2005, Recife, Brazil.

40. Lomauro C J., Bakshi A S., Labuza T P. Ealuation of food moisture sorption isotherm equations. Part I: fruit, vegetable and meat products. Lebensmittel – Wisseenschaft & Technologie 1985; 18 111-117.

41. Ferreira C D., Pena R S. Comportamento higroscópio da farinha de pupunha (Bactris gasipaes). Ciência e Tecnologia de Alimentos 2003; 23 251-255.

42. Rodrigues M I., Iemma A F. Planejamento de Experimentos e Otimização de Processos: uma estratégia sequencial de planejamentos. Campinas: Casa do Pão Editora; 2005.

43. Berovic M., Ostroversnik H. Production of Aspergillus niger pectolytic enzymes by solid state bioprocessing of apple pomace. Journal of Biotechnology 1997; 53(1) 47-53.

44. Bari M R., Alizadeh M., Farbeh F. Optimizing endopectinases production from date pomace by Aspergillus niger PC5 response surface methodology. Food and Bioproducts Processing 2010; 88(1) 67-72.

45. Gomes E., Guez M A U., Martin N., Silva R. Enzimas termoestáveis: fonte, produção e aplicação industrial. Química Nova 2007; 30 136-145.

46. Zheng Z., Shetty K. Cranberry processing waste for solid state fungal inoculats production. Process Biochemistry 2000; 33(3) 323-329.

47. Venturini Filho W G. Bebidas Não alcoólicas: Ciência e Tecnologia. São Paulo: Blucher, 2010.

48. Maller A., Damásio A R L., Silva T M., Jorge J A., Terezin H F., Polizeli M L T M. Biotechnological potential of agro-industrial wastes as a carbon source to thermostable polygalacturonase production in Aspergillus niveus. Enzyme Research 2011; 2011 1-6.

19

Chapter

IMPACT OF MUCILAGE EXTRACTION ON THE FUNCTIONAL CHARACTERISTICS OF CHIA MEALS

1. INTRODUCTION: AN OVERVIEW

Chia (Salvia hispanica L.) is an annual herbaceous plant that belongs to the Lamiaceae family, which is native to southern Mexico and northern Guatemala. The Salvia hispanica fruit consists of four nutlets, similar to an indehiscent achene, which contain a single seed. These nutlets are commonly called "seeds" [1]. Chia seed, together with corn, beans, and amaranth were important crops for pre-Columbian civilizations in America, including the Mayan and Aztec populations [2, 3]. With time its use was abandoned, but by at the end of the last century there was a resurgence of interest in chia due to its nutritional value [4]. Chia is considered an alternative crop to diversify and stabilize the economy of Northwestern Argentina [5]. The plant produces numerous small white and dark seeds that mature in autumn [6]. These seeds contain about 30% oil, and they mainly consist of unsaturated fatty acids [4, 7]. Chia seeds are a natural source of omega-3 fatty acids, antioxidants, proteins, vitamins, minerals and dietary fiber [5, 7, 8].

Chia meal (residue of the seeds after oil extraction) is a good source of proteins (19-23%) [9], dietary fiber (33.9-39.9%) [10], and compounds with antioxidant activity [7]. It also exhibits some interesting functional properties for its use in the food industry [11]. Functional properties are generally associated with the presence of proteins [12] and also of dietary fiber [13-16].

Dietary fiber (DF) consists of a heterogeneous mixture of compounds that are classified according to their physical properties and effects of their intake into: soluble dietary fiber (SDF) and insoluble dietary fiber (IDF) [17], referring to the solubility of fibers in water. Plant secretions such as pectins and gums, components such as mucilage, and chelating agents such as phytates are sources of SDF; whereas cellulose, lignin, some fractions of hemicellulose, phenolic compounds such as tannins and lipid structures such as waxes, suberins and cutins constitute IDF [18].

The functional properties of food components can be defined as any physicochemical property that affects and/or modifies some of its characteristics and that contributes to the quality of the final product. Knowledge about of functional properties such as color, particle size, water holding, absorption and adsorption capacity, as well as those linked to the affinity for lipid components is very useful for the food industry, because during the processing some modifications can occur that must be taken into account according to the usage of the final product and its marketing conditions [19]. For example, water-holding capacity (WHC) is related to the freshness and softness effect present in bakery products, and the oil-holding capacity (OHC) is related to the un-fatty effect in fried food when it is low and to the juiciness and texture in meat products when it is high [18, 20, 21].

In addition to the characteristics mentioned above, it is important to consider the physiological effects of the DF intake. Given the capacity of SDF to form gels, it increases the viscosity of the bolus in the gastrointestinal tract, slowing the intestinal transit, making digestion and the absorption of nutrients more efficient, providing more of a feeling of satiety. Soluble fiber are fermentable fibers that can be microbiologically decomposed in the colon, producing gases such as carbon dioxide, hydrogen and methane, and short-chain fatty acids (acetic, propionic and butyric) which are absorbed and used as energy sources. Some of the most important beneficial effects of SDF is that it regulates blood sugar and lower cholesterol levels. On the other hand, IDF is responsible for adding bulk to the stool, speeding the passage of stool through the intestine by promoting peristalsis, alleviating constipation and other gastrointestinal disorders [22, 23]. Both types of fiber may also reduce the risk of obesity, hypertension, appendicitis, and other disorders [24]. The beneficial effects noted above show the important role that DF play in human intake, and that is why a daily intake of 25-30 g is recommended, with a good SDF/IDF balance (a minimum of 30% SDF and 70% IDF, optimum 50/50 ratio) in order to benefit from both fractions of fiber [25, 26].

Chia mucilage (SDF), a complex carbohydrate of high molecular weight, is an important component of the seed due to its physiological role. The mucilage is secreted when the seed comes into contact with water, generating high-viscosity solutions [27, 28]. Many studies have examined the functional properties of different types of gums (Linux usitatissimum, Opuntia Picus indica, Alyssum homolocarpum, Psyllium plantago) [29-32]. However, little information has been reported on the functionality of chia seed mucilage as a stabilizing or thickening agent of food products.

The aim of this study was to conduct a comparative assessment of the functional qualities of chia meals (Salvia hispanica L.) derived from seeds with and without mucilage.

2. MATERIALS AND METHODS

2.1 Seeds

Chia seeds were obtained from commercial sources in Salta, Argentina (25º S and 65.5º W). They were cleaned manually by removing the foreign matter such as stones, dirt and broken seeds. They were packed in hermetic plastic vessels and stored at 5ºC until further use.

2.2 Meal without mucilage (Msm)

Meal without mucilage refers to the residue obtained after the oil extraction process using a Soxhlet apparatus following the IUPAC Standard Method [33] (n-hexane under reflux, 8 h, 90 °C,) of seeds that had previously had the mucilage extracted.

2.2.1 Mucilage extraction

The mucilage was extracted of chia seeds previously soaked in water (1:4) for 4 hours at room temperature. This mixture was distributed into plastic trays and covered with aluminum foil and frozen at -80°C, lyophilized, and the mucilage was removed by a sieving process (20 sieve mesh ASTM, 840 μm) (3 sections of 15 min each).

2.3 Meal with mucilage (Ms)

The data corresponding to the meal with mucilage that was considered in the analysis of the results corresponds to that reported by Capitani et al. [11]. The meal was obtained after oil solvent extraction (n-hexane) in a Soxhlet apparatus (Buenos Aires, Argentina) by thermal cycles at 80ºC for 8 h, following the IUPAC Standard Method [33], of chia seeds previously ground in a laboratory grinder (Moulinex, horizontal blade grinder, Buenos Aires, Argentina).

Meals (Msm and Ms) were homogenized and stored in plastic vessels at 5ºC until further use.

2.4 Scanning Electron Microscopy (SEM)

The whole seeds and seeds after mucilage removal were adhered to a cover slip, coated with a thin gold film (600 Å) in a sputter coater (Pelco 91000) and observed in a scanning electron microscope (LEO model EVO 40) at 5 kV. Longitudinal sections were sliced with a razor blade, after being plunged into liquid nitrogen to ensure the maintenance of their internal structure, and analyzed by microscopy using the same procedure and magnification ranges between x136 and x5000.

2.5 Characterization of meals

2.5.1 Proximate composition

Moisture, crude fiber and ash content were determined according to AOCS recommended practices Ba 2a-38, Ba 6-84 and Ba 5a-49, respectively [34]. Oil and nitrogen content (N) were determined following IUPAC Standard Method [33] and AOAC Method [35], respectively. Protein content was calculated as nitrogen x 6.25. Carbohydrate content was estimated by calculating the nitrogen-free extract (NFE) by difference using Eq. (1).

$$NFE{:}100 - (oil + protein + crudefiber + ash) \quad (1)$$

2.5.2 Total, soluble and insoluble dietary fiber

Total dietary fiber (TDF), soluble dietary fiber (SDF) and insoluble dietary fiber (IDF) were determined according the enzymatic gravimetric method [36].

2.5.3 Neutral Detergent fiber (NDF), Acid Detergent fiber (ADF), lignin, cellulose and hemicellulose

The vegetable cell was separated into two parts (Van Soest method): cell content (highly digestible) and cell wall (partially digestible). The cell wall was analyzed and its components (cellulose, hemicellulose and lignin) were determined. The technique makes use of acidic and neutral detergent [35, 37].

2.5.4 Antioxidant activity

The extraction of phenolic compounds was carried out according to the method of Re et al. [38]. Ten mL ethanol were added to 1 g sample, then it was homogenized in Vortex for 2 min, decanted and filtered (0.45 μm nylon paper). The supernatant was transferred into a flask and evaporated using a rotavapor apparatus (BUCHI R124, Germany) to concentrate the sample. It was then redissolved in 1000 μL ethanol.

A spectrophotometric method was used to determine the antioxidant activity using a Hitachi U-1900 UVeVIS spectrophotometer (Japan). The antioxidant activity was quantified by a dying assay of the radical cation ABTS+ measuring ABTS+ reduction as the percentage of absorption inhibition at 734 nm, just 6 min later. The radical cation ABTS and potassium persulfate were obtained from Sigma Aldrich. Chlorogenic acid was used as standard antioxidant. Results were expressed as μmol/L Trolox g/sample, considering that chlorogenic acid diminishes twice the amount of absorption than Trolox [39].

2.5.5 Functional properties

2.5.5.1 Water-Holding (WHC) and Oil-Holding Capacity (OHC)

Water and oil holding capacities were determined according to the method of Chau et al. [40]. Briefly, 1 g (dry base (d.b.)) sample was weighed and then stirred into 10 mL distilled water or corn oil (density 0.92 g/mL, Arcor). These suspensions were centrifuged at 2200 x g for 30 min (Rolco Centrifuge Refrigerate, Model CR-5850, 22 cm radius, Buenos Aires, Argentina) and the supernatant volumes were measured. Water-holding capacity was expressed as gram water held per gram sample, and oil-holding capacity as gram oil held per gram sample.

2.5.5.2 Water Absorption Capacity (WAbC)

This property was determined according to the AACC method 88-04 [41]. Approximate water absorption capacity was first determined by weighing out 2 g (d.b.) sample, adding water until saturation (approx. 35 mL) and centrifuging at 2000 x g for 10 min in a Rolco Model CR-5850, 22-cm radius centrifuge (Buenos Aires, Argentina). Approximate water absorption capacity was calculated by dividing the increase in sample weight (g), by initial weight, quantifying the water needed to complete the original sample weight (2 g d.b.) to 15 g. Water absorption capacity (WAbC) was then determined by placing samples in four tubes, adding different quantities of water (1.5 and 0.5 mL water above original weight, and 1.5 and 0.5 mL water below; one in each tube), agitating vigorously, and centrifuging the samples at 2000 x g for 10 min in a Rolco Model CR 5850. The supernatant was discarded and the residue weighed. Average water absorbed was calculated, and WAbC was determined and expressed as gram water absorbed per gram sample.

2.5.5.3 Organic Molecule Absorption Capacity (OMAC)

This capacity was determined according to the method of Zambrano et al. [19]. A three gram (d b.) sample was placed in excess quantity corn oil (approx. 25 mL) for 24 h at room temperature, and then centrifuged at 2000 x g for 15 min in a Rolco Model CR-5850. OMAC was expressed as the absorbed hydrophobic component and calculated in terms of sample weight gain (g oil/sample g).

2.5.5.4 Emulsifying Activity (EA) and Emulsion Stability (ES)

These properties were evaluated according to the method [36] of Chau et al. [40]. Briefly, 100 mL 2 g/100 mL suspension was homogenized using an Ultra-Turrax T25 disperser (Janke & Kunkel, IKA-Labortechnik, Germany) at 7800 rpm for 2 min. Then, 100 mL corn oil (density 0.92 g/mL, Arcor) were added and homogenized at 15,000 rpm for 2 min. Emulsions were centrifuged in a 15 mL graduated centrifuge tube at 455 x g for 10 min, and then emulsion volume was measured. The EA was expressed as the remaining volume of the centrifuged emulsion corresponding to 100 mL of initial emulsion. The emulsion stability was determined by heating the emulsions to 80ºC for 30 min, cooling them to room temperature and then centrifuging the samples at 455 g for 10 min. ES was expressed as the remaining volume of the centrifuged emulsion corresponding to 100 mL of initial emulsion. On the other hand, all emulsions were evaluated by optical characterization using a Vertical Scan Analyzer (QuickSCAN, Beckman Coulter, Fullerton, USA). The QuickSCAN head scans the entire length of the sample (approximately 65 mm), collecting backscattering (BS) data every 40μm. Thus, it is possible to obtain curves showing the percentage of backscattering light flux, relative to external standards, as a function of the sample height in mm [42]. Coalescence kinetics were determined by measuring the mean values of BS as a function of time in the 25-30 mm zone (Backscattering % 25-30 mm).

2.6 Statistical analysis

The results obtained were analyzed using ANOVA and Tukey's test ($p \leq 0.05$), using Infostat software [43].

3. RESULTS AND DISCUSSION

The nutlet of Salvia hispanica consists of the seed and a pericarp surrounding the seed. The true seed, in turn, consists of a coat (testa), the endosperm and the embryo, consisting mainly of two cotyledons [1]. Basically, the pericarp of the chia seed is similar to that other Nepetoideaes because it shows cuticle, exocarp, mesocarp, layers of sclereids and endocarp. The cells of the mesocarp and exocarp are parenchimatic. Figure 1 shows a scanning electron microscopy of a Salvia hispanica nutlet. In the exocarp there often are cells that produce mucilage when the nutlets get wet (Figure 1D).

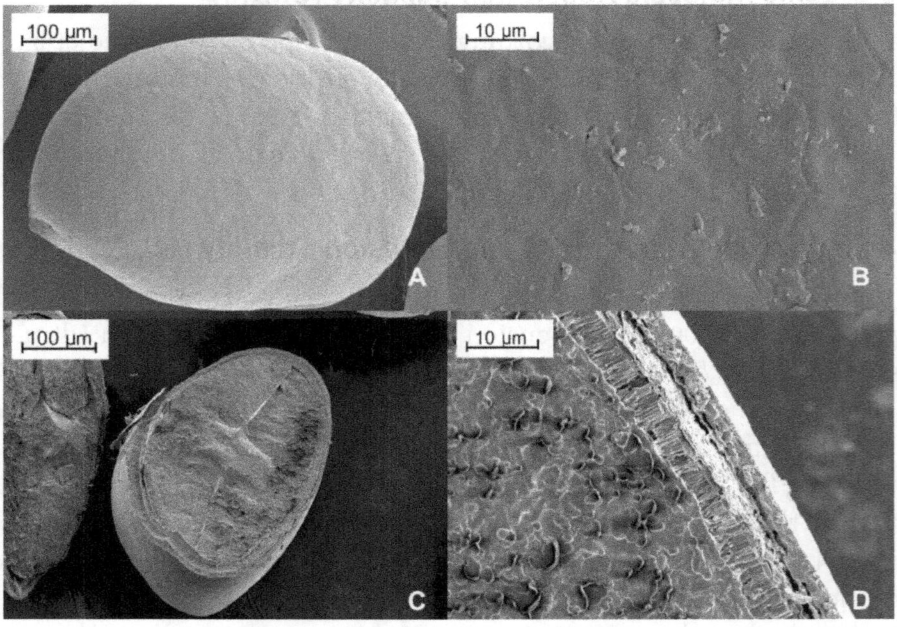

Figure 1. SEM microscopy of Salvia hispanica nutlets. (A) Whole nutlet, lateral view (x150), (B) pericarp surface (x5000), (C) broken nutlet, longitudinal section (x149), (D) broken nutlet (x1500).

Figure 2 shows SEM microscopy of Salvia hispanica L. seeds after mucilage extraction. In these images it can be observed that the mixocarpy phenomenon occurs in the outer layers (cuticle and exocarp). After removing the mucilage, the nutlet surface is characterized by small hill-like eminences, spaced, that cover the entire surface, corresponding to the mesocarp cells. Chia seeds presented a similar structure to that of two mucilaginous species (Carrichtera annua and Anastatica hierochuntica), which could be associated with the presence of concentric aggregates of glucuronic acid [44]. The retention of the mucilage close to the seed can be due to the association of the mucilage with the columella (a secondary wall cell produced after mucilage secretion) and portions of the cell wall [45].

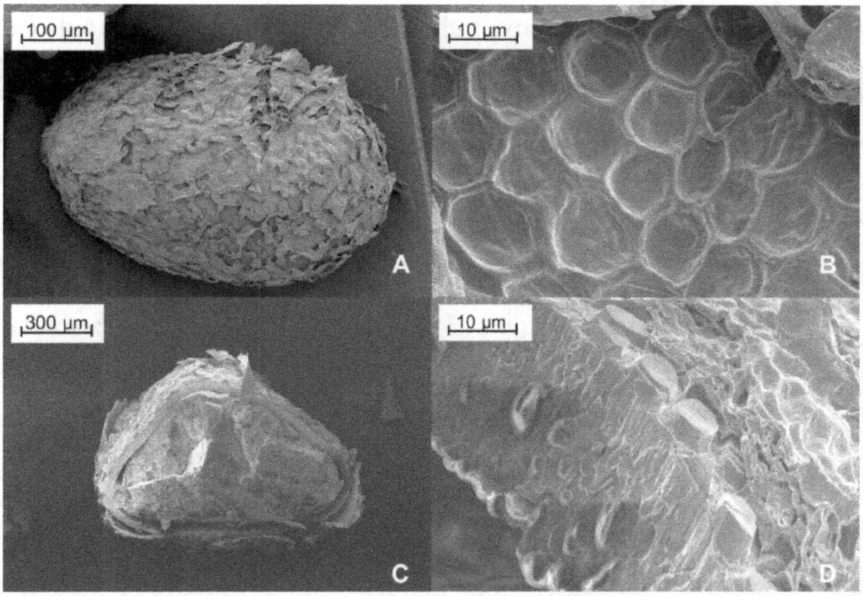

Figure 2. SEM microscopy of Salvia hispanica L. nutlets after mucilage extraction. (A) Whole nutlet (x136), (B) nutlet surface (x5000), (C) broken nutlet (x150), (D) broken nutlet (x3500).

The proximate composition of chia meals with and without mucilage is presented in Table 1. Both meals were characterized by a high protein content, higher than that reported for sunflower meals of different origin (20.6-23.1%) [46] and canola meals (36.1-40.0%) [47, 48], and within the range of the values reported for linseed meals (38.9-43.3%) [48, 49].

On the other hand, both types of meals presented a high crude fiber content, with values higher than those reported for sesame, soybean, linseed and canola meals, 5.8%, 3.5%, 5.27% and 11.54%, respectively [50, 48].

In Table 2 it is possible to observe that both types of meals presented a high content of TDF, consisting mainly of IDF. Even though the value of SDF was relatively low, this could be attributed to the fact that, during the determination of this fiber, some components were not quantified because they cannot precipitate during the treatment with ethanol, and thus SDF was underestimated [26]. It is worth noting that the meal obtained from seeds that previously had their mucilage extracted (Msm) exhibited a statistically higher content of IDF ($p<0.05$) than that for Ms, at the expense of a significant decrease in its SDF content. These results are consistent with the data obtained from the analysis of NDF, consisting of cellulose, hemicellulose and lignin (structural polysaccharides that contribute to the IDF fraction), which was statistically higher ($p<0.05$) in Msm (Table 3). Regarding Ms, it presented a better IDF/SDF balance, with a 89/11 ratio.

Table 1. Proximate composition of chia (Salvia hispanica L.) meals (% d.b.)

Component	Msm	Ms #
Moisture	10.66 ± 0.04 a	10.47 ± 0.16 a
Protein*	42.43 ± 0.71 a	41.36 ± 0.28 a
Crude fiber	27.75 ± 0.97 a	27.57 ± 0.07 a
Ash	7.82 ± 0.13 a	7.24 ± 0.15 a
Oil	0.22 ± 0.25 a	0.21 ±0.08 a
NFE	24.17 ± 0.76 a	23.62 ± 0.94 a
# Capitani et al. [11] Mean value (n = 3)		
Values followed by different letters differ significantly (p ≤ 0.05), according to Tukey's test. *Factor: 6.25; NFE: nitrogen-free extract.		

Table 2. Total (TDF), soluble (SDF) and insoluble (IDF) dietary fiber of chia (Salvia hispanica L.) meals (% d.b.)

Component	Msm	Ms #
TDF	47.13 ± 0.17 a	46.06 ± 0.86 a
IDF	45.62 ± 0.37 b	41.13 ± 0.47 a
SDF	1.51 ± 0.24 a	4.93 ± 0.65 b
# Capitani et al. [11] Mean value (n = 3)		
Values followed by different letters differ significantly (p ≤ 0.05), according to Tukey's test.		

Table 3. Fiber composition of chia meals analyzed according to the method of Van Soest (% d.b.)

Component	Msm	Ms #
NDF	63.6 ± 2.1 [b]	53.9 ± 0.3 [a]
ADF	30.8 ± 1.2 [a]	38.1 ± 1.2 [b]
Lignin	6.9 ± 0.5 [a]	4.5 ± 0.7 [a]

Cellulose	23.1 ± 0.9 ^{a}	34.6 ± 1.3 ^{b}
Hemicellulose	33.6 ± 1.0 ^{b}	14.8 ± 1.2 ^{a}
# Capitani et al. [11] Mean value (n = 3)		
Values followed by different letters differ significantly (p ≤ 0.05), according to Tukey's test.		

The antioxidant activity of the two chia meals compared with other types of meals is shown in Table 4. Both for Ms and Msm, the activity was high, without a significant difference between them (p>0.05). These values were higher than those found for wheat bran and sorghum and barley whole grain meals. But they were significantly lower than those found for chia meal obtained as a byproduct of cold-pressing oil extraction. The latter could be attributed to the fact that the meal obtained by pressing shows a higher percentage of residual oil (11.39% d.b.), which contains tocopherols, a class of compound with natural antioxidant activity [11].

Table 4. Antioxidant activity of chia (Salvia hispanica L.) meals compared with other meals

Sample	Trolox equivalent antioxidant coefficient (TEAC, μmol/g)
Msm	187.4 ± 33.21 a
Ms1	226.6 ± 4.13 a
Chia meal from oil pressing extraction1	557.2 ± 28.18 b
Wheat bran 2	48.5
Sorghum meal 3	51.7
Barley meal 3	14.9
1 Capitani et al., [11] 2 Iqbal et al. [57] 3 Ragaee et at. [58] Mean value (n = 3)	
Values followed by different letters differ significantly (p ≤ 0.05), according to Tukey's test.	

Regarding the functional properties, the meal with mucilage (Ms) exhibited a statistically higher absorption and water holding capacity (p<0.05) than that of the meal without mucilage (Msm) (Table 5). This behavior can be associated with the presence of mucilage in Ms, which acts as soluble dietary fiber, capable of holding water inside its matrix [51]. The WAbC of both meals was higher than that observed for canola and soybean meals (3.90 g/g and 3.28 g/g, respectively) and similar to that of linseed meal (6.03 g/g) [48].

Both types of chia meals presented a low absorption of organic molecules and oil-holding capacity, being significantly higher in Msm. These differences could be explained in terms of the particle size and the cellulose content of the meal [52, 53]. The determination of OHC is important because it is related to the capacity that food components have to hold oil, affecting their flavor and mouthfeel [54]. The OMAC is associated with the interaction of the fiber with fats, bile acids, cholesterol, drugs, and toxic and carcinogenic compounds at the intestinal level. Due to their low OHC levels, both types of chia meals could be considered an important ingredient in the manufacture of fried products due to their low fatty mouthfeel contribution. It is noteworthy that the OHC values were higher than those reported for the Jessenia polycarpa fruit meal [55] and similar to those of the fibrous residue of Canavalia ensiformis and barley [53, 56] and those reported by Khattab and Arntfield [48] for linseed and canola meals (2.01 g/g and 2.09 g/g, respectively).

Table 5. Functional properties of chia (Salvia hispanica L.) meals

Property	Msm	Ms #
WHC (g/g)	5.25 ± 0.39 a	10.64 ± 0.60 b
WAbC (g/g)	4.79 ± 0.49 a	6.45 ± 0.41 b
OHC (g/g)	2.94 ± 0.14 b	2.03 ± 0.08 a
OMAC (g/g)	2.22 ± 0.01 b	1.64 ± 0.02 a
# Capitani et al. [11] Mean value (n = 3)		
Values followed by different letters differ significantly ($p \leq 0.05$), according to Tukey's test.		

In Figure 3 it can be noted that the differences observed in the emulsifying activity were not significant, although the stability of the emulsion prepared with Ms was statistically higher than that with Msm ($p<0.05$). This effect can be associated with the capacity of mucilage to act as a thickening agent due to its ability to increase the viscosity of the aqueous phase in an O/W emulsion, thus hindering movement of the oil droplets of the dispersed phase [59]. This property is similar in the linseed mucilage, which has a strong thickening capacity, favorably affecting the water-holding capacity and the emulsifying properties of defatted linseed flour [60].

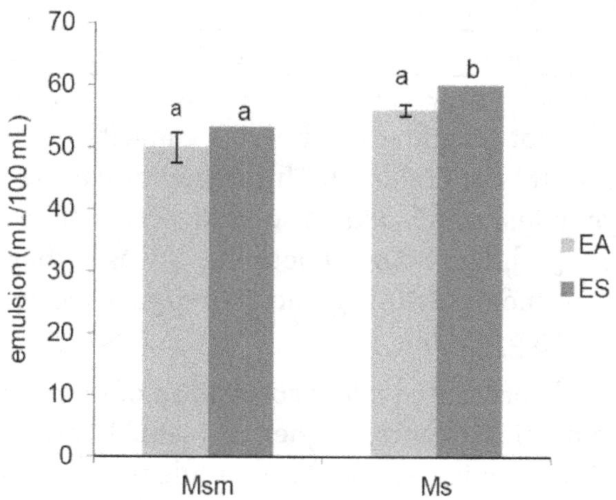

Figure 3. Activity and stability of O/W emulsions (50:50 p/p) with chia meal with and without mucilage. Values followed by different letters differ significantly (Tukey's test, p ≤0,05)

The behavior of the meals studied with respect to emulsion stability, examined by their optical characterization with a vertical scan analyzer (QuickScan), is shown in Figure 4. Both meals presented a high initial emulsifying capacity (66.3 and 62.2 %BS. for Ms and Msm, respectively), which remained approximately constant for all the time span studied (60 min) for Ms. However, the emulsion stability of the meal without mucilage decreased markedly by the end of the 60 min (50.4 %BS).

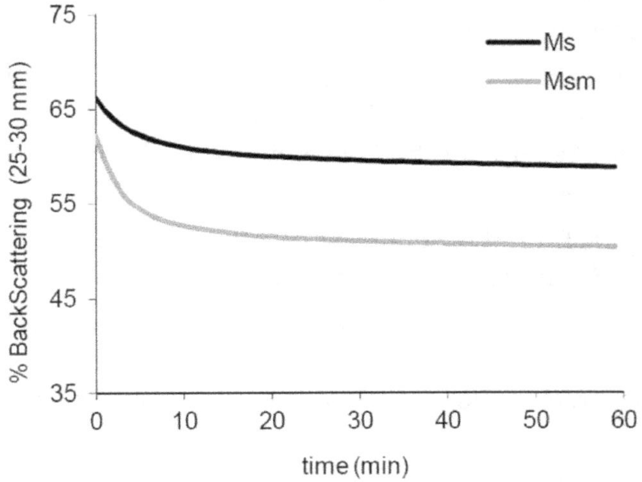

Figure 4. Destabilization kinetics of O/W emulsions (50:50 p/p) with chia meal

4. CONCISE SUMMARY

The findings indicate that both varieties of chia meals (with and without mucilage) have noteworthy functional features for the food business, potentially applicable in the production of bakery items, powdered beverages, yogurts, ice creams, sauces, and creams. They propose the prospective utilization of two chia byproducts: the residual meal derived from the oil extraction of whole chia seeds and a byproduct resulting from mucilage extraction. Concerning the creation of stable emulsions, the meal containing mucilage is advised due to mucilage's function as a thickening agent. From a physiological perspective, mucilage emerges as a potentially valuable food ingredient due to its health benefits, as it can form high-viscosity gels that slow intestinal transit, enhance satiety, and aid in the prevention of conditions such as obesity, colon cancer, hypercholesterolemia, and diabetes.

REFERENCES

1. Ixtaina VY, Nolasco SM, Tomás MC. Characterization of chia (Salvia hispanica L.) white and dark seeds and oils. Ed/ MC. Tomás/, Advances in Fats and Oils Research, Transwordld Research Network Kerala, India, pp. 135-147. 2010.

2. Ayerza R, Coates W. Ground chia seed and chia oil effects on plasma lipids and fatty acids in the rat. Nutrition Reasearch 2005; 25 995-1003.

3. Álvarez-Chávez LM, Valdivia-López MA, Alberto-Juárez ML, Tecante A. Chemical characterization of the lipid fraction of mexican chia seed (Salvia hispanica L.). International Journal of Food Properties 2008; 11 687-697.

4. Bushway A, Belya. Chia seed as a source of oil, polysaccharide and protein. Journal of Food Science 1981; 46 1349-1356.

5. Coates W, Ayerza R. Production potencial of chia in northwestern Argentina. Industrial Crops and Products 1996; 5 229-233.

6. Sahagun B. 1950-1982, Florentine codex: general history of the things of New Spain. In: Monographs of the School of America Research. Anderson AJO and Dibble CE, eds. Salt Lake City: University of Utah Press (Originally written 1575-1577 or 1578-1580).

7. Taga MS, Miller EE, Pratt DE. Chia seeds as a source of natural lipid antioxidants. Journal of the American Oil Chemists Society 1984; 61 928-931.

8. Ixtaina VY, Martínez ML, Spotorno V, Mateo CM, Maestri DM, Diehl BW, Nolasco SM, Tomás MC. Characterization of chia seed oils obtained by pressing and solvent extraction. Journal of Food Composition and Analysis 2011; 24(2) 166-174.

9. Craig R, Sons M. Application for approval of whole chia (Salvia hispanica L.) seed and groubd whole chia as novel food ingredients. Advisory committee for novel food and proceses. Ireland, Company David Armstrong, 2004. pp. 1-29.

10. Reyes-Caudillo E, Tecante A, Valdivia-López MA. Dietary fiber content and antioxidant activity of phenolic compounds present in Mexican chia (Salvia hispanica L.) seeds. Journal of Food Chemistry 2008; 107 656-663.

11. Capitani MI, Spotorno V, Nolasco SM, Tomás MC. Physicochemical and functional characterization of by-products from chia (Salvia hispanica L.) seeds of Argentina. Lebensmittel Wissenschaft and Technology - Food Science Technology 2012; 45 94-102.

12. Badui S. Química de los Alimentos. Longman / Alhambra; 1999.

13. Chau C, Huang Y. Comparison of the chemical composition and physicochemical properties of different fibers prepared from the peel of Citrus sinensis L. cv. Liucheng. Journal of Agriculture and Food Chemistry 2003; 51 2615-2618.

14. Wong K, Cheung P. Dietary fibers from mushroom sclerotia: 1. Preparation and physicochemical and functional properties. Journal and Agriculture Food Chemistry 2005; 53 9395-9400.

15. Yoshimoto M, Yamakawa O, Tanoe H. Potencial chemopreventive properties and varietal difference of dietary fiber from sweetpotato (Ipomea batatas L.) root. Japan Agricultural Research Quarterly 2005; 39 37-43.

16. Raghavendra S, Ramachandra S, Rastogi N, Raghavarao K, Kumar S, Tharanathan R. Grinding characteristics and hydration properties of coconut residue: A source of dietary fiber. Journal of Food Enginering 2006; 72 281-286.

17. Gorinstein S, Zachwieja Z, Folta M, Barton H, Piotrowicz J, Sembré M, Weisz M, Trakhtenberg S, Martín-Belloso O. Comparative content of dietary fiber, total phenolics, and minerals in persimmons and apples. Journal of Agriculture Food Chemistry 2001; 49 952-957.

18. Sánchez GBS. Caracterización Fisicoquímica y Funcional de la Fibra dietética del fruto del Nispero (Eriobotrya japonica) y de la cáscara de mango Obo (Mangifera indica L.). Tesis de Grado, Universidad Tecnológica de la Mixteca, Oaxaca, México. 2005.

19. Zambrano-Zaragoza ML, Meléndez R, Gallardo Y. Propiedades funcionales y metodología para su evaluación en fibra dietética. Ed/ FM. Lajolo, F Saura-Calixto, E Witting and Wenzel de Menezes/, Fibra dietética en Iberoamérica: Tecnología y Salud. Obtención, caracterización, efecto fisiológico y aplicación en alimentos. Brasil: Livraría LTDA. pp. 195-209. 2001.

20. Peraza GM. Caracterización de los residues fibrosos de Canavalia ensiformis L. y Phaseolus lunatus L. y su incorporación a un product alimenticio. Tesis de Maestría. Facultad de Ingeniería Química, Universidad Autónoma de Yucatán, México. 2000.

21. Ator CJ. Caracterización química de la fibra dietetic de subproductos obtenidos de la leguminosa Vigna Unguiculata L. Walps. Tesis de Maestría, Facultad de Ingeniería Química, Universidad Autónoma de Yucatán, Mérida, Yucatán, México, 2000.

22. Larrauri JA, Goñi I, Martín-Carrón N, Rupérez P, Saura-Calixto F. Measurement of health-promoting properties in fruti dietary fibres: antioxidant capacity, fermentability and glucosa retardation index. Journal of Agriculture and Food Chemistry 1996; 71 515-519.

23. McDougall GJ, Morrison IM, Stewart D, Hillman RJ. Plant cell walls as dietary fibre: range, structure, processing and function. Journal of Science and Food Agriculture 1996; 70 133-150.

24. Rehinan Z, Rashid M, Shah WH. Insoluble dietary fibre components of food legumes as affected by soaking and cooking processes, Journal of Food Chemistry 2004; 85 245-249.

25. Martín-Belloso O, Gorinstein S, Grijelmo-Miguel N. Characterization of peach dietary fibre concentrate as a food ingredient. Food Chemistry 1999; 65 175-181.

26. Saura-Calixto F, García-Alonso A. Metodología para el análisis de fibra y carbohidratos. Ed/ FM. Lajolo, F Saura-Calixto, E Witting and Wenzel de Menezes/, Fibra dietética en Iberoamérica: Tecnología y Salud. Obtención, caracterización, efecto fisiológico y aplicación en alimentos. Brasil: Livraría LTDA. pp. 17-25. 2001.

27. Lin KY, Daniel JR. Structure of chia seed polysaccharide exudates. Carbohydrate Polymer 1994; 23 13-18.

28. Marin Flores FM, Acevedo MJ, Tamez RM, Nevero MJ, Garay AL WO/2008/0044908 Method for obtaining mucilage from Salvia hispánica L. Word Internacional Property Organization. 2008.

29. Mazza G, Biliaderis CG. Functional properties of flax seed mucilage. Journal of Food Science 1989; 54 1302-1305.

30. Sáenz C, Sepúlveda E, Matsuhiro B. Opuntia spp mucilage´s: a functional component with industrial perspectivas. Journal of Arid Environments 2004; 57 275-290.

31. Koocheki A, Mortazavi SA, Shahidi F, Razavi SMA, Taherian AR. Rheological properties of mucilage extracted from Alyssum homolocarpum seed as a new source of thickening agent. Journal of Food Engineering 2009; 91, 490-496.

32. Cheng Z, Blackford J, Wang Q, Yu L. Acid treatment to improve psyllium functionality. Journal of Functional Foods 2009; 1 44-49.

33. IUPAC (1992), International Union of Pure and Aplplied Chemistry – Standard Methods for the Analysis of Oils, Fats and Derivates. 7 edn. Paquot, C. and Hautffene, A., Blackwell Scientific Publications, Inc., Oxford.

34. AOCS (1998), Official and Recommended Practices of the American Oil Chemists' Society, 5th edn., edited by D. Firestone, AOCS Press, Champaign Illinois.

35. AOAC (1997), In W. Horwitz (Ed.), Official methods of analysis (17th ed.), Washington, D.C., USA, Association of Official Analytical Chemists.

36. Prosky L, Asp N, Schweizer T, Devries S, Furda I. Determination of insoluble, soluble and total dietary fiber in food and food products: interlaboratory study. Journal of the Association of Official Analytical Chemists 1988; 71 1017-1023.

37. Guiragossian VY, Van Scoyoc SW, Auxtell JD. Chemical and Biological Methods for Grain and Forage Sorghum, Departament of Agronomy Internacional Programs in Agricultura, Purdue University, West Lafayette, Indiana, USA. 1979.

38. Re R, Pellegrini N, Proteggente A, Pannala A, Yang M, Rice-Evans C. Antioxidant activity applying an improved ABTS radical cation decolorization assay. Free Radical Biological Medicine 1999; 26 (9/10) 1231-1237.

39. Walker RB, Everette JD Comparative reaction rates of various antioxidants with ABTS radical cation. Journal of Agriculture and Food Chemists 2009; 57 1156-1161.

40. Chau C, Cheung K, Wong Y. Functional properties of protein concentrate from three Chinese indigenous legume seeds, Journal of Agricultural and Food Chemistry 1997; 45 2500-2503.

41. AACC (1984), Official methods of análisis (12th ed.). St. Paul, MN, USA: American Association of Cereal Chemists.

42. Pan LG, Tomás MC, Añón MC. Effect of sunflower lecithins on the stability of water in oil (W/O) and oil in water (O/W) emulsions. Journal of Surfactant and Detergents 2002; 5(2) 135-143.

43. Infostat Group, Infostat version 2004, Facultad de Ciencias Agrarias, Universidad Nacional de Córdoba, Argentina 2004.

44. Gutterman Y, Shem-Tov S, Mucilaginous seed coat structure of Carrichtera annua and Anastatica hierochuntica from the Negev Desert highlands of Israel, and its adhesion to the soil crust. Journal of the Arid Environment 1997; 35 695-705.

45. Western T, Skinner D, Haughn G. Differentiation of mucilage secretory cells of the Arabidopsis seed coat. Plant Physiological 2000; 122 345-355.

46. Pérez EE, Carelli AA, Crapiste GH. Chemical characterization of oils and meals from wild sunflower (Helianthus petiolaris Nutt). Journal of the American Oil Chemists` Society 2004; 81 1-5.

47. Sun M, Xu L, Saldaña DA. Comparison of canola meals obtained with conventional methods and supercritical CO2 with and without ethanol. Journal of the American Oil Chemists` Society 2008; 85 667-675.

48. Khattab RY, Arntfield SD. Functional properties of raw and processed canola meal, LWT – Food Science and Technology 2009; 42 1119-1124.

49. Mueller K, Eisner P, Yoshie-Stark Y, Nakada R, Kirchhoff E. Functional properties and chemical composition of fractionated brown and yellow linseed meal (Linum usitatissimum L.). Journal of Food Engineering 2010; 98 453-460.

50. Egbekun MK, Ehieze MU. Proximate composition and functional properties of fullfat and defatted beniseed (Sesamum indicum L.) flour. Plant Foods for Human Nutrition 1997; 51 35-41.

51. Figuerola F, Hurtado ML, Estévez AM, Chiffelle I, Asenjo F. Fibre concentrates from apple pomace and citrus peel as potential fibre sources for food enrichment. Food Chemistry 2005; 91 395-401.

52. López G, Ros G, Rincón F, Periago M, Martínez M, Ortuño J. Relationship between physical and hydration properties of soluble and insoluble fiber of artichoke. Journal of the Agricultural and Food Chemistry 1996; 44 277-2778.

53. Betancur-Ancona D, Peraza-Mercado G, Moguer-Ordoñez Y, Fuertes-Blanco S. Physicochemical characterization of lima vean (Phaseolus lunatus) and jack bean (Canavalia ensiformis) fibrous residues, Food Chemistry 2004; 84 287-295.

54. Kinsella JE. Functional properties of protein foods, Critical Review of Food Science and Nutrition 1976; 1 219-229.

55. Belén D, Alemán R, Alvarez F, Álvarez M. Evaluación de algunas propiedades funcionales y reológicas de harinas de coroba (Jessenia polycarpa Karst). Revista de la Facultad de Agronomía (LUZ, Venezuela) 2004; 21 161-170.

56. Mongeau R, Brassard R. Insoluble dietary fibre from breakfast cereals and brands, bile salt binding and water holding capacity in relation to particle size. Cereal Chemistry 1982; 59(5) 413-415.

57. Iqbal S, Bhanger M, Anwar F. Antioxidant properties and components of some commercially available varieties of rice bran in Pakistan. Food Chemistry 2005; 93 265-272.

58. Ragaee S, Abdel-Aal M, Norman M. Antioxidant activity and nutrient composition of selected cereals for food use. Food Chemistry 2006; 98 32-38.

59. McClements DJ. Food emulsions: Principles, practice, and techniques. Boca Raton, FL: CRC Press. 1999.

60. Dev DK, Quensel E. Functional and microestructural characteristics of linseed (Linum usitatissimum L.) flour and a protein isolate, Lebensmittel Wissenschaft and Technology - Food Science Technology 1986; 19 331-337.

20 Chapter

THE REDESIGN OF PROCESSES' DEVELOPMENT IN FOOD PRODUCTION ORGANIZATIONS USING QUALITY ENGINEERING METHODS AND TOOLS

1. INTRODUCTION:AN OVERVIEW

Various trends and difficulties, including heightened competition, emerging technology and laws, quality demands, and evolving customer preferences, have compelled the food business to adapt. In reaction to these emerging problems, food companies are enhancing competitiveness by restructuring, rethinking current processes, and expanding efforts for market share through product differentiation and/or the creation of new food products [1]. To enhance performance, companies in the food industry must implement organizational restructuring and process redesign. Automation of production and other processes to reduce reliance on human resources and shift activities to self-service facilities for customers and partners; Enhancement of logistical infrastructure and systems; Energy conservation strategies utilizing innovative technology and materials, novel manufacturing techniques, and the application of best practices; Political and regulatory advancements (food safety and other regulations); Technological innovations (biotechnology, information and communication technology, radio-frequency identification, robots, sensors, e-business); Comprehend globalization, market dynamics, and consumer trends.

This chapter examines process development due to its influence on food quality, safety, and sustainability. [4, 5, 6]. A redesign of process development may be achieved by several methodologies, strategies, and technologies [7, 8, 9, 10]. This chapter employs Business Process Management (BPM) alongside quality engineering methodologies and instruments. Several critical inquiries will be examined regarding the reconfiguration of process development in food producing entities utilizing quality engineering instruments and methodologies.

The first question that will be raised is the quality of processes in the food industry. The analysis will start from a typical process map for companies in the food processing industry in Serbia. After that the production process, as one of the most important processes in food processing will be decomposed. For each sub process appropriate metrics will be defined as the road of evaluation for the quality of the sub processes and quality of the goal of the process itself. Redesign of process development will be analyzed by comparing different approaches and by consideration of process redesign as the process. Different quality engineering methods and tools in the food industry will be compared according to the frequency of their implementation in the Serbian food industry as well as a correlation between the application of different quality engineering methods and tools, and profit in the companies. As an extension of the general ranking idea presented on the ranking and definition of goals in the production process, a fuzzy approach for evaluation of the importance of entities in supply chains in the food industry is presented. The general idea is to present an application of a mathematical tool in a situation that is very common in the food industry where conditions have been constantly changing so the observed values could not be stochastically described and where there is not a sufficient amount of data for statistical analysis. In other words, the application of fuzzy sets on evaluation of the importance of entities in the supply chain will be presented. A strategic map as a strategic part of the BSC (Balanced Score Card) framework is presented as one of the quality engineering methods. The presented strategic map started from the Kaplan – Norton model but it was adjusted in order to meet the needs of food processing companies in Serbia. Relations between entities (from all four perspectives) are defined as the result of research among Serbian companies. In the final part of the chapter the process framework for food processing companies is presented. The questionnaire used for the research is presented as well as gathered data from 53 Serbian companies. The collected data served as input for the modeling and evaluations previously addressed.

The chapter's primary focus is to present an overview of the redesign of process development, commencing with analysis (decomposition of processes), followed by redesign, and the implementation of contemporary quality engineering tools and methods (including their frequency of usage and impact, as well as the integration of select tools). It also encompasses theoretical and mathematical frameworks for ranking quality objectives (theory of fuzzy sets) and culminates in the establishment of a process framework, while maintaining a robust connection to data collected from the food industry.

2. QUALITY OF PROCESSES IN THE FOOD INDUSTRY

The food industry contains a number of completely different processes which require a wide range of measuring instruments. On the one hand, the quality of processes has market goals in brand development, demand management and new product introduction, while embracing food security and quality requirements. On the other hand, quality of processes in the food industry depends on many factors such as customer demands, key performance indicators, the process map, technology level, management level etc. In order to provide quality analysis, and redesign and improvement of the process the first step is an analysis of the process map of a typical organization in the food industry. A typical process map for an organization in the food industry is presented in figure 1.

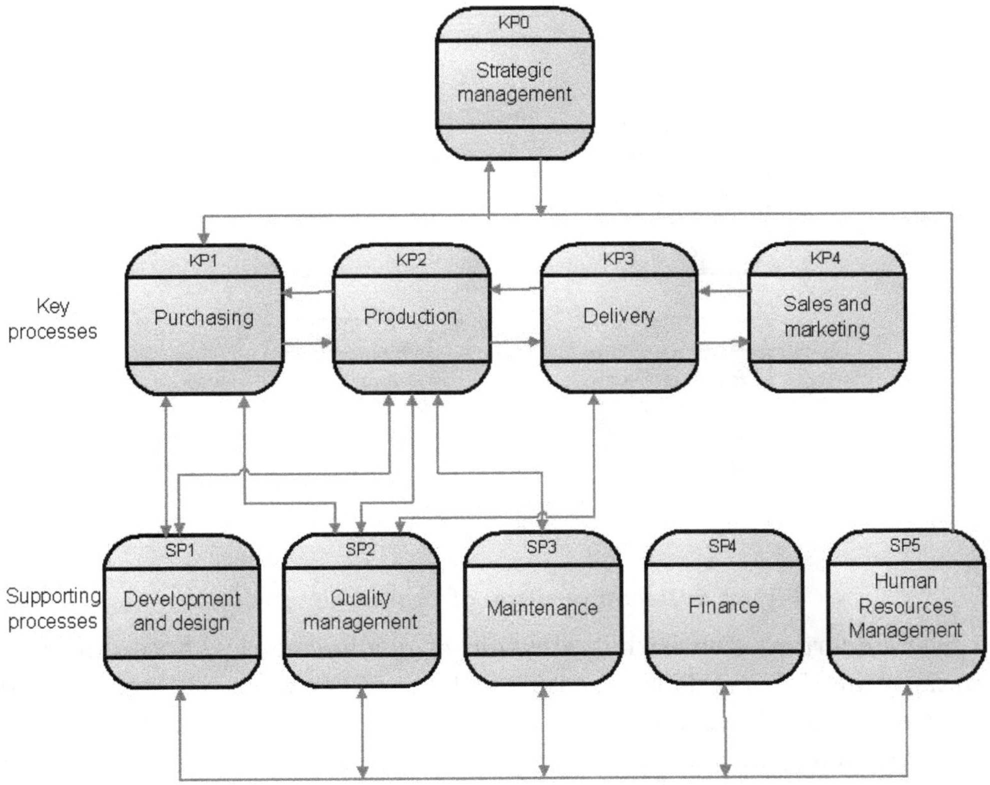

Figure 1. Typical process map for organizations in the food industry

Besides strategic management, each component process is functioning at a tactical and operative level and has quality metrics. In this chapter the major focus will be placed on the production process (KP2), all other decompositions could be performed using the same pattern, with appropriate quality metrics. In further analysis the production process could be decomposed into the sub processes presented in figure 2.

The production process (according to figure 2) consists of four sub processes: Logistic process, Process realization, Process planning and scheduling and Production process control.

In further analysis the logistic sub process (KP 2.1) could be decomposed into the following sub processes: Definitive logistic strategy; Plan inbound material flow; Operate outbound warehousing; Operate transportation and Manage reverse logistics.

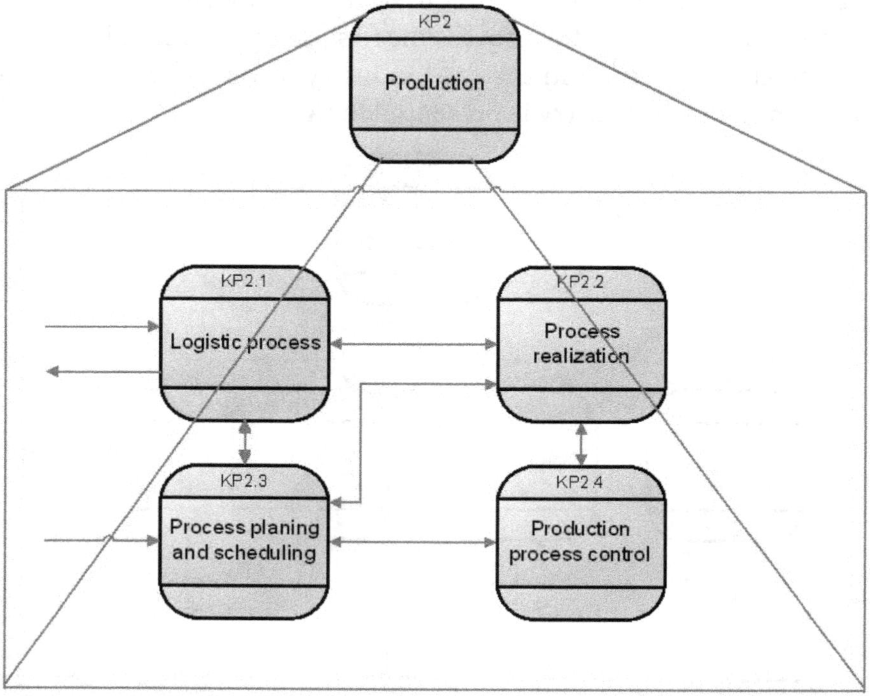

Figure 2. Decomposition of production process

The next important step is the definition of quality metrics. According to the definition a metric is a verifiable measure stated in either quantitative or qualitative terms. Quality metric data may be used to: spot trends in performance, compare alternatives and predict performance. Organizations need to collect information for a particular quality metric in order to evaluate and improve their processes. For further analysis of the logistic process the following quality metrics are presented in Table 1.

The second sub process of KP2, Production process realization (KP 2.2) is decomposed into sub processes: Preparing workers for obligatory measures; Preparing working places; Realization of working activities and Work reporting. The accompanied quality metrics are presented in table 2.

Production process planning and scheduling (KP 2.3) is decomposed into the following sub processes: Manage demand for products; Create material requirement plan (MRP) and Schedule production. The accompanied quality metrics are presented in table 3.

Table 1. Quality metrics of "Logistic process" KP 2.1

Logistic strategy realization %	Realization of inbound plan of material flow	Costs of warehousing / plan *100	Realization of plan of outbound transportation	Score
95-100	>100	<50	>100	10
85-95	90-100	50-60	90-100	9
75-85	80-90	60-70	80-90	8
65-75	70-80	70-80	70-80	7
55-65	60-70	80-90	60-70	6
45-55	50-60	90-100	50-60	5
35-45	40-50	100-110	40-50	4
25-35	30-40	110-120	30-40	3
<25	<30	>120	<30	2
0.25	0.25	0.25	0.25	weight

Table 2. Quality metrics of "Production process realization" KP 2.2

Level of preparing workers	Flexibility of working plans	Level of plan fulfillment *100%	Quantitative Waste %	Value of waste %	Score
10	10	>100	<0.5	<0.5	10
9	9	90-100	0.5-1.0	0.5-1.0	9
8	8	80-90	1.0-2.0	1.0-2.0	8
7	7	70-80	2.0-3.0	2.0-3.0	7
6	6	60-70	3.0-4.0	3.0-4.0	6
5	5	50-60	4.0-5.0	4.0-5.0	5
4	4	40-50	5.0-6.0	5.0-6.0	4
3	3	30-40	6.0-7.0	6.0-7.0	3
2	2	20-30	7.0-8.0	7.0-8.0	2
1	1	<20	>8.0	>8.0	1
0.2	0.1	0.2	0.25	0.25	weight

Table 3. Quality metrics of "Production process planning and scheduling" KP 2.3

Accuracy of demand %	Accuracy of MRP %	Work in progress/ production %	On time delivery (OTD) %	Score
85-95	85-95	5-10	90-95	9
75-85	75-85	10-15	85-90	8
65-75	65-75	15-20	80-85	7
55-65	55-65	20-25	75-80	6
45-55	45-55	25-30	70-75	5
35-45	35-45	30-35	65-70	4
25-35	25-35	35-40	60-65	3
15-25	15-25	40-45	55-60	2
<15	<15	>45	<55	1
0.15	0.25	0.3	0.3	weight

Table 4. Quality metrics of "Production process control" KP 2.4

Effectiveness of control of inputs %	Effectiveness of process control %	Effectiveness of control of outputs %	Level of control of measurement devices %	Score
>95	>95	>95	10	10
85-95	85-95	90-95	9	9
75-85	75-85	85-90	8	8
65-75	65-75	80-85	7	7
55-65	55-65	75-80	6	6
45-55	45-55	70-75	5	5
35-45	35-45	65-70	4	4
25-35	25-35	60-65	3	3
15-25	15-25	55-60	2	2
<15	<15	<55	1	1
0.25	0.25	0.3	0.2	weight

Production process control (KP 2.4) could be decomposed into the following sub processes: Control of inputs; Process control, Control of outputs, Non conformance product control and Control of measurement devices. The accompanied quality metrics are presented in table 4.

Quality systems focus on the quality of what the organization in the food industry produces, the factors which will cause the organization to achieve its goals, the factors which might prevent it satisfying customers and the factors which might prevent it from being productive, innovative and profitable. To control, assure and improve quality there is a need to focus on certain goals, in the case of goals of the production process (KP2) of a typical company in the food industry we can define goals as the constituent of the previous sub process goals (figure 3).

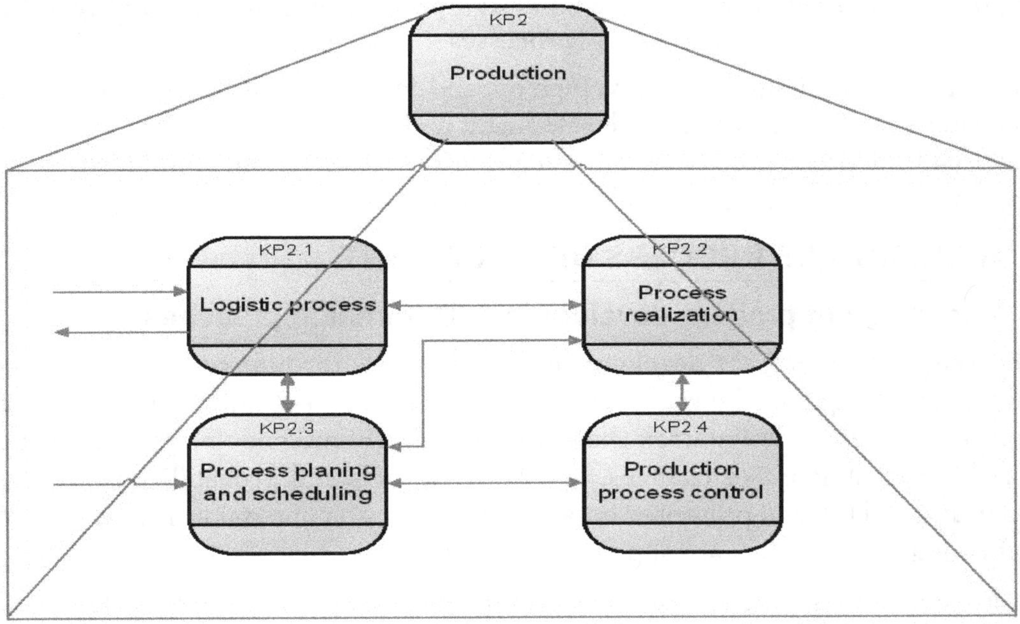

Figure 3. Quality goals of Production process KP2

By decomposition of the production process, analysis of sub processes and definition of metrics for each sub process it is possible to control, assure and improve the quality of a process in companies from the food industry sector. Definition of scores and weights in the metrics of each sub process is performed according to the authors' experience and available literature. In order to clearly demonstrate the idea all values are presented as deterministic ones. Of course some of them could be expressed by linguistic expressions rather than precious numbers but that issue will be elaborated in further text. The key processes in the food industry were analyzed and compared, and the results of research are presented in figure 4., depicting the existence of the gap between the quality of development and design process, and other processes. The analysis was performed on selected companies from Serbia.

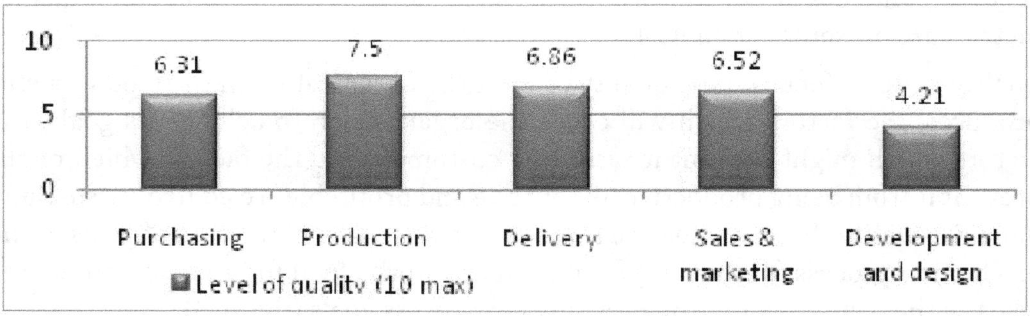

Figure 4. Quality of development and design process compared to quality of key processes

The gap between quality of the development and design process and other processes indicates that the quality of the development and design process is lower than the quality of the other processes so the focus in process redesign and improvement should be on the redesign of process development and design.

3. REDESIGN OF PROCESS DEVELOPMENT

3.1 Redesign of process development: Different approaches

The redesign of process development is connected to different methodologies of process change. In the redesign of process development and in the process of business process redesign itself there are a number of methodologies which cover the different amount of changes, results, used tools and probability of success. According to [11], and presented in table 5 a comparison of possible methodologies indicating whether they are applicable in the food industry.

According to table 5 it is clear that Continuous Process Improvement (CPI) as a never ending effort to discover, and eliminate the main causes of problems, is most likely to succeed in companies from the food industry.

Process improvement and BPR & lean are less likely to succeed but they cover a larger number of changes in the companies. Beside a redesign of process development can be viewed as a process. According to [11], the redesign of process development is divided into 5 phases (4 different entities) (Fig.5) with 10 steps.

Table 5. Comparison of change methodologies (adapted from [11])

No.	Change methodology	Amount of change	Score of change	Used tools	Probability of success	Applicable in food industry
1	BPR & lean	-reduction more than 50% of time, costs and quality	-cross func. teams or functional teams	• -process maps • -design principles • -benchmarking • -best practices • -lean tools	-less than 40%	yes
2	Process improvement	-reduction more than 20% of time, costs and quality	-cross teams or functional teams	• -process maps • -design principles • -LE • -six sigma • -lean tools	-more than 70%	yes
3	Continuous process improvement	-small reduction more than 20% of time, costs and quality	-one person or one sub process		-more than 90%	yes

There are five phases in this model:

- Analysis Phase — Identify areas of opportunity and target specific problems.
- Design Phase — Generate solutions and identify the required resources to implement the chosen solution with approval of senior management.
- Development Phase — Formulate a detailed procedure for implementing the approved solution with staff and customers.
- Implementation Phase — Execution of the solution and implementation of the redesign.
- Evaluation Phase — Build metrics, measurement tools, monitor implementation, and evaluate measurements for continuous improvement.

Process redesign as a process could be presented in the 10 steps, according to figure 5. The first five steps are common for most companies and the other steps could be defined according to the specific problem and in some cases using different quality methods and tools.

The first step covers:

- Meeting with senior management for the purpose of discussing barriers to process,
- Improvement, problem of eventual job losses and crafting of the kick-off-speech,
- Meeting with affected process managers and employees.

The second step is very important for success of the complete process. It is realized according to the team engineering approach [12, 13] with specified roles of team members: project manager, project principal, process improvement team, facilitator, and expert in ICT.

Senior management has the important task of directing team work in the first meetings and by the definition of statements and roles which will define the procedures for the team work.

The third step is performed by team work, starting from analysis of the existing state of processes, best practice and creating the as-is state of processes.

The fourth step covers the interviews with customers, according to: Customer request and needs; Ranking the criteria; Needed performance according to each criteria and Competitor position according to ranking and criteria.

The use of benchmarks and the development of new forms of benchmarking for best practice is well established in the food industry, although less so in other areas of the food chain. Benchmarking and best practice analysis is performed in step five through analysis of food industry competitions, using information from: industry trade associations (trade chambers etc.), industry studies, consultants' reviews, distributors, former employees, competitors themselves, published documents, indirect information sources from competitors, government sources, customers, supplies, and reverse – engineering products.

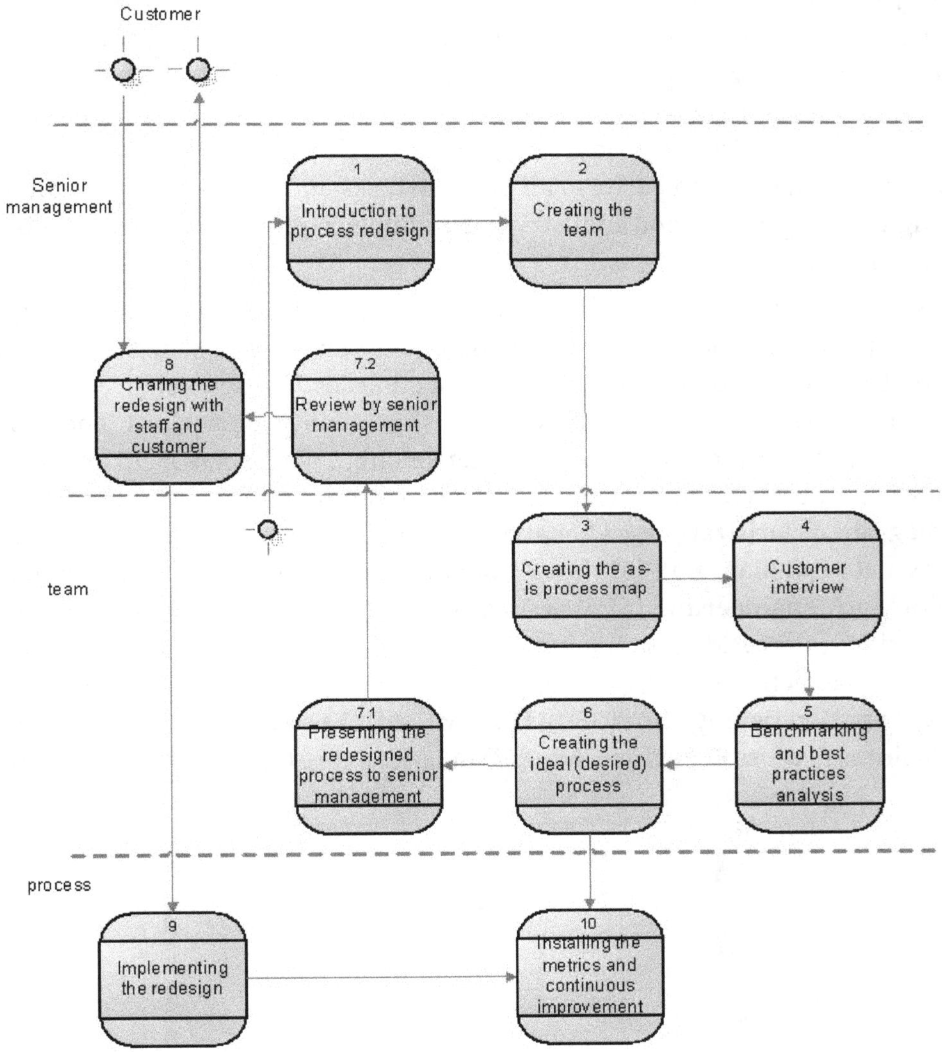

Figure 5. Process redesign as process

Other steps are performed according to appropriate redesign methodology adapted for each specific organization from the food industry. In addition, different quality engineering tools and methods could be employed in the following steps. It is important to emphasize that ICT could be used as a support during process redesign on the one hand, and on the other hand ICT and Business Process Management cover various aspects such as process control and supply chain management.

3.2 Quality engineering methods and tools in the food industry

Quality engineering methods and tools have an important role in the food industry because customers and markets demand proven high quality products and protection against low quality and unsafe products. On the other hand, the food industry needs to attain the quality that meets international standards for quality products. As it was mentioned the food industry needs to cope with the challenge of modern technological production methods, know them and assimilate them in quality assurance areas using new innovative hi-tech sensory and measurement instruments, supervise the production process, so that the designated quality level is always met. In order to perform all of these tasks and in order to meet all of the existing challenges the food industry has a number of engineering methods and tools available.

We analyzed companies from Serbia (listed in the following text) and compared results with world practice. The following quality engineering methods and tools used by the food industry have been analyzed according to their frequency and success rate: (1) Ichicawa's seven basic tools for quality (flow chart, check sheet, histograms, scatter plots, control chart, cause-effect diagram, Pareto analysis); (2) The seven new tools for improvement (affinity diagram, interrelationship diagram, tree diagram, prioritization grid, matrix diagram, process decision program chart, activity network diagram); (3) Cost of Quality (CoQ); (4) Project management; (5) Simultaneous engineering; (6) Statistical process control; (7) Reliability and risk engineering; (8) World class manufacturing (WCM); (9) Six-sigma; (10) Lean six-sigma; (11) Taguchi method; (12) Zero defect (ZD); (13) Design of experiments; (14) Quality Function Deployment (QFD); (15) FMEA; (16) FMECA; (17) Just in Time (JiT); (18) Business Process Reengineering; (19) Balance Score Cards (BSC); and (20) other.

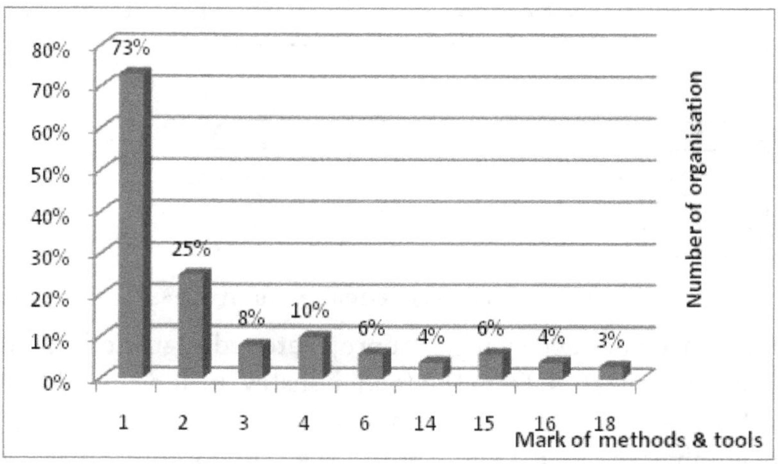

Figure 6. Distribution of quality methods' & tools' application

The results of the usage and frequency of the specific quality engineering methods and tools listed above, in the case of the Serbian food industry, is presented in Figure 6.

According to the results of the research, presented in figure 6, the dominant method or tool in food companies in Serbia position have Ichicawa's seven basic tools for quality (flow chart, check sheet, histograms, scatter plots, control chart, cause-effect diagram, Pareto analysis). It is also obvious that Serbian companies in general do not employ (at sufficient level) more advanced, new methods needed for higher quality process improvement.

Project management and FMEA are also popular in Serbian food processing companies. The second question is an analysis of profit compared with the quality tools and methods. According to research there is a high positive correlation between an increase of profit and application of quality tools and methods and it is presented in figure 7.

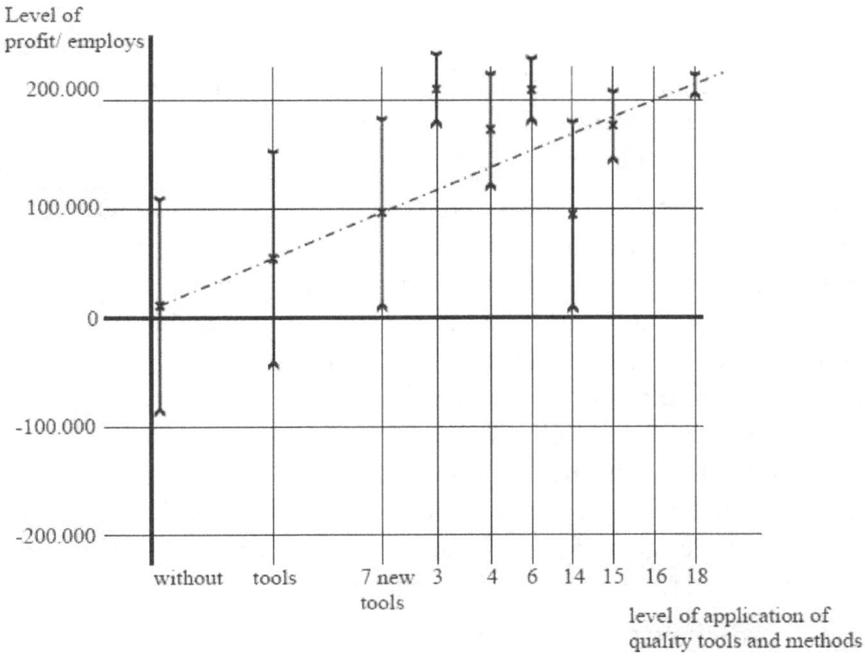

Figure 7. Impact of level of application of quality tools and methods on profit/ employee

According to figure 7 the level of profit (per employee) increases with implementation of different quality methods and tools. The largest increase in profit can be seen in the application of: Cost of Quality (CoQ), Statistical process control, Business Process Reengineering. The increase in the profit with the application of seven new tools is larger than with the application of the basic seven tools for quality. It is interesting that the basic seven tools for quality are the most commonly used but their contribution to profit increase is the lowest compared with other frequently used quality methods and tools. Another result is that the seven new tools do not contribute more significantly to the increase of profit due to the lower level of knowledge of employees connected with the new approaches.

Finally, although rather "old" QFD proved itself as a very useful and efficient tool. Generally, with an increase of the level of application of quality tools and methods profit/employees increases in the analyzed organizations.

3.3 Fuzzy approach for evaluation of the importance of entities in supply chains in the food industry

The process of logistics and the food supply chain is very important for all companies from the food industry. In food supply chains, definition of weight / importance of different entities (processes, sub processes and goals) in the presence of uncertainties is one of the important goals for the management team. A similar problem has been met and emphasized in the definition of quality metrics and evaluation of scores and weights in section 2. The solution to this problem must be placed on the whole of the supply chain because of its critical effect on efficiency.

It is realistic to assume, that decision makers express their evaluations more easy and more precisely by using linguistic expressions than numbers. The number and the type of linguistic expression used for a description of importance are defined by the management team and depend on the size of supply chain in food companies. Different mathematical approaches such as probability theory, fuzzy sets, rough set theory and others enable quantification of linguistic expressions. The development of fuzzy set theory enables the elimination of uncertainties and imprecision caused by lack of good evidence. In the fuzzy approach the uncertainties and imprecision caused are described by linguistic variables. They can be modeled by fuzzy sets with a different shape (triangle, trapezoid, but in some cases with Gaussian distribution, discrete fuzzy numbers) of membership functions.

The fuzzy approach has been used in cases: (1) where conditions have been constantly changing so the observed value could not be stochastically described (2) where there is no sufficient amount of data for statistical analysis. In other words, fuzzy sets theory could simulate human way of thinking in the process of decision making under imprecise, approximative and unclear data.

According to [14] the advantages of fuzzy sets theory can be presented in the following: it is conceptually clear, flexible, covers different non-linear functions of different complexity; tolerant on imprecise data; includes expert opinions and viewpoints; based on natural language; enables better communication between experts and managers.

Estimation of the relative importance of the processes and sub processes on the level of the supply chain $p \in P$ is defined by a pair-wise comparison matrix whose elements are defined as the relative importance of process/sub-process i/j compared to process/sub-process i ' / j '', i, i, ' = 1, .., I ; j, j ' = 1, .., Ji where I and, Ji are the total number of analyzed processes and sub processes i, i (1,..,I, respectively). A number of authors consider this approach better compared to direct estimation. Elements of the pair-wise comparison matrix are linguistic expressions which are modeled by triangular fuzzy numbers [14, 15]. The domains of these fuzzy numbers are defined on interval [1-5]. Value 1, or value 5 define that analyzed entity i/j compared to entity i ' / j '', i, i, ' = 1, .., I ; j, j ' = 1, .., Ji has the same or extremely higher importance retrospectively. These triangular fuzzy numbers are defined as: Very low importance - ~ ~ ~ R1 =(x; 1, 1, 2) Low importance - R2 =(x; 1, 1, 3), Medium importance – R3 =(x; 1, 3, 5), High ~ ~ importance -R4 =(x; 3, 5, 5) and Very high importance-R5 =(x; 4, 5 , 5).

The weights vector can be calculated by applying fuzzy extent analysis [16]. The weights vector of processes is denoted as W =(w1, .., wi, ..., wI)and sub processes Wi =(wi1, .., wij, ..., wi J i). The relative importance of processes wi and sub-processes ware ordinal numbers. In literature, there are many papers in which the weights vector is given by applying extent analysis [17].

The importance of the goals which are defined on the level of each sub process are defined by direct estimation made by a management team. According to literature, it can be concluded that this approach of estimation of importance of an entity is justified when the number of entities is less than five.

The importance of each goal k, k=1,,.,Kij on the level of sub process j, j=1,.., Ji could be described using five linguistic expressions which are modeled by triangular fuzzy numbers ~ vijk =(y; lijk , mijk , uijk). The total number of goals on the level of each sub process is denoted as Kij. The value of domains is defined on a standard measurement scale [18]. These triangular fuzzy numbers are defined in the following way: very low value-(y; 1, 1, 3), low value- (y; 1, 3, 5), medium value-(y; 3, 5, 7), high value-(y; 5, 7, 9) and very high value- (y; 8, 9, 9). Since the goals could be beneficial or costly it is necessary to perform ~ normalization of fuzzy values vijk, i = 1, .., I ; j = 1, .., Ji; k = 1, .., Kij. In this case the procedure of linear normalization is used [16]. Normalized values of goals weights are marked as ~ r ijk, i = 1, .., I ; j = 1, .., J i, k = 1, .., Kij. In further analysis only one goal with a critical effect on the management of sub-process j, j=1,.., Ji is considered.

For the management team carrying out the analysis, the following tasks are important: (1) to determine the rank of the process in a company (2) to determine the rank of a sub process on the process level in a company, (3) to determine the rank of sub processes with respect to the importance of goals and the importance of the considered sub process (4) to determine the rank of processes with respect to the importance of the goals, the relative importance of sub- processes of process i, i=1,..Im and the relative importance of process i, i=1,.,I, and (5) calculate the degree of belief that the sub process, or process which is on second place in the rank could be on first place; answers to these questions are given by comparing triangular fuzzy ~ ~ numbers cij, d i , i = 1, .., I ; j = 1, .., Ji, respectively.

The algorithm for analysis of the relative importance of processes, sub-processes and goals in a company p ∈ P is formally given as follows.

Step 1. Input fuzzy matrix W = [wii '], i, i ' = 1, .., I ; i ≠ i '; p = 1, ..., P

Step 2. Calculate weight vectorW =(w1, .., wi, ..., wI); rank the processes by placing on first place the process with highest wi.

Step 3. Calculate weight vectorW i =(wi1, .., wij, ..., wi J i); rank sub processes on the level of each process i, i=1,..,I by placing on the first place the sub process with the highest value wij.

Step 4. Transform all linguistic expressions which are modeled by triangular fuzzy numbers vijk into r ijk =(z; L ijk , Mijk , Uijk)by applying linear normalization procedure [16].

Step 5. Calculate the weighted normalized aggregated relative importance of sub-process j:

$$\tilde{c}_{ij} = \frac{1}{K_{ij}} \cdot \sum_{k=1}^{K_{ij}} w_{ij} \cdot \tilde{r}_{ijk} \; i=1, .., I; \; j=1, .., J^{i}, \; k=1, .., K_{ij},$$

Step 6. Calculate the weighted normalized aggregated relative importance of process i:

$$\tilde{c}_{i} = \frac{1}{J_{i}} \cdot \sum_{j=1}^{J_{i}} w_{i} \cdot \tilde{c}_{ij}, \; i=1, .., I; \; j=1, .., J^{i}$$

Step 7. Rank sub-processes and processes according to decreasing order, cij and d i, respectively and define level of belief that sub-process j, j=1,...,Ji, or process i, i=1,..,I could have the highest importance with respect to the importance of all goals and the importance of sub process j, j=1,...,Ji, or process i, i=1,..,I [19].

The developed procedure is illustrated with an example with real-life data from the authors' research.

The pair-wise comparison matrix of relative importance of processes is:

$$\begin{bmatrix} 1,1,1 & 1/\tilde{R}_2 & 1/\tilde{R}_2 & 1/\tilde{R}_3 & 1/\tilde{R}_2 & 1,1,1 & 1,1,1 \\ \tilde{R}_2 & 1,1,1 & 1,1,1 & 1/\tilde{R}_2 & 1,1,1 & \tilde{R}_2 & \tilde{R}_3 \\ \tilde{R}_2 & 1,1,1 & 1,1,1 & \tilde{R}_2 & 1,1,1 & \tilde{R}_2 & \tilde{R}_3 \\ \tilde{R}_3 & \tilde{R}_2 & \tilde{R}_2 & 1,1,1 & \tilde{R}_2 & \tilde{R}_3 & \tilde{R}_4 \\ \tilde{R}_2 & 1,1,1 & 1,1,1 & 1/\tilde{R}_2 & 1,1,1 & \tilde{R}_2 & \tilde{R}_3 \\ 1,1,1 & 1/\tilde{R}_2 & 1/\tilde{R}_2 & 1/\tilde{R}_3 & 1/\tilde{R}_2 & 1,1,1 & \tilde{R}_2 \\ 1/\tilde{R}_1 & 1/\tilde{R}_3 & 1/\tilde{R}_3 & 1/\tilde{R}_4 & 1/\tilde{R}_3 & 1/\tilde{R}_2 & 1,1,1 \end{bmatrix} \quad (3)$$

The weight vector of processes is:

W =(0.015 0.072 0.114 0.114 0.171 0.171 0.171 0.171)

The most important processes in considered food company are: Project execution (i=5), Process execution (i=6), Quality assurance (i=7) and Support processes (i=8).

The pairwise comparison matrix of relative importance of subprocesses under process i=1 is:

$$\begin{bmatrix} 1,1,1 & 1/\tilde{R}_1 & 1/\tilde{R}_3 & 1/\tilde{R}_3 & 1/\tilde{R}_4 & 1/\tilde{R}_4 & 1/\tilde{R}_4 & 1/\tilde{R}_4 \\ \tilde{R}_1 & 1,1,1 & 1/\tilde{R}_2 & 1/\tilde{R}_2 & 1/\tilde{R}_3 & 1/\tilde{R}_3 & 1/\tilde{R}_3 & 1/\tilde{R}_3 \\ \tilde{R}_3 & \tilde{R}_2 & 1,1,1 & 1,1,1 & 1/\tilde{R}_3 & 1/\tilde{R}_3 & 1/\tilde{R}_3 & 1/\tilde{R}_3 \\ \tilde{R}_3 & \tilde{R}_2 & 1,1,1 & 1,1,1 & 1/\tilde{R}_3 & 1/\tilde{R}_3 & 1/\tilde{R}_3 & 1/\tilde{R}_3 \\ \tilde{R}_4 & \tilde{R}_3 & \tilde{R}_3 & \tilde{R}_3 & 1,1,1 & 1,1,1 & 1,1,1 & 1,1,1 \\ \tilde{R}_4 & \tilde{R}_3 & \tilde{R}_3 & \tilde{R}_3 & 1,1,1 & 1,1,1 & 1,1,1 & 1,1,1 \\ \tilde{R}_4 & \tilde{R}_3 & \tilde{R}_3 & \tilde{R}_3 & 1,1,1 & 1,1,1 & 1,1,1 & 1,1,1 \\ \tilde{R}_4 & \tilde{R}_3 & \tilde{R}_3 & \tilde{R}_3 & 1,1,1 & 1,1,1 & 1,1,1 & 1,1,1 \end{bmatrix} \quad (2)$$

The weight vector of sub-processes on level process i=1 is:

W 1 =(0.295 0.497 0.208)

Sub process under process is "Strategic choice" (j=2).

The pairwise comparison matrix of relative importance of subprocesses under process i=2 is:

$$\begin{bmatrix} 1,1,1 & 1/\tilde{R}_2 & 1/\tilde{R}_2 & 1/\tilde{R}_3 & 1/\tilde{R}_2 & 1,1,1 & 1,1,1 \\ \tilde{R}_2 & 1,1,1 & 1,1,1 & 1/\tilde{R}_2 & 1,1,1 & \tilde{R}_2 & \tilde{R}_3 \\ \tilde{R}_2 & 1,1,1 & 1,1,1 & \tilde{R}_2 & 1,1,1 & \tilde{R}_2 & \tilde{R}_3 \\ \tilde{R}_3 & \tilde{R}_2 & \tilde{R}_2 & 1,1,1 & \tilde{R}_2 & \tilde{R}_3 & \tilde{R}_4 \\ \tilde{R}_2 & 1,1,1 & 1,1,1 & 1/\tilde{R}_2 & 1,1,1 & \tilde{R}_2 & \tilde{R}_3 \\ 1,1,1 & 1/\tilde{R}_2 & 1/\tilde{R}_2 & 1/\tilde{R}_3 & 1/\tilde{R}_2 & 1,1,1 & \tilde{R}_2 \\ 1/\tilde{R}_1 & 1/\tilde{R}_3 & 1/\tilde{R}_3 & 1/\tilde{R}_4 & 1/\tilde{R}_3 & 1/\tilde{R}_2 & 1,1,1 \end{bmatrix} \quad (3)$$

The weight vector of sub-processes on level process i=2 is:

W 2 =(0.095 0.165 0.178 0.229 0.165 0.107 0.061)

Sub process under process is "Level of leadership transformation" (j=4).

The pairwise comparison matrix of relative importance of subprocesses under process i=3 is:

$$\begin{bmatrix} 1,1,1 & \tilde{R}_2 & \tilde{R}_3 & \tilde{R}_3 \\ 1/\tilde{R}_2 & 1,1,1 & \tilde{R}_2 & \tilde{R}_2 \\ 1/\tilde{R}_3 & 1/\tilde{R}_2 & 1,1,1 & 1,1,1 \\ 1/\tilde{R}_3 & 1/\tilde{R}_2 & 1,1,1 & 1,1,1 \end{bmatrix} \quad (4)$$

The weight vector of sub-processes on level process i=3 is:

W 3 =(0.39 0.275 0.168 0.168)

Subprocess under process is "Roles and responsibilities" (j=1).

The pairwise comparison matrix of relative importance of subprocesses under process i=4 is:

$$\begin{bmatrix} 1,1,1 & 1/\tilde{R}_2 & 1,1,1 & 1/\tilde{R}_2 & 1,1,1 & 1/\tilde{R}_2 & 1/\tilde{R}_3 \\ \tilde{R}_2 & 1,1,1 & \tilde{R}_2 & 1,1,1 & \tilde{R}_2 & 1,1,1 & 1/\tilde{R}_2 \\ 1,1,1 & 1/\tilde{R}_2 & 1,1,1 & 1/\tilde{R}_2 & 1,1,1 & 1/\tilde{R}_2 & 1/\tilde{R}_3 \\ \tilde{R}_2 & 1/\tilde{R}_2 & \tilde{R}_2 & 1,1,1 & 1/\tilde{R}_2 & 1,1,1 & 1/\tilde{R}_2 \\ 1,1,1 & 1/\tilde{R}_2 & 1,1,1 & \tilde{R}_2 & 1,1,1 & 1/\tilde{R}_2 & 1/\tilde{R}_3 \\ \tilde{R}_2 & 1,1,1 & \tilde{R}_2 & 1,1,1 & \tilde{R}_2 & 1,1,1 & 1/\tilde{R}_2 \\ \tilde{R}_3 & \tilde{R}_2 & \tilde{R}_3 & \tilde{R}_2 & \tilde{R}_3 & \tilde{R}_2 & 1,1,1 \end{bmatrix} \quad (5)$$

The weight vector of sub-processes on level process i=4 is:

W 4 =(0.1 0.155 0.1 0.144 0.123 0.155 0.222)

Sub-process under process i=4 is "Process significance". (j=7).

Processes i=5 and i=6 could be decomposed on four sub processes each. The relative importance of sub processes under process i=5 and i=6 are equal. Such that w5 j = w6 j = 0.25, j = 1, 2, 3, 4.

The pairwise comparison matrix of relative importance of subprocesses under process i=7 is:

$$\begin{bmatrix} 1,1,1 & 1/\tilde{R}_3 & 1/\tilde{R}_4 & 1/\tilde{R}_4 & 1/\tilde{R}_3 \\ \tilde{R}_3 & 1,1,1 & 1/\tilde{R}_3 & 1/\tilde{R}_3 & 1,1,1 \\ \tilde{R}_4 & \tilde{R}_3 & 1,1,1 & 1,1,1 & \tilde{R}_2 \\ \tilde{R}_4 & \tilde{R}_3 & 1,1,1 & 1,1,1 & \tilde{R}_2 \\ \tilde{R}_3 & 1,1,1 & 1/\tilde{R}_2 & 1/\tilde{R}_2 & 1,1,1 \end{bmatrix} \quad (6)$$

The weight vector of sub-processes on level process i=7 is:

W 7 =(0.027 0.186 0.29 0.29 0.205)

Sub process under process i=7 is "Level of accomplishment of quality goals".(j=3).

Process i=8 could be decomposed on eight sub processes. According to the fuzzy rating of the management team, all sub processes have equal of the relative importance, so that w8 j = 0.143, j = 1, ..., 8.

Estimation of the importance of goals with critical effect management of the sub processes are given in Table 6. By applying the Algorithm (Step 5 to Step 7) rank of sub processes with respect to the relative importance of the defined goals and the relative importance of the sub processes and the rank of the processes with respect to the relative importance of the sub processes and the relative importance of processes is determined. The calculated ranks are presented in Table 7 and Table 8.

According to the calculated values of importance of the sub processes with respect to their relative importance and the relative importance of the goals of each sub process (see Table 2) the following analysis can be made: The sub process which has the highest importance: (1) Strategic alignment (i=1) is Strategic choice (j=2), (2) Process development (i=4) is Process significance (j=7) and (3) Quality assurance of goals (j=3) and documentation of the process (j=4). Based on the calculated degree of belief it is clear that the other sub processes, i=1, i=4 and i=7 have very low importance compared to the first ranked sub process.

Table 6. Importance of goals on the level of each sub process with the critical effect on management of those sub processes

ijk, i=1,..,I; j=1,...,Ji k=1,.., Kij	~ vijk	~ r ijk	ijk, i=1,..,I; j=1,...,Ji k=1,.., Kij	~ vijk	~ r ijk
111	low value	(0.11,0.11,0.33)	511	high value	(0.56,0.78,1)
122	medium value	(0.33,0.56,0.78)	522	high value	(0.56,0.78,1)
133	very low value	(0.11,0.11,0.33)	533	high value	(0.56,0.78,1)
222	medium value	(0.33,0.56,0.78)	544	high value	(0.56,0.78,1)
233	high value	(0.56,0.78,1)	611	high value	(0.56,0.78,1)
244	medium value	(0.33,0.56,0.78)	622	very high value	(0.89,1,1)
255	medium value	(0.33,0.56,0.78)	633	very high value	(0.89,1,1)
266	high value	(0.56,0.78,1)	644	very high value	(0.89,1,1)
277	medium value	(0.33,0.56,0.78)	711	high value	(0.56,0.78,1)
311	medium value	(0.33,0.56,0.78)	722	very high value	(0.89,1,1)
322	medium value	(0.33,0.56,0.78)	733	very high value	(0.89,1,1)
333	high value	(0.56,0.78,1)	744	very high value	(0.89,1,1)
344	very high value	(0.89,1,1)	755	very high value	(0.11,0.11,0.12)
411	high value	(0.56,0.78,1)	811	high value	(0.56,0.78,1)
422	high value	(0.56,0.78,1)	822	high value	(0.56,0.78,1)
433	very high value	(0.89,1,1)	833	high value	(0.56,0.78,1)
444	high value	(0.56,0.78,1)	844	high value	(0.56,0.78,1)
455	high value	(0.56,0.78,1)	855	high value	(0.56,0.78,1)
466	high value	(0.56,0.78,1)	866	medium value	(0.33,0.56,0.78)
477	very high value	(0.89,1,1)	877	very high value	(0.89,1,1)

Table 7. Rank of sub-process with respect to the importance and goals of the sub process

ij, i=1,..,I; j=1,,Ji	~ cij	Rank	Degree of belief	ij, i=1,..,I; j=1,..Ji	~ cij	Rank	Degree of belief
11	(0.032,0.097,0.165)	2	0.005	51	(0.14,0.195,0.25)	1	
12	(0.164,0.278,0.388)	1		52	(0.14,0.195,0.25)	1	
13	(0.023,0.023,0.068)	3		53	(0.14,0.195,0.25)	1	
21	(0.031,0.053,0.074)	5		54	(0.14,0.195,0.25)	1	
22	(0.092,0.129,0.165)	1		61	(0.14,0.195,0.25)	2	0.34
23	(0.059,0.1,0.139)	3		62	(0.222,0.25,0.25)	1	
24	(0.076,0.128,0.179)	2	0.99	63	(0.222,0.25,0.25)	1	
25	(0.092,0.129,0.165)	1		64	(0.222,0.25,0.25)	1	
26	(0.035,0.05,0.083)	4		71	(0.015,0.021,0.027)	4	
27	(0.02,0.034,0.048)	6		72	(0.165,0.185,0.185)	2	0.00
31	(0.129,0.218,0.304)	1		73	(0.258,0.29,0.29)	1	
32	(0.091,0.154,0.214)	3		74	(0.258,0.29,0.29)	1	
33	(0.094,0.131,0.168)	4		75	(0.023,0.023,0.026)	3	
34	(0.15,0.168,0.168)	2	0.44	81	(0.08,0.112,0.143)	2	0.34
41	(0.056,0.078,0.1)	6		82	(0.08,0.112,0.143)	2	0.34
42	(0.087,0.121,0.155)	2		83	(0.08,0.112,0.143)	2	0.34
43	(0.089,0.1,0.1)	4		84	(0.08,0.112,0.143)	2	0.34
44	(0.081,0.112,0.144)	3		85	(0.08,0.112,0.143)	2	0.34
45	(0.069,0.096,0.123)	5		86	(0.055,0.08,0.112)	3	
46	(0.087,0.121,0.155)	2	0.00	87	(0.127,0.143,0.143)	1	
47	(0.196,0.22,0.23)	1					

Under the following processes, the sub processes which have the most importance are: (1) Process governance (i=3)- Rules and responsibilities (j=1), (2) Process execution (i=6)- Planned and achieved goals (j=2), Resource utilization (j=3), and Human resources (j=4), and (3) Support processes (i=8)- ICT support (j=7). Based on the degree of beliefs, the management team can conclude that the sub processes which are placed on second place in the rank under treated processes have less importance compared to the sub processes which are placed on first place. However, in making operational decisions, the importance of the sub processes which are on second place should be considered. The most important sub-process under Process leadership (i=2) is Level of leadership transformation (j=5). Since the degree of belief for the sub process Level of trust and communication in an organization (j=4) is 0.99, it is clear that these two sub-processes have equal relative importance.

Process execution is the most important process in the analyzed food processing companies according to analysis of the importance of all sub processes, the importance of the defined goals of the sub process and the importance of the process. The degree of belief that the process which is denoted as Project execution has the highest importance of 0.8. According to the calculated result, the mentioned process has high importance for the specific food company, so the management team must have this in mind in making strategic decisions.

Table 8. Rank of process with respect to its importance and importance of goals of the sub-process of each process

i i=1,..,I	~di	Rank	Degree of belief that process has the highest importance
1	(0.001,0.0.002,0.003)	4	
2	(0.004,0.006,0.009)	7	
3	(0.013,0.019,0.024)	5	
4	(0.011,0.014,0.016)	6	
5	(0.024,0.033,0.043)	2	0.8
6	(0.1015,0.04,0.043)	1	
7	(0.025,0.028,0.028)	3	
8	(0.014,0.019,0.024)	5	

The presented model is used for the development of a very usable software solution that enables calculation of the importance of each goal, process and sub process [20, 21].

3.4. Strategy map

A strategy map describes how an organization can create sustained value for its shareholders, customers and communities.

The strategy map is developed based on the Kaplan and Norton model. A Strategy map describes how the organization creates value by connecting strategy objectives in an explicit cause and effect relationship in the four BSC objectives (financial, customer, processes, learning and growth). Strategy map is a strategic part of the BSC (Balanced Score Card) framework to describe strategies for value creation.

- Financial perspective is recognized in competitiveness.
- Customer perspective is identified: (1) quality, (2) safety, and (3) image.
- Internal perspective's eight processes: (P1) strategy alignment, (P2) process leadership, (P3) process governance, (P4) process development, (P5) project execution, (P6) process execution, (P7) quality assurance process, and (P8) supporting processes.
- Perspective or growth and learning: (1) technology and (2) people capability.

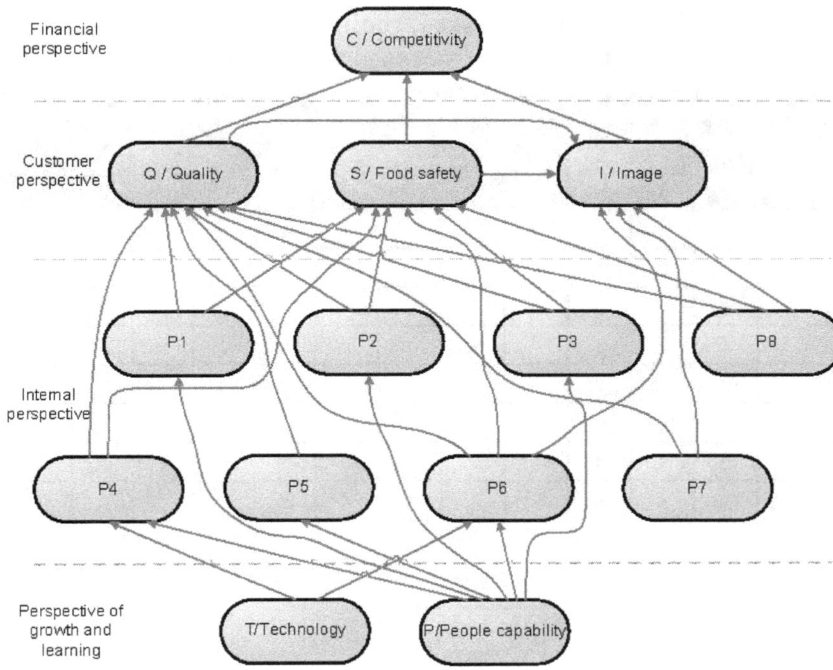

Figure 8. Strategy map

A strategy map for the food industry is presented in figure 8. The level of each component process is identified using a questioner for companies in the Serbian food industry. Relations among processes are defined using a method with 3 iterations.

A strategic map should describe a strategy and present the strategy to management and employees, and in the same way connect stakeholders, customer management, process management, quality management, core capabilities, innovations, human resources, ICT, organizational design / redesign and learning.

4. RESEARCH RESULTS: CASE STUDY OF THE SERBIAN FOOD INDUSTRY

4.1 Proposed model: Process framework for companies in the food industry

In this section we will provide the process framework for companies from the food industry sector. All processes are divided into the following categories: leadership processes, core processes and support processes. As it was shown in the section above "Project execution" process (P5) has the highest importance.

The support processes contain the following: CRM (Customer Relation Management), Supply management, Human resources management, ICT support, Maintenance, Marketing, Sales, Finances, Quality management and Other management systems.

Table 9. Cross reference of processes, their importance and importance of their sub processes and goals

Process Pi	Estimation of process importance	Estimation of importance of sub process in the frame of specific process							Estimation of importance of sub process goals						
		PP1	PP2	PP3	PP4	PP5	PP6	PP7	PP1	PP2	PP3	PP4	PP5	PP6	PP7
P1	5	7	8	6					5	6	4				
P2	6	6	7	7	8	7	6	5	5	6	7	6	6	7	6
P3	7	8	7	6	6				6	6	7	8			
P4	7	6	7	6	7	6	7	8	7	7	8	7	7	7	8
P5	8	7	7	7	7				7	7	7	7			
P6	8	7	6	7	6				7	8	8	8			
P7	8	7	8	9	9	8			7	8	9	9	8		
P8	8	5	6	6	6	5	5	6	7	7	7	7	7	6	8

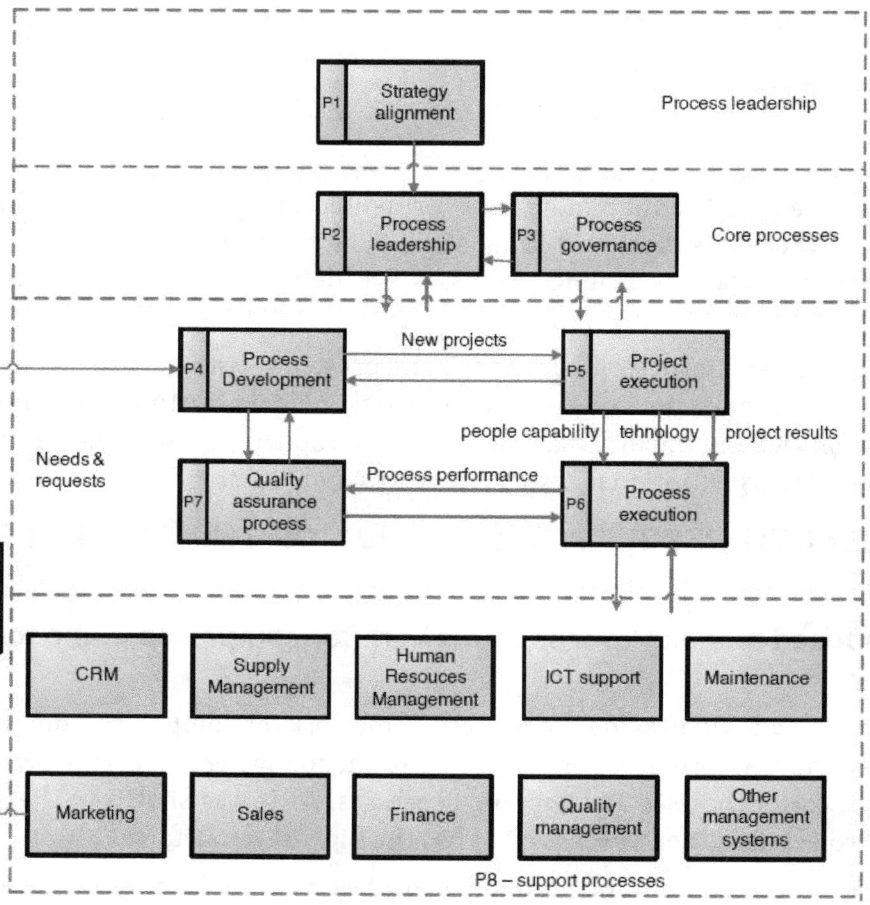

Figure 9. Process framework for companies in the food industry

Each process presented in figure 9 has its own importance as a whole. It is clear that each process could be decomposed on the accompanied sub processes.

Table 10. Questionnaire for Serbian companies

Questionnaire	
Questionnaire M	Questionnaire M
1.Questionnaire for strategy alignment	**6.Questionnaire for process execution**
1. Estimation of strategic alignment of processes; 2. Estimation of strategic choices; 3. Estimation of process architecture.	1. Level of achieving of process goals 2. Gap between planned and achieved 3. goals 4. Level of resource utilization (for process 5. needs): 6. Human; 7. Equipment; 8. ICT; 9. Knowledge.
2.Questionnaire for Process Leadership	
1. Level of transformational leadership; 2. Level of transactional leadership; 3. Level of trust in leadership; 4. Level of trust and communication in organization; 5. Level of business process awareness; 6. Level of process innovation; 7. Level of promotion of manager success.	**6.1Questionnaire for process performance**
3.Questionnaire for Process governance	1. Evaluation of increase of process awareness; 2. Evaluation of definition and establishment of rewards; 3. Evaluation of understanding or responsibilities for the process; 4. Evaluation of process metrics; 5. Evaluation of performance monitor; 6. Evaluation of management of the process; 7. Evaluation of continuous improvement; 8. Evaluation of communications
1. Estimation of roles selection and responsibilities; 2. Estimation of roles selection; 3. Estimation of evaluation and control of management including estimation of the risk of process; 4. Evaluation of implementation of contemporary methods and tools in business processes.	
4.Questionnaire for Process development	**7.Questionnaire for Quality assurance of processes**
1. Evaluation of concept of desired process; 2. Evaluation of relevance of process goals and indicators; 3. Evaluation of achievement of the process goals; 4. Evaluation of documentation of processes; 5. Evaluation of processes investigation;	1. Level of effectiveness of processes – On time Delivery of products 2. Level of quality of inputs in process 3. Level of quality of working procedures 4. Level of achieving the quality goals 5. Level of process coast/planned process costs

6. Evaluation of analysis of processes; 7. Evaluation of process significance; 8. Evaluation of process flexibility; 9. Evaluation of Process agility; 10. Level of inclusion and complexity of process demands .	
5.Questionnaire for Project execution	**8.Evaluation of support processes**
1. Evaluation of „right" projects; 2. Evaluation of establishment of project organization; 3. Evaluation of portfolio management and control; 4. Evaluation of project management frameworks. **5.1Evaluation – Technology** 1. Evaluation of technology level of production 2. Evaluation of level of quality approach (WCM, Lean, 6 sigma, IMS) 3. Evaluation of implementation of system and process approach (reengineering, time, electivity) 4. Evaluation of business decision making, business intelligence 5. Evaluation of human resources 6. Evaluation of internal capability. 7. Evaluation of Center of Business Innovation (CBI) 8. Evaluation of CBI engagement model.	1. Evaluation of marketing process; 2. Evaluation of sale;. 3. Evaluation of customer relations 4. management (CRM); 5. Evaluation of supply chain management (SCM); 6. Evaluation of finances; 7. Evaluation of human resources management; 8. Evaluation of ICT support; 9. Evaluation of implementation of quality management; 10. Evaluation of implementation of quality management; 11. Evaluation of other management systems **9.Evaluation of entities in customers perspective** 1. Evaluation of product quality 2. Evaluation of quality of organization 3. Evaluation of product safety 4. Evaluation of the brand, image of organization **10.Evaluation of entities in financial perspective** 1. Level of competitiveness compared to EU market

Total number of 53 companies were analyzed and results are presented in table 10.

Each sub process has importance in the frame of the specific process as well as the importance of its goal. The overall data gathered as the result of research in Serbian companies in the food industry is presented in table 9. The presented data could be combined with linguistically expressed opinion and used for ranking and simulation of quality goals according to the approach presented using fuzzy sets. Data were gathered using the questionnaire presented as table 10.

Table 11. Gathered data from Serbian companies

No.	Company	Stand.	1	2	3	4	5	5.1	5.2	6	6.1	7	8	9	10
	Company 01	22000	8	6	7	7	5	8	8	1	7	9	8	6	9
	Company 02	22000	5	8	5	4	7	6	4	1	6	7	7	6	5
	Company 03	22000	7	6	7	5	8	6	4	1	7	5	9	7	2
	Company 04	22000	7	6	7	4	5	6	5	1	6	6	7	5	6
	Company 05	22000	6	8	5	5	7	8	7	1	7	6	9	8	6
	Company 06	CAC	4	7	2	2	5	5	1	1	5	2	6	4	3
	Company 07	CAC	5	7	4	4	6	8	4	1	6	5	8	8	3
	Company 08	CAC	7	8	6	4	6	8	5	1	8	7	9	7	5
	Company 09	CAC	3	6	2	2	6	4	1	1	3	2	4	4	2
	Company 10	CAC	5	6	5	3	5	4	4	1	4	4	6	4	5
	Company 11	CAC	6	7	6	5	5	6	6	1	5	7	7	5	8
	Company 12	CAC	4	7	4	3	5	6	2	1	5	5	5	5	4
	Company 13	CAC	4	5	4	2	5	6	5	1	5	4	6	6	4
	Company 14	CAC	4	5	3	3	7	5	4	1	6	6	6	6	2
	Company 15	CAC	5	6	5	3	6	8	1	1	5	5	6	7	4
	Company 16	CAC	4	4	3	3	4	4	3	1	5	4	4	3	4
	Company 17	CAC	4	3	2	2	4	3	1	1	4	1	3	3	3
	Company 18		5	4	5	4	5	8	6	0	7	6	7	8	5
	Company 19		5	3	5	3	5	8	6	1	7	6	7	8	2
	Company 20		5	3	5	3	5	8	6	1	8	5	8	8	4
	Company 21		4	2	4	3	4	9	7	0	6	5	6	7	7
	Company 22		6	5	5	4	4	8	6	0	5	7	6	5	9
	Company 23		4	3	4	2	6	7	5	0	5	4	5	5	3
	Company 24		4	6	4	2	6	4	4	0	4	2	6	5	2
	Company 25		4	4	4	2	6	4	4	0	5	2	7	7	2
	Company 26		5	5	5	4	4	6	5	0	5	4	6	6	4
	Company 27		5	5	5	4	6	7	5	0	6	6	5	5	6
	Company 28		4	5	4	2	5	7	4	0	4	5	5	5	4
	Company 29		4	2	4	3	5	5	4	0	5	4	6	7	5
	Company 30		6	2	7	7	5	3	7	0	5	8	7	5	3
	Company 31		4	5	4	3	4	5	5	0	4	4	5	5	3
	Company 32		4	5	4	3	4	4	4	0	5	4	4	4	3

	Company 33		5	5	4	3	4	5	4	0	4	5	6	6	4
	Company 34		5	5	4	3	5	4	5	0	5	4	6	6	3
	Company 35		4	5	4	3	4	4	4	0	5	4	5	5	3
	Company 36		5	4	5	3	5	6	5	0	6	6	7	5	3
	Company 37		4	4	3	2	4	4	5	0	6	5	6	5	3
	Company 38		4	4	4	2	4	5	4	0	4	4	7	8	2
	Company 39		5	5	5	2	5	3	4	0	4	4	5	6	4
	Company 40		4	4	3	2	5	3	4	0	3	5	4	5	3
	Company 41		4	4	2	1	6	5	4	0	5	4	5	5	3
	Company 42		3	5	4	2	4	4	4	0	4	3	6	5	3
	Company 43		4	4	4	2	4	5	4	0	6	4	6	5	4
	Company 44		5	4	6	6	5	6	6	0	6	6	6	5	4
	Company 45		2	6	1	1	2	2	1	0	2	1	4	6	2
	Company 46		5	5	5	2	4	5	6	0	4	2	4	6	4
	Company 47		3	5	3	2	3	6	2	0	3	2	5	6	3
	Company 48		4	5	2	2	3	5	4	0	4	1	4	7	2
	Company 49		8	4	6	7	4	4	3	0	3	6	7	6	8
	Company 50		6	5	6	6	5	6	7	1	6	5	7	8	5
	Company 51		5	7	7	7	7	7	7	1	7	6	7	7	4
	Company 52		4	6	7	7	6	6	5	1	7	6	8	7	4
	Company 53		6	4	6	6	5	7	3	0	5	5	8	8	7

Legend: CAC – Codex Alimentarius – Food Hygiene, Recommended International Code of Practice General Principles of Food Hygiene CAC/RCP – 1969, Rev. 4-2003.

In table 11, the numbers in the columns correspond to the Questionnaire presented in table 10 (The names of companies are left and replaced by "company x" due to protection of companies data.). Structure of the sample is: 25 organizations with less than 10 employees, 21 organizations with 10-50 employees, 3 organizations with 50-250 employees, 3 organizations with 250-500 employees and 1 organization with more than 500 employees.

Based on the expert opinion of consultants working with organizations in the sample the relation among processes and other entities are determined. The form of relation is: RI, O where I - goal (performance) of entity and O – goal (performance source entity) of destination entity starting values of constants are determined by investigation of the organization in the sample. Presented relations describe the importance of relations presented in the strategy map.

RT, P4:	GP4= GPot + 0.11*ΔT	RP1, S:	GS= GS01 + 0.03*ΔP1
RT, P6:	GP6= GPop + 0.15*ΔP	RP2, S:	GS= GS02 + 0.04*ΔP2
RP1, P:	GP1= GP01 + 0.25*ΔP	RP3, S:	GS= GS03 + 0.08*ΔP3
RP2, P:	GP2= GP02 + 0.28*ΔP	RP4, S:	GS= GS04 + 0.18*ΔP4
RP3, P:	GP3= GP03 + 0.24*ΔP	RP6, S:	GS= GS06 + 0.28*ΔP6
RP4, P:	GP4= GP04 + 0.27*ΔP	RP7, S:	GS= GS07 + 0.15*ΔP7
RP5, P: RP6, P:	GP5= GP05 + 0.18*ΔP GP6= GP06 + 0.16*ΔP	RP8, S:	GS= GS08 + 0.20*ΔP8
RP1, Q:	GQ= GQ01 + 0.30*ΔP1	R S, I:	GI= GS0 + 0.38*ΔI
RP2, Q:	GQ= GQ02 + 0.32*ΔP2	RP6, I:	GI= GI06 + 0.32*ΔP6
RP3, Q:	GQ= GQ03 + 0.29*ΔP3	RP7, I:	GI= GI07 + 0.30*ΔP7
RP4, Q: RP5, Q: RP6, Q:	GQ= GQ04 + 0.35*ΔP4 GQ= GQ05 + 0.18*ΔP5 GQ= GQ06 + 0.17*ΔP6	RP8, I: RQ, C:	GI= GI08 + 0.35*ΔP8 GC= GQ0 + 0.39*ΔQ
RP7, Q:	GQ= GQ07 + 0.24*ΔP7	RS, C :	GC= GS0 + 0.28*ΔS
RP8, Q:	GQ= GQ08 + 0.17*ΔP8	RI, C :	GC= GI0 + 0.15*ΔI

5. CONCISE SUMMARY

The food business is a sector undergoing swift, multifaceted transformations. This is an expanding industrial sector that is gaining significance in relation to various trends and challenges: population expansion, heightened demand for healthy food, food safety standards, and diverse client requirements.

The food business must respond in various ways to mitigate obstacles, diminish risks, and leverage its strengths by utilizing available opportunities. Changes in the food industry will trend towards the automation of manufacturing processes and technical advancements, utilizing the full capabilities of biotechnology, information and communication technology, RFID, robotics, sensors, and e-business. Additional modifications will focus on optimizing logistical infrastructure and enhancing energy efficiency. However, organizations in the food industry must also comprehend globalization trends, market advancements, and fluctuations in consumer demands. These modifications will enhance the food industry's success in global marketplaces. These adjustments may yield only limited and partial outcomes, failing to realize the full potential of the proposed modifications unless corporations undertake organizational restructuring and process redesign. Modifying and embracing organizational adjustments, together with the restructuring and redesign of processes, are prerequisites for all subsequent alterations. Conversely, firms typically neglect organizational difficulties and process redesign in favor of other change initiatives, such as the adoption of new and emerging technologies.

This chapter examined many concerns and challenges related to the redesign of process development in food production companies through the application of quality engineering tools and methodologies. The initial step involves analyzing current processes, decomposing them, and implementing quality metrics to assess both process quality and goal quality. A standard process map for a food processing company in Serbia is illustrated, with the production process segmented into subprocesses. All indications for quality measurements were presented as numerical values to properly illustrate the notion.

Following decomposition, the subsequent step is the redesign of process development. Various change approaches are evaluated based on the extent of change, change score, utilized instruments, likelihood of success, and their relevance to the food business. The notion of continuous process improvement is introduced, and the process redesign has five phases with ten steps. A comparative analysis of several quality engineering methods and instruments utilized in the food sector was conducted, focusing on their frequency of use and correlation with profit enhancement in Serbian enterprises. The overarching result was that organizations continue to employ traditional, established processes and tools, notwithstanding their limited influence on profit enhancement relative to contemporary tools and procedures. A total of 20 distinct quality engineering methodologies

Tools were examined among 53 Serbian enterprises within the food industry sector. Given the multitude of challenges, ranging from prioritizing quality objectives to assessing the significance of processes or entities, as well as evaluating methods or tools, there is a necessity for a comprehensive approach to address these concerns. All these issues could be resolved with the application of the fuzzy technique. This text introduces a fuzzy approach for assessing the significance of entities inside supply chains in the food sector, extending the general ranking concept related to goal definition in the production process. A strategic map, as an integral component of the Balanced Scorecard (BSC) framework, serves as a model for food processing enterprises. This map's unique contributions are the relationships among entities, described from all four perspectives, as a result of research conducted in Serbian firms. The framework of processes for companies in the food industry was presented. The overall contribution of the discussed topics—decomposition, redesign, evaluation of quality tools and methodologies, fuzzy ranking, and the strategic map—derives from research conducted in 53 Serbian enterprises. The questionnaire for Serbian companies and the research findings are presented. A significant addition is that all decompositions, redesigns, modeling, simulations, and computations were conducted using actual data obtained from Serbian enterprises.

An advantageous expansion of this research could be a comparative analysis of data with that from the EU; nevertheless, literature suggests that the variation would likely be little.

REFERENCES

1. Trends and drivers of change in the food and beverage industry in Europe: Mapping report, www.eurofound.eu.int
2. Nikhil Daxini, Use BPM to Assist in New Product Development, TRIZ Journal, http:// www.triz-journal.com/content/c081201a.asp
3. Aikens, H., Quality Inspired Management: The Key to Sustainability, Prentice Hall, 2011.
4. Hur, M., The influence of total quality management practices on the transformation of how organizations works, Total Management and Business Intelligence, Vol. 20, Numbers 7-8, p. 847 – 863, ISSN: 1478 – 3363, 2009.
5. Foster, T., Managing Quality: Integrating the supply chain, Prentice Hall, ISBN: 978-0-13-507819-8, 2010.
6. Summers, D., Quality Management: Creating and Sustaining Organizational Effectiveness, Pearson International Edition, 2009.
7. Foster, T., Managing Quality: An Integrative Approach, Pearson & Prentice Hall, 2004, ISBN: 0-13-123018-3
8. Juran, J., Managerial Breakthrough, Mc Grow Hill, New York, 1995.
9. Rao et all., Total Quality Management: A Cross Functional Perspective, John Willey & Sons, New York, 1996.
10. Goetch, D., Davis, S., Introduction to Total Quality, Prentice Hall, Upper Saddle River, New Jersey, 1997.
11. Jeston, J., Nelis, J., Business Process Management, Elsevier, Amsterdam, 2008.
12. Born, G., Process Management to Quality Improvement, John Willey & Sons, New York, 1996.
13. Davenport T.H., Process Innovation Reengineering Work Through Information Technology, Boston, MA, 1993.
14. Zimmermann, H.J., Fuzzy set Theory and its applications. Kluwer Nijhoff Publising: Boston, 1996.
15. Klir, G.J., Folger, T.A., Fuzzy Sets, Uncertainty and Information, Pretence-Hall, Englewood, Cliffs, NJ., 1988.

16. Chang, D.Y., Applications of the extent analysis method on fuzzy AHP, European Journal of Operational Research, 95, 649-655, 1996.
17. Tadić, D., Milanović, D., Misita, M., Tadić, B., New integrated approach to the problem of ranking and supplier selection under uncertainties, Proceedings of the Institution of Mechanical Engineers. Part B: Journal of Engineering manufacture, doi: 10.1243/09544054JEM2105.
18. Saaty, T.L., How to make a decision: The Analytic Hierarchy Process, European Journal of Operational Research, 48, 9-26, 1990.
19. Shih, H.S., Shyur, H.J., Lee, E.S., 2007. An extension of TOPSIS for group decision making. Mathematical and Computer Modelling 45 (7/8), 801-813.
20. Marić, A, Arsovski, S., Mastilović, J.: Contribution to the improvement of products quality in baking industry, International Journal for Quality research, Vol.3, No. 3, pp. 209-216, 2009
21. D. Tadić, M. Stefanović, D. Milanović, Fuzzy Approach in Evaluation of Operations in Food Production, International Journal for Quality Research, Vol1, no 2, pp 97 - 104, 2007.

21 Chapter

CALCULUS PRINCIPLES FOR MECHANICAL PRESSES IN THE OIL INDUSTRY

1. INTRODUCTION: AN OVERVIEW

The oil products sector manufactures both edible and inedible oils. Approximately two-thirds of total oil products consist of edible oils, utilized directly in food or in the production of margarine, mayonnaise, baked goods, cooking fats, and preserves, among others. One-third of the total volume of produced oil comprises technical oils utilized in the manufacture of diverse products, including detergents, paint, glycerin, fatty acids, varnish, medicines, and cosmetics (Banu, 1999).

Vegetable oils represent one of the most ancient categories of recognized chemical substances. Numerous references and indications exist on the utilization of these oils during the Stone Age and Bronze Age (Willems, 2007).

The primary input for the vegetable oil business is oilseeds, a crucial element of contemporary agriculture. Oilseeds offer highly nutritious nourishment for humans, and oil crops and their derivatives constitute a significant commercial commodity. Vegetable oils provide vitamins, calories, and vital fatty acids for the human diet at a comparatively moderate expense. Following the processing of oilseeds, the residual cake, a solid byproduct, serves as a significant protein source for animal feed (Bargale, 1997).

Fats are present in both plant and animal tissues, as well as in the secretions of animal glands, such as milk. Plant fatty matter is predominantly concentrated in seeds, kernels, germs, fruits, and tubers, serving as a reserve of compounds utilized by the plant as an energy source. The fat level in these plant sections is quite varied, often below 5% in most specimens.

A diverse array of raw materials exists for the oil business. In the vegetable kingdom, there are over 100 oilseed species; however, only 40 of them are suitable for oil extraction.

The other plants are unprofitable, as they have low oil content of the seeds or as they require a difficult expression process. The most important oleaginous plants are: sunflower, soya, rape, cotton, poppy, almond, sesame, nut, palm, coconut, olive, flax, castor (Banu, 1999).

Separation of oil from oilseeds is an important processing operation. The process employed has a direct effect on the quality and quantity of protein and oil obtained from oilseeds (Bargale, 1997). The terms “expression” and “extraction” are used frequently when discussing about vegetable oil separation. Expression is the process of mechanically pressing liquid out of liquid-containing solids. Extraction is the process of separating a liquid from a liquid-solid system with the use of a solvent (Khan & Hanna, 1983). There has been some confusion in the literature between the operations of “expression” and “extraction”. The latter word has been used quite loosely to designate either operation (Gurnham & Mason, 1946). This tendency has been so extensive that the distinction between the two terms appears to be disappearing from the literature. The term ”extraction” is also used for mechanical oil expression (Biris et al., 2009/a).

Worldwide, for extraction of oil from seeds, fruits and nuts, four basic methods are used, as shown in the following figure.

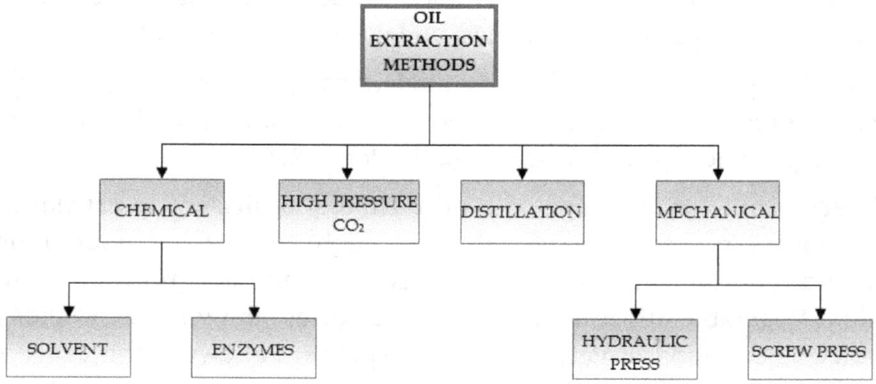

Figure 1. Oil extraction methods (Sari, 2006)

Chemical extraction method is based on the use of enzymes or solvent to extract the oil from the raw material. Solvent extraction method uses a solvent (which is, in generally, a hexane, meaning a petroleum distillate) mixed with ground seeds. Seeds are grounded to maximize the contact area of the seed with the solvent; thus the oil yield is higher. After the mixing process, the obtained mixture is heated up to 100°C to separate the oil from the solvent. The other chemical extraction, enzymatic extraction, is adopted by powerfull vegetable oil companies, as the process produces many high value products. For this extraction methods, seeds are cooked and put into water. Next, enzymes are added as they digest the solid material. The basic difference of this type of extraction method from the solvent type is that the residual enzymes in the oil are separated by the use of a liquid-liquid centrifuge (Sari, 2006).

The extraction using high pressure carbon dioxide (i.e. supercritical fluid extraction, SFE) embodies several features of conventional solvent extraction, but it has important features of its own (Bulley et al., 1984). This extraction method consists in seeds mixing with a liquid form of high pressure carbon dioxide. Oil is dissolved in the carbon dioxide and when the pressure is released, the carbon dioxide becomes a gas, leaving the oil behind.

Steam distillation is the method used for the extraction of 93% from the essential oils, the rest of 7% being extracted with other method (Masango, 2005). Hot steam releases the aromatic molecules from the plant material, by forcing the open of the pockets in which the oils are kept in the plant material. The molecules of these volatile oils will exit from the plant material and evaporate into the steam. The steam mixed with the essential oil is furter passed through a cooling system to condense the steam, which forms a liquid from which the essential oil and water is then separated.

The mechanical process is another method for oil extraction. Mechanical expression of oil requires the application of pressure to force oil out of the oil bearing material (Ogunsina et al., 2008). Various types of machines can be used for compression: screw presses, hydraulic presses, roll presses and mills, collapsible-plate and frame-filter presses, disc mills, interlocking-finger juice extractors, juice reamers (Khan & Hanna, 1983). From this variety of machines, there are available for expression processing two of them, the hydraulic press and screw press mechanisms. Hydraulic press, based on the principle of the hydraulic ram, originates in England and it was first patented in 1795 by Joseph Bromah (Dunning, 1953). The first screw press was developed by V.D. Anderson in the United States, in 1900. Due to the advantages it presents (continuous operation, high working capacity, run without high shocks and vibrations, working pressures which can be easily adjusted, etc.) screw presses quickly replaced hydraulic presses (Biris et al., 2009/b). In the figure below is presented a model of screw press developed by De Smet Rosedowns company, from United Kingdom.

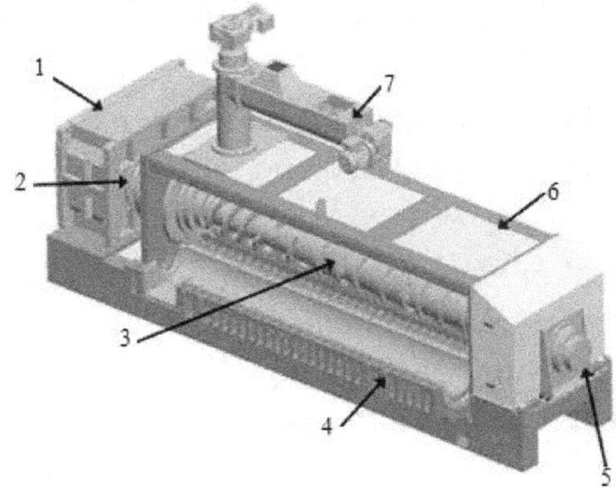

Figure 2. De Smet Rosedowns screw oil press 1 – gearbox; 2 – thrust housing; 3 – worm assembly; 4 – drainage cage; 5 – discharge end bearing; 6 – frame; 7 – feeder.

Mechanical pressing and solvent extraction are the most commonly used methods for commercial oil extraction. Oils obtained by mechanical pressing has high quality, but it can be recovered up to 90-95% of available oil. Solvent extraction has the advantage of the high yield that can be obtained (up to 99%), and as for disadvantages, oil quality is lower (Karaj & Müller, 2009). This quality reduction is produced by the extensive solvent recovery processes that are necessary and the fact that the solvent co-extracts undesired components from the cell walls (Willems et al., 2008).

Pressing technology of oleaginous material meal occurs under the influence of compression forces in mechanical presses. First, the oil from particles surface is separated – retained by surface forces of the molecular field, by the channels formed between particles. For a certain pressure, deformation and strong compression of particles begin, producing the elimination of oil from the capilars of particles. At a certain point, the space between particles gets so small that the oil film is subjected to the retaining forces exerted by particles surfaces, the oil can not be removed, the particle breaks in several places, the particle surfaces are in contact, and the briquetting begins, namely the forming of broken (cakes).

Increasing the pressure on meal particles must be done gradually, as for the harsh increase, fine meal particles will block the outler of oil from the capilaries, thus reducing the general pressing yield.

Pressing process cand be assimilated to the process of capillary filtration (fluid flow throuth capillary), expressed by the equation:

$$V = \frac{\pi \cdot p \cdot d \cdot t}{128 \cdot \eta \cdot l} \left[\mathrm{m}^3\right] \qquad (1)$$

where: V – volume of separated liquid (passing through capillaries), [m3]; p – applied pressure, [N/m2]; d – diameter of cappilary channel, [m]; η - dinamic viscosity of oil, [Pa s]; l – lenght of capillary channel which must be passed by the separated oil, [m]; t – time of applied pressure, [s].

From the above equation it results that the process of oil separation can be positively influenced if the values of p, d and t are increasing and if values of l and η will decrease.

Pressure p in mechanical presses is created by an helical conveyer (worm) which rotates in a closed space (pressing chamber). Gradual increase of pressure is done by the decrease of free volume of pressing chamber from one stage to another (by increasing the shaft diameter and decreasing the chamber diameter) and by reducing the pitch of worm. Pressing force is influenced by the resistance at the exit of material from the pressing chamber, as the amterial is forced to pass through a space with variable section.

Pressing time t must be high enough to allow the proper oil flow. A prolongation of the pressing time does not lead to significant increase of oil extraction efficiency, but leads to the sensitive decrease of press productivity.

Pressing time can be determined as the sum of the pressing times in each section (compression stage) of the press:

$$t = \sum_{i=1}^{n} t_{si} \tag{2}$$

where: n is the number of pressing stages, and the pressing time in a certain section is given by:

$$t_s = \frac{V_{ls} \cdot c_s}{Q_v \cdot (1 - \beta_s)} \tag{3}$$

where: Vls – volume of free space in the pressing section, [m3]; cs – pressing degree of meal in the respective section ; Qv – volumic feed flow of the press with meal, [m3/s]; βs – correction coefficient related to the quantity of meal removed from the press with the oil, until the analyzed section.

Pressing time also depends on the design and functional characteristics of the press and it ranges between 40-200 seconds. Pressing time depends on the shaft speed, cake (brocken) thickness at the press exit, and physico-chemical characteristics of the meal. Pressing time is inversely proportional to shaft speed and to cake thichness. As cake thickness is greater, the pressure in pressing chamber decreases, and pressing time decreases due to the fact that the material passes easier through the pressing chamber. At high pressure presses, by reducing the thickness of sunflower brocken from 11 mm to 4 mm the pressing time increases from 93 s to 106 s.

Parameters η, l and d are influenced by the preparation oparations of meal thus:

- oil viscosity η decreases by meal heating in roasting operation;
- capillars lenght l can be lowered by advanced destruction of cellular structure by grinding and, partially, during roasting, as well as by the reduction of passing distance of oil to the outlet hole from the pressing chamber (material layer in the pressing chamber must have small thickness).

In the technological diagram of pressing followed by solvent extraction, the presses can be of various types:

- for moderate preliminary pressing, which ensure the separation of 75-80 % oil and 18-22

% remaining oil in brocken;

- for advanced preliminary pressing, which ensure 12-14 % remaining oil in brocken.

To obtain oil just by pressing, without solvent extraction, there are:

- presses with a single pressing stage, mechanical presses for final pressing, which realize separation with maximum 3-6 % remaining oil in brocken;
- presses with two pressing stages, the first is used for moderate pressing, and the second stage for final pressing.

Depending on the press type, pressure applied on the meal gets to 250-280 daN/cm2 – for preliminary pressing, respectively 400-2000 daN/cm2 – for final or single pressing.

In Romanian oil factories, processing oleaginous seeds with high oil content is done after the diagram of pressing-extraction, so by using preliminary pressing equipments (moderate or advanced).

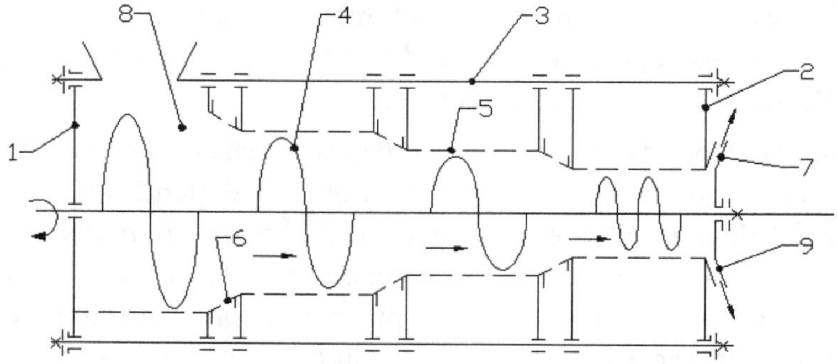

Figure 3. The operating principle of the continuous mechanical press 1 –front plate; 2 –back plate; 3 –clamping columns; 4 –screw; 5 –cylindrical strainer (barrel); 6 –tapered strainer (barrel); 7 –tapered exhaust end for pressure adjustment of pressing chamber; 8 –feed area of meal; 9 – evacuation area of cake from the pressing chamber.

Mechanical oil expression involves the release of oil from the seed interior into the interparticle voids on application of pressure. Filling of the interparticle voids leads to a buildup of pore pressure, thereby to the development of pressure gradient in the voids. As a result, oil flows through the porous medium and is finally expressed through the porous retaining envelope (Ajibola et al., 2002). The efficiency of expression is influenced by: the porosity of the cake, yield stress of the solid phase, the compressive force applied and viscosity of the expressed liquid (Clifford, 1973). The pressing process has been studied by several authors and they have found the following parameters influencing oil expression: applied pressure, moisture content, heating temperature and heating time, particle size (Adeeko & Ajibola, 1990; Khan & Hanna, 1983). Thus, increasing parameters such as heating temperature, heating time and applied pressure while reducing oilseed moisture to a certain degree will result in the increase of oil yield. A significant influence on oil yield has the postheating moisture content of some oilseeds (Ajibola et al., 1993). Effects of oilseeds heat treatment are: rupture of the oil bearing cells of the seed, coagulate the protein in the meal, adjust the moisture level of the meal to optimum level for oil expression. Lower the viscosity and increase the fluidity of the oil to be expelled and destroy mould and bacteria thereby facilitating oil expression from material (Adeeko & Ajibola, 1990). Optimum heating temperature for oilseeds is found in the range of 90-110°C for an average retention time of 20 minutes (FAO, 1989).

Oil expression is accompanied by compression and consolidation process brought about by the reduction in the volume of the compressed material (figure 4).

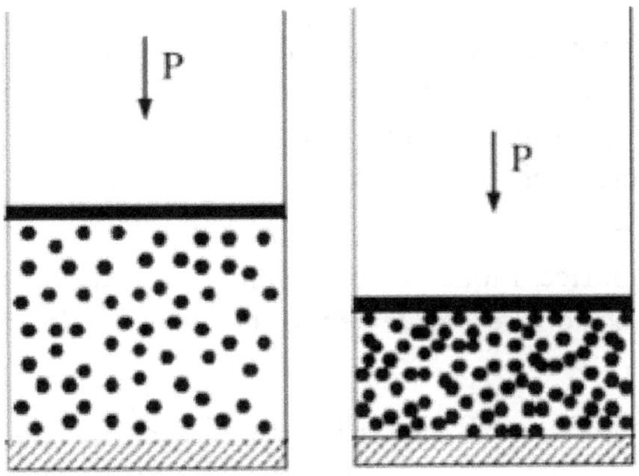

Figure 4. Volume reduction of the compressed material (Venter et al., 2007)

Thus, reduction of the total void space ocurrs, causing oil elimination (Sivalla et al., 1991). Value of the applied pressure at the point that oil leaves the interparticle voids is viewed as the oil point pressure, namely the minimum pressure that must be applied before oil expression begins. Applied pressures below this point are regarded as effort required to mobilize oil from the seed cells to the surface (Sukumaran & Singh, 1989; Mrema, 1979).

The general theoretical description of expression is based on consolidation theories originally developed for soil mechanics (Terzaghi, 1954). There are several studies on the modelling of oilseed expression, resulting in the development of empirical models, Terzaghi-type models and models mased on the nature of the cell structure of the oilseeds (Venter, 2007).

Processes and phenomena that occur during the pressing process of oleaginous materials are very complex.This paper contains a theoretical model on the power necessary to operate an oil press. The necessary components to power the press are: the power required for material transport along the pressing chamber, the power required to press the oleaginous material, the power required to overcome the friction between the screw spire and the material, the power required to push the material from the press through the exhaust cylinder head. The paper presents some diagrams showing the influence of various constructive and functional parameters on the pressing process.

Theoretical elements of a functional calculus, and of the power necessary to operate the press are rather poor and based especially on simple formulas containing some correction coefficients, whose value is empirically obtained from experiments. This is due to the complexity of the processes and phenomena taking place during pressing, such as: material transport, proper pressing, overcoming the frictions between auger and material, pushing the material through the slot at the end of the pressing chamber.

Mechanical work for material pressing it results from the expression of the equivalent stress, which occurs in the pressing chamber after applying on the cross surface of the pressing chamber an equivalent pressing force, so that it may be produced a reduction of the volume occupied by the material, from the initial value to the final value.

Based on the mathematical model developed in this study, was found the variation of the component parts of the power necessary for operation in the case of a press from the oil industry.

The main data taken into consideration for modelling are: the variable auger rotational speed (n=15-40 min-1), the variable pressure inside the pressing chamber (p=50 105-200 105 Pa), the diameter of the pressing chamber (D=200 mm), the diameter of the auger shaft (d=100 mm).

This work presents a mathematical model that enables the precise determination of functional parameters and the requisite power for operating presses in the food sector.

The data in this research indicate that the power required for effective pressing is the greatest, followed by the power needed to overcome the friction between the auger spiral and the material subjected to pressing. The power requirements for material movement via the exhaust area and the pressing chamber are far lower than those of the presses and can be considered negligible.

This paper is beneficial for students pursuing bachelor's degrees, as well as for professors and researchers involved in the design and development of mechanical oil presses.

2. THEORETICAL ELEMENTS

2.1 Functional calculus elements

Pressure ratio is the reduction of the material subjected to pressing and it can be calculated using equation [4], where Vi is the initial volume of the material, [m3] and Vf is the final volume, [m3].

$$\varepsilon = \frac{V_i - V_f}{V_i} \quad (4)$$

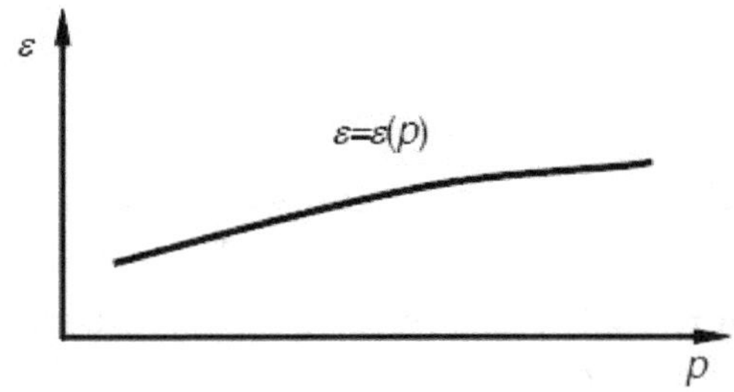

Figure 5. Variation of the pressure ratio with the pressure

The value of pressure ratio is directly proportional to the working pressure of the press, and its variation is shown in figure 5.

Press volume flow rate can be evaluated by using the relation [5]:

$$Q_v = V_{te} \cdot (1-\varepsilon) \cdot n \cdot k \cdot 60 \quad \left[\mathrm{m}^3\right] \quad (5)$$

where: Vte – is the theoretical volume of the material displaced by the auger spire during a complete rotation, in the exhaust area [cu.m.]; n – auger rotative speed, [rpm]; k – coefficient taking into account the material flowing back through the spire extremities, as well as the incomplete feed with material, (k=0,2 * 0,35).

The theoretical volume of the material displaced by the auger spire is calculated using the following equation:

$$V_{te} = \frac{\pi}{4} \cdot (D^2 - d^2) \cdot (s - \delta) \quad \left[\mathrm{m}^3\right] \quad (6)$$

where: s – the auger spire pitch [m]; δ – thickness of the auger spire, [m]; D – outer diameter of the auger spire, [m]; d – inner diameter of the auger spire (of the auger shaft), [m].

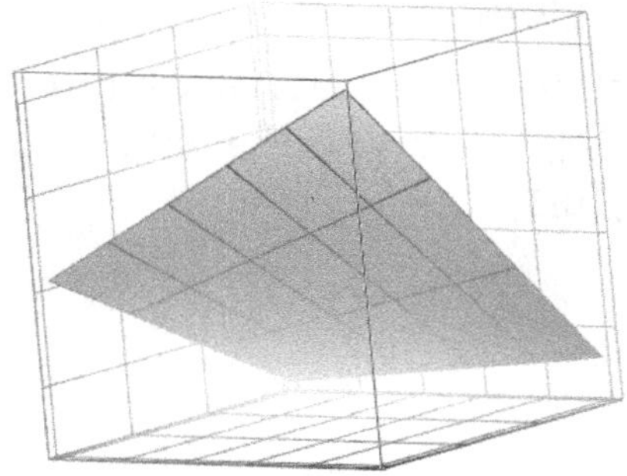

Figure 6. Variation of the press flow rate depending on the auger rotative speed and the spire pitch

By replacing in equation (5) the expression of the theoretical volume given by equation (6), it results the expression of the press volume flow rate under the form (see Figure 6):

$$Q_v = \frac{\pi}{4} \cdot (D^2 - d^2) \cdot (s - \delta) \cdot (1 - \varepsilon) \cdot n \cdot k \cdot 60 \quad \left[\mathrm{m^3/h}\right] \tag{7}$$

Calculus of the power necessary to operate the press

The power necessary to operate the press can be evaluated by using the equation:

$$P_p = \frac{P_{tr} + P_{pres} + P_{fr} + P_{cap}}{\eta_{tm}} \quad [\mathrm{kW}] \tag{8}$$

where: Ptr – represents the necessary power to transport the material from feeding chamber to exhaust head, [kW]; Ppres – necessary power for pressing the material, [kW]; Pfr – necessary power for overcoming the frictions between the auger spire and the material, [kW]; Pcap – necessary power for pushing the material through the exhaust space in the press, [kW]; ηtm – mechanical transmission yield (output).

Necessary power for material transport

Taking into account the calculus equations of the slow helical conveyors, it can be written the expression of the necessary power for proper transport of the material along the auger:

$$P_{tr} = \frac{F_r \cdot v}{1000} \quad [\mathrm{kW}] \tag{9}$$

where: Fr – represents the resistant force to the material advancing along the press auger, [N]; v – mean speed by which the material moves along the press auger, [m/s].

The resistant force Fr is given, on one part, by the phenomenon of outer friction between the material and the walls of the pressing chamber, and, on the other part, by the phenomenon of outer friction of the material subdued to pressing. The value of this force can be calculated by the expression:

$$F_r = q \cdot l \cdot g \quad [\mathrm{N}] \tag{10}$$

where: g – represents the gravity acceleration, [m/sq.s]; q – the linear load (mass per linear meter of material) in the press, [kg/m]; l – length of pressing chamber, [m].

The expression of the linear load, q, can be written:

$$q = S \cdot \psi \cdot \gamma = \frac{\pi \cdot (D^2 - d^2)}{4} \cdot \psi \cdot \gamma \quad [\mathrm{kg/m}] \tag{11}$$

where: ψ – represents the coefficient of admission for the press section; S – area of the cross section of the pressing chamber, [sq.m.].

By replacing into the relation (10) it results:

$$F_r = \frac{\pi \cdot (D^2 - d^2)}{4} \cdot \psi \cdot \gamma \cdot l \cdot g \quad [\mathrm{N}] \tag{12}$$

It results the necessary power for transporting the material along the pressing chamber:

$$P_{tr} = \frac{F_r \cdot v}{1000} = \frac{\pi \cdot (D^2 - d^2) \cdot \Psi \cdot \gamma \cdot l \cdot g \cdot v}{4 \cdot 1000} = \frac{\pi \cdot (D^2 - d^2) \cdot \Psi \cdot \gamma \cdot l \cdot g \cdot s \cdot n}{4 \cdot 1000 \cdot 60} \quad [\mathrm{kW}] \tag{13}$$

Necessary power for pressing the material

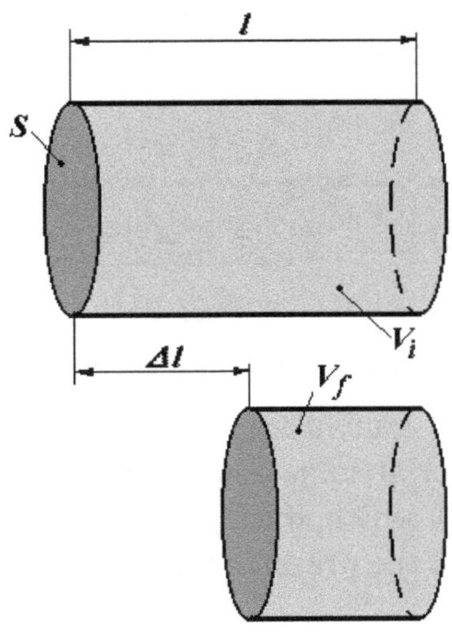

Figure 7. Variation of the material volume in the pressing process

The mechanical work for material pressing (Lpres) results from the expression of the equivalent tension (stress) σ, which appears inside the pressing chamber as a result of applying on the cross surface (S) of the pressing chamber an equivalent pressing force (Fpres), so that it may be produced a reduction of the volume occupied by the material, from the initial value, Vi to the final value, Vf, as it is also shown in figure 7. Hence, the equivalent tension (stress) in the pressing chamber can be written as:

$$\sigma = \frac{F_{pres}}{S} = \frac{F_{pres} \cdot \Delta l}{S \cdot \Delta l} = \frac{L_{pres}}{\Delta V} = \frac{L_{pres}}{V_i - V_f} \quad [\text{Pa}] \tag{14}$$

Taking into account equation (4), it results:

$$V_f = V_i \cdot (1 - \varepsilon) \tag{15}$$

and from the equations (14) and (15) is obtained the expression of the mechanical work for the material pressing:

$$L_{pres} = \sigma \cdot (V_i - V_f) = \sigma \cdot [V_i - V_i \cdot (1 - \varepsilon)] = \sigma \cdot \varepsilon \cdot V_i \quad [\text{J}] \tag{16}$$

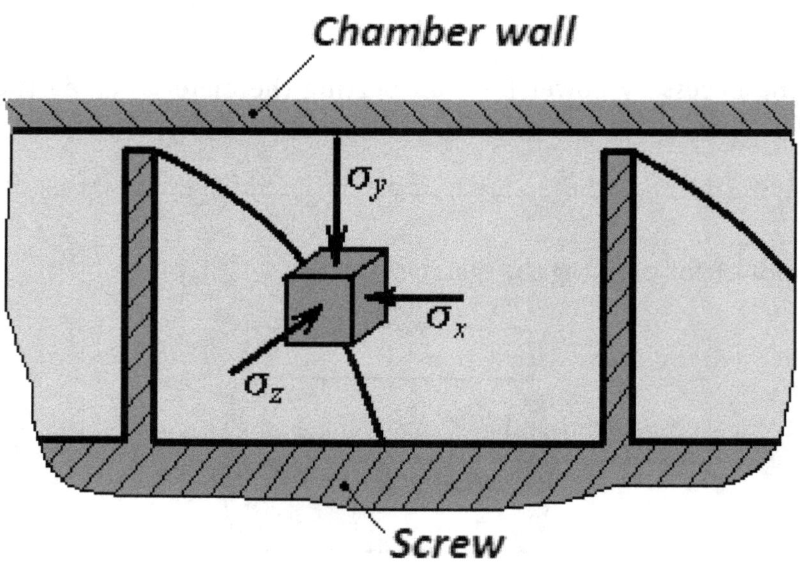

Figure 8. Elementary volume of material subdued to the pressing process

To determine the value of equivalent tension (stress) σ, it is considered an elementary volume of material subjected to pressing, uniformly loaded on each section, as shown in figure 8, which, during the pressing process will move only on the longitudinal direction of the press (direction x). Under these conditions it can by written:

$$\begin{cases} \sigma_y = \sigma_z \\ \sigma_x = p \end{cases} \tag{17}$$

where: p [Pa] - represents the pressure performed by the auger, which is exerted on the material.

It is considered that the tensions (stresses) on the direction y and z occur due to the pressure oriented to the direction of material displacement, respectively:

$$\sigma_y = \sigma_z = \beta \cdot \sigma_x \tag{18}$$

where: β - represents the coefficient of the side pressure. Taking into account equation (18), it results:

$$\sigma_x + \sigma_y + \sigma_z = p \cdot (1 + 2 \cdot \beta) \tag{19}$$

As the materials subjected to the pressing process in the food industry also contains a certain percentage of liquid substance (oil, must, etc.), it can be considered that the hydrostatic pressure law remains valid, respectively:

$$\sigma = \frac{\sigma_x + \sigma_y + \sigma_z}{3} = \frac{p \cdot (1 + 2 \cdot \beta)}{3} \tag{20}$$

Thus, it results the expression of the mechanical work necessary for pressing the material:

$$L_{pres} = \frac{1 + 2 \cdot \beta}{3} \cdot p \cdot \varepsilon \cdot V_i \quad [\mathrm{J}] \tag{21}$$

respectively, the expression of the necessary power for pressing the material:

where: Fpres – represents the pressing force [N]; vpres – the pressing speed, [m/s]; Δt – the time interval when the reducing of the material volume is performed from the initial value Vi to the final value Vf, [s]. The value of this time interval can be calculated depending on the rotative speed [rpm] of the press auger, respectively:

$$P_{pres} = \frac{F_{pres} \cdot v_{pres}}{1000} = \frac{F_{pres} \cdot \frac{\Delta l}{\Delta t}}{1000} = \frac{F_{pres} \cdot \Delta l}{1000 \cdot \Delta t} = \frac{L_{pres}}{1000 \cdot \Delta t} \quad [\mathrm{kW}] \tag{22}$$

Taking into account equations (21), (22) and (23), the expression of the necessary power for pressing the material is obtained:

$$\Delta t = \frac{60}{n} \tag{23}$$

In reality, for presses in food industry, the value of the pressure, p, is not kept constant along the auger, having a variation which can be as that seen in figure 9.

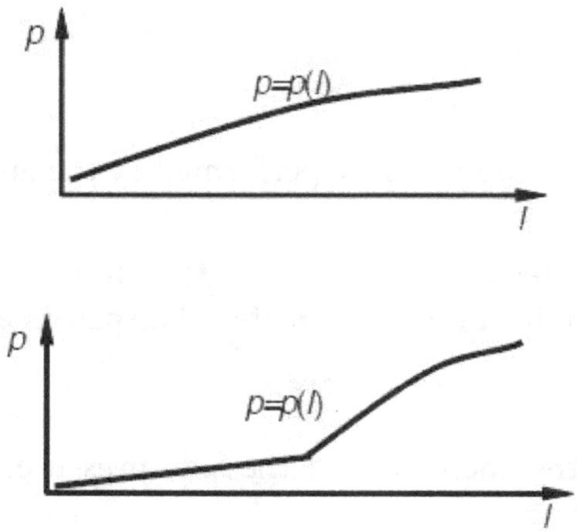

Figure 9. Pressure variation along the pressing chamber

- Power necessary to overcome the frictions between the auger spire and material

To calculate the necessary power for overcoming the frictions between the auger spire and the material, first it must be calculated the friction torque (moment), which occurs on the spire surface when it comes into contact with the material. For the calculus of this friction torque (moment) it is first taken into consideration an elementary ring, dr, situated on the auger spire on the radius r (Figure 10) and for the auger length suitable to a pitch, s it is determined the normal force exerted on the elementary ring:

$$dN = p \cdot dS = p \cdot 2 \cdot \pi \cdot r \cdot dr \tag{25}$$

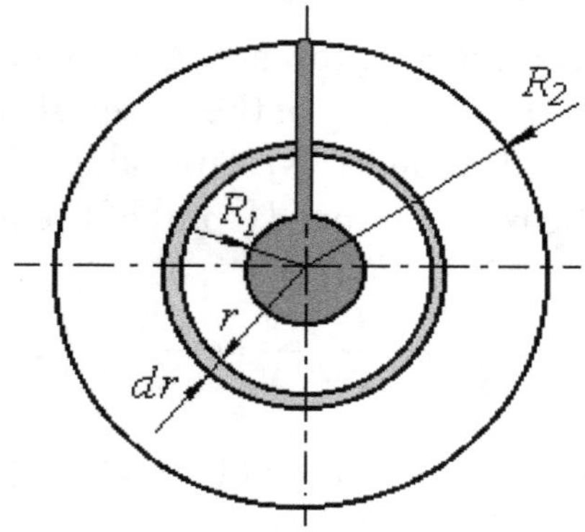

Figure 10. Elementary ring on the auger spire

The value of the friction force which occurs on the surface of the elementary ring is calculated by the following equation:

$$dF_f = \mu \cdot dN = \mu \cdot p \cdot 2 \cdot \pi \cdot r \cdot dr \tag{26}$$

The expression of the friction torque (moment) at the surface of the elementary ring is:

$$dM_f = r \cdot dF_f = \mu \cdot p \cdot 2 \cdot \pi \cdot r^2 \cdot dr \quad (27)$$

which, by integration, for the whole active cross surface of the auger spire, suitable to a length equal to a pitch, s, leads to the equation:

$$M_f = \int_{R_1}^{R_2} \mu \cdot p \cdot 2 \cdot \pi \cdot r^2 \cdot dr = 2 \cdot \mu \cdot p \cdot \pi \cdot \int_{R_1}^{R_2} r^2 \cdot dr = 2 \cdot \pi \cdot \mu \cdot p \cdot \frac{r^3}{3}\Big|_{R_1}^{R_2} \quad (28)$$

respectively:

$$M_f = 2 \cdot \pi \cdot \mu \cdot p \cdot \frac{R_2^3 - R_1^3}{3} \ [\text{Nm}] \quad (29)$$

It results the expression of the necessary power for overcoming the friction between the auger spire and the material:

$$P_{fr} = \frac{M_f \cdot n}{9550} = \frac{2 \cdot \pi \cdot \mu \cdot p \cdot \left(R_2^3 - R_1^3\right) \cdot n}{3 \cdot 9550} \ [\text{kW}] \quad (30)$$

- Power necessary for pushing the material through the exhaust space

For pushing the material through the exhaust space from the end of the pressing chamber the power consumed is given by:

$$P_{cap} = \frac{F_c \cdot v_{cap}}{1000} = \frac{F_c \cdot \frac{\Delta l_c}{\Delta t}}{1000} = \frac{F_c \cdot \Delta l_c}{1000 \cdot \Delta t} = \frac{L_c}{1000 \cdot \Delta t} = \frac{L_c \cdot n}{1000 \cdot 60} \ [\text{kW}] \quad (31)$$

where: Fc – resistant force to material pushing through, the head of the pressing chamber, [N]; vcap – the material speed through the head of the pressing chamber, [m/s]; lc – length of exhaust canal, [m].

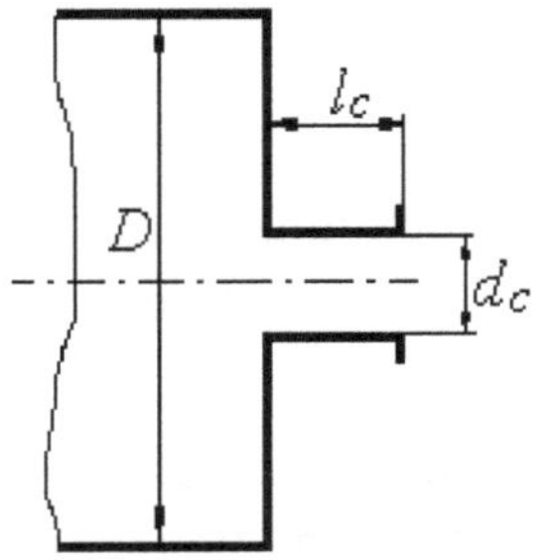

Figure 11. End of the pressing chamber

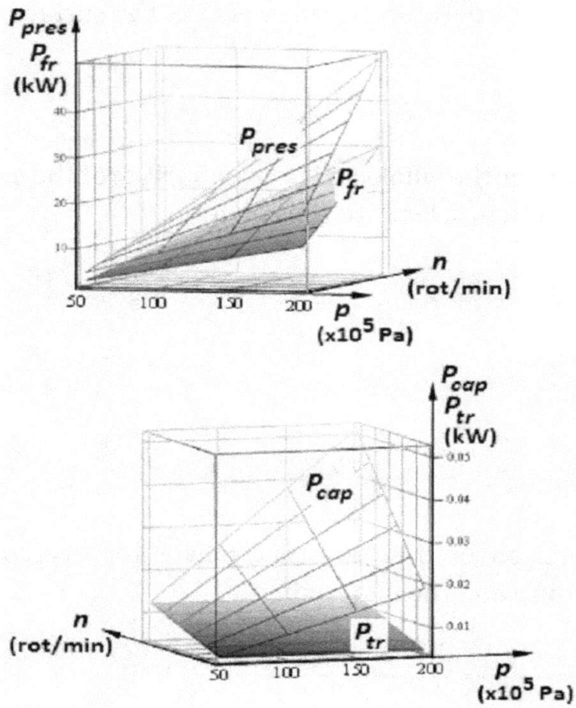

Figure 12. Power variation depending on pressure and auger rotational speed

The necessary mechanical work for pushing the material through the exhaust space (Fig. 11) for the end of the pressing chamber Lc is calculated using the following equation:

$$L_c = F_c \cdot l_c = p \cdot A_c \cdot l_c = p \cdot \frac{\pi \cdot d_c^2}{4} \cdot l_c \quad [\mathrm{J}] \tag{32}$$

It results the expression for the calculus of power Pcap:

$$P_{cap} = \frac{p \cdot \pi \cdot d_c^2 \cdot l_c \cdot n}{4 \cdot 1000 \cdot 60} \quad [\mathrm{kW}] \tag{33}$$

3. APPLICATION

Using the mathematical model developed in this study, figure 12 shows the variation of the component parts of the power necessary for operation of a press from the oil industry.

The main data taken into consideration for modelling are: the variable auger rotational speed (n=1540 min-1), the variable pressure inside the pressing chamber (p=50 105200 105 Pa), the diameter of the pressing chamber (D=200 mm), the diameter of the auger shaft (d=100 mm).

4. CONCISE SUMMARY

This work presents a mathematical model that enables the precise determination of functional parameters and the requisite power for operating presses in the food sector. The data in this research indicate that the power required for effective pressing is the greatest, followed by the power needed to overcome the friction between the auger spiral and the material subjected to pressing. The power requirements for material movement via the exhaust area and the pressing chamber are far lower than those of the presses and can be considered negligible. This paper is beneficial for students pursuing bachelor's degrees, as well as for professors and researchers involved in the design and development of mechanical oil presses.

REFERENCES

1. Adeeko, K. A., & Ajibola, O. O. (1990). Processing factors affecting yield and quality of mechanically expressed groundnut oil, Journal of Agricultural Engineering Research, 45, pg. , 31-43.
2. Ajibola, O. O., Okunade, D. A., & Owolarafe, O. K. (2002). Oil point pressure of soybean, Journal of Food Process Engineering, 25, pg. , 407-416.
3. Ajibola, O. O., Owolarafe, O. K., Fasina, O. O., & Adeeko, K. A. (1993). Expression of oil from sesame seeds, Canadian Agricultural Engineering, 35, pg. , 83-88.
4. Banu, C. (1999). Manualul inginerului din industria alimentară, vol. I și II, Editura Tehnică, București.
5. Bargale, P. C. (1997). Mechanical oil expression from selected oilseeds under uniaxial compression, Doctoral Thesis, University of Saskatchewan, Canada.
6. Biriş, S.Şt., Manea, M., Maican, E., Ungureanu, N., & Toma, L. M. (2009). a), Studies regarding the use of Finite Elements Method for the working process modelling of the oil presses, Proceeding of the 37 International Symposium „Actual Tasks on Agricultural Engineering", Croația, Opatija, february, pg. 485-496., 10-13.
7. Biris, S., St, , Manea, M., Paraschiv, G., Vlăduț, V., & Bungescu, S. (2009). b), Power calculus elements for oil presses, Proceeding of the 37th International Symposium „Actual Tasks on Agricultural Engineering", Croația, Opatija, february, pg. 475-484., 10-13.
8. Bulley, N. R., Fattori, M., Meisen, A., & Moyls, L. (1984). Supercritical fluid extraction of vegetable oilseeds, Journal of the American Oil Chemists' Society, 61, pg. , 1362-1363.
9. Clifford, W. C. (1973). Theory of filtration, Chemical Engineers' Handbook, 5, pg. , 19-69.

10. Dunning, J. W. (1953). History and latest developments in expeller and screw press operation on cottonseed, Journal of the American Oil Chemists' Society, 30, pg., 486-492.
11. FAO- Food and Agriculture Organization(1989). Food Composition Table for Use in Africa. US Department of Health, Education and Welfare Press, Washington D.C.
12. Gurnham, F. C., & Masson, H. J. (1946). Expression of liquids from fibrous materials, Industrial and Engineering Chemistry, 38, pg. , 1309-1315.
13. Karaj, S., & Müller, J. (2009). Optimization of mechanical extraction of Jatropha curcas seeds, Landtechnik 64, pg. (3), 164-167.
14. Khan, L. M., & Hanna, M. A. (1983). Expression of oil from oilseeds- A review, Journal of Agricultural Engineering Research, 28, pg. , 495-503.
15. Masango, P. (2005). Cleaner production of essential oils by steam distillation, Journal of Cleaner Production, 13, pg. , 833-839.
16. Mrema, G. C. (1979). Mechanisms of mechanical oil expression from rapeseed and cashew, Doctoral Thesis, National University of Ireland, Dublin.
17. Ogunsina, B. S., Owolarafe, O. K., & Olatunde, G. A. (2008). Oil point pressure of cashew (Anacardium occidentale) kernels, International Agrophysics, 22, pg. , 53-59.
18. Sari, P. (2006). Preliminary design and construction of a prototype canola seed oil extraction machine, Doctoral Thesis, Middle East Technical University, Ankara, Turkey.
19. Sivalla, K., Bhole, N. C., & Mujherjee, R. K. (1991). Effects of moisture on rice bran oil expression, Journal of Agricultural Engineering Research, 50, pg. , 81-91.
20. Sukumaran, C. R., & Singh, B. (1989). Compression of a bed of rapeseeds- the oil point, Journal of Agricultural Engineering Research, 42, pg. , 77-84.
21. Terzaghi, K. (1954). Theoretische Bodenmechanik, Berlin: Springer.
22. Venter, M. J., Kuipers, N. J. M., & de Haan, A. B. (2007). Modelling and experimental evaluation of high-pressure expressionn of cocoa nibs, Journal of Food Engineering, 80, pg. , 1157-1170.
23. Willems, P. (2007). Gas Assisted Mechanical Expression Of Oilseeds, Doctoral Thesis, University of Twente, Netherland.
24. Willems, P., Kuipers, N. J. M., & De Haan, A. B. (2008). Hydraulic pressing of oilseeds: Experimental determination and modelling of yield and pressing rates, Journal of Food Engineering, 89, pg. , 8-16.

22
Chapter

IMMUNOREGULATION OF THE GASTROINTESTINAL TRACT AND THE CHALLENGES POSED BY NANOTECHNOLOGY IN FOOD

1. INTRODUCTION: AN OVRVIEW

1.1. Nanoparticles: Physicochemical Properties and Applications in Food

Nanoparticles are fundamental three-dimensional entities ranging from 1 to 100 nanometers (nm) in size, displaying distinctive physicochemical properties that underpin their applications and provide unique problems in the creation of novel uses [1, 2]. Due to their diminutive size, nanoparticles can interact with human physiology at the subcellular level, presenting numerous potential applications in nutrient and medication delivery, immunization therapy, and tissue regeneration. Chemical modification of nanoparticles can facilitate certain physiological uses by enhancing blood circulation parameters, hence prolonging their residence duration in tissues, or by targeting specific tissues utilizing ligands. The applications of nanotechnology in food are as intricate and diverse as the formulations that can be developed using this technology. Current technology applications affecting food encompass the production of food packaging, which integrates antibacterial compounds like silver,[2] detection particles such as gold, flavor improvement, and the delivery of dietary supplements and nutraceuticals[2-5]. Although their potential or current application offers significant advantages, it is essential to possess a comprehensive grasp of the physiology of nanoparticle absorption and the implications of their containment or integration into the mammalian physiological and cellular environment.

1.2 Food packaging

Historically, food packaging has typically consisted of conventional materials such as paper or metal-based materials. The use of polymeric formulations improved the ability to retain moisture and provided a gas barrier, thus extending food shelf life. Typically, the Food and Drug Administration requires that the manufacturer of food contact material comply with the regulatory requirements for each individual substance that comprises the entire formulation of the food contact material [6]. These food contact materials have typically included paper, metallic-based items, and polymeric compounds such as polyethylene terephthalate (PET), polypropylene, polyethylene, polystyrene, and others. Recently the formulation of nanocomposites has improved the ability to produce food contact surfaces that are superior with respect to their heating and gas barrier resistance characteristics [2, 7, 8]. Typically, a combination of previously approved compounds and nano-material has been used for the construction of the newer nanocomposite materials that strive to enhance the storage and preservation of foods. Nanocomposites are described as a combination of inorganic nanomaterial and a continuous phase consisting of synthetic polymers [9]. Nanoclay composites consist of magnesium aluminum silicate nanoparticles (bentonite or montmorillonite), and have proven to be a superior gas barrier for the preservation of foods [7]. The production of sustainable, biodegradable polylactide (PLA)-based polymers present the potential to reduce the dependence upon petrochemical based polymers by using alternative renewable sources to produce packaging materials with qualities comparable to presently used products. The combination of PLA with montmorillonite (MMT) nanocomposite [10-12] has been reported to produce a short term packaging material with good $O2$ gas permeability, and can be converted into $CO2$ and $H2O$ through decomposition by microorganisms [10].

Silver, which has long been recognized for its antimicrobial characteristics [13], has been among the inorganic constituents incorporated into nanocomposite materials. Prior to the advent of nanotechnology, silver had long been used as an ingredient within dental composite material [14], integrated into wound dressings [15-17] and other medical devices approved by the FDA, and is recognized as a biocide by the EPA [18]. Analysis of the antimicrobial effects of silver ion on gram positive and gram negative cell walled microorganisms demonstrated similar effects [19, 20]. Exposure of microbial organisms to silver results in the retraction of the cytoplasm from the cell wall, condensation of the DNA into electron-dense granules, and there is an accumulation of silver ions into the cytoplasm. The damage, as inferred in these studies, is due to the inability to replicate at the DNA level [19]. Additional denaturant effects attributed to the silver ion include its ability to attach to sulfhydryl groups, amino groups, and the terminal phosphate and carboxyl groups of bacterial proteins [13], essentially inactivating the enzymes involved with electron transport and metabolism. Of the electron transfer functions, cytochrome reductase and cytochrome oxidase are targeted [21]. While interest in silver's use as an antimicrobial has increased due to the observed rise in hospital and community-acquired antibiotic resistances, it is important to note that a growing microbial resistance to silver has also been reported [22]. Antimicrobial properties are similarly attributed to silver (Ag) nanoparticles [20, 23-27], and this property has spurred the inclusion of this material into a wide array of products within the foods sector including packaging and service containers, and bottles, or used as a measure to prevent or control surface contamination by Escherichia coli and Staphylococcus aureus [28]. Indeed, the incorporation of silver into MMT composite preparations was shown to inhibit the growth of Escherichia coli 0157:H7, Staphylococcus aureus, and Klebsiella pneumonia on agar at levels that were 35% of the levels achieved by cefotaxime and chloramphenicol [29].

1.2 Dietary supplements and nutraceutical delivery

The encapsulation of dietary vitamins and other nutritional supplements as a nanoparticle has gained considerable interest as a means to increase the shelf life of such materials, and to improve delivery and release within the body. The engineered particles provide potential strategies with which to overcome the impermeability of the mucosal epithelium, and offer a possible means of circumventing the degradation of the nutrient by harsh degradative gastrointestinal conditions. Several candidate materials, used successfully for the delivery of drugs and vaccines, have been examined for their ability to encapsulate nutrients. Finally, compounds such as polysaccharides and proteins that are already in use within commercial food applications are attractive candidates for the production of new nanocomposite packaging and encapsulation material, as several are generally regarded as safe and are biodegradable.

Poly (D,L)-lactic co glycolic acid (PLGA) nanoparticles are widely used for the encapsulation and delivery of drugs due to their reported biocompatibility and lack of overt toxicity. The physicochemical properties of the PLGA particles are affected by specific formulation and processing parameters, such as drug and polymer concentration, solvent volume, polymer molecular weight, the type of emulsifier used in the processing and its concentration, and the aqueous-to-organic phase ratio [30, 31]. Thus, PLGA nanoparticles have been shown to adequately encapsulate hydrophobic and hydrophilic molecules albeit the latter present some challenges with respect to a lowered load efficiency, and many PLGA-encapsulated delivery systems have been designed for a wide variety of macromolecules including drugs, biologically active cytokines, and peptides [31-33].

Chitosan, an N-deacetylated derivative of chitin, has been analyzed for use in nutrient delivery due to its wide acceptance in drug delivery, and is generally regarded as non-toxic and biocompatible. Chitosan ((1 • 4)-2-amino-2-deoxy-β-D-glucan) is a naturally occurring cationic polysaccharide found in the shells of shrimp, lobsters, and crab that has an intrinsic ability to bind mucin. The bioadhesive property of chitosan permits organ-specific delivery, and surface modification of the polysaccharide particle has been successfully used to alter organ delivery [34]. Chitosan has been demonstrated to induce increased permeability in Caco-2 monolayers across tight junctions as measured by changes in the measured transepithelial electrical resistance and in a 14C-mannose absorption assay [35, 36]. The improved absorption across cell layers due to the opening of tight junctions is thought to be the result of ionic interactions between the cell membrane and chitosan polysaccharide. While these characteristics favor the polysaccharide's use as a delivery method for a variety of compounds, it is necessary to incorporate anionic alginate to prevent burst release of the encapsulated material due to protonation in an acidic environment. The results obtained by encapsulation of Vitamin A within dual layered chitosan-alginate nanospheres have been reported to be successful [37, 38]. In this instance, a high encapsulation efficiency and improved storage stability was achieved using double-layered microcapsules that incorporated chitosan, alginate, calcium chloride and Tween 20. The production of combined Chitosan/ PLGA spherical particles have also been reported for the encapsulation of Vitamin A [38].With this construct, the microspheres (averaged 283 nm) demonstrated stability within an acidic environment and a lowered release rate into the gastric environment when compared to particles composed solely of PLGA. Thus, release of the target nutrient would be mainly in the small intestine where the vitamin would be absorbed. Interestingly, the material was visualized in the intestinal villi, and in the endothelium of rabbit GI.

Whey protein, derived from dairy, is recognized for its natural ability to form films and gels [7]. Whey nanospheres containing alginate have demonstrated the controlled release of an encapsulated nutrient, riboflavin, when tested in simulated gastric juices [39]. In this instance, 94 nm whey nanoparticles were constructed using an emulsification and cold gelation method, which averts the use of toxic solvents, and modification of the alginate concentration provides some control over degradation of the particle by pepsin in their assay. The encapsulation of viable probiotic yeast cells has been reported using whey –alginate microspheres produced by a cold gelation extrusion technique [40]. The encapsulation of a hydrophobic, fat-soluble nutrient can be achieved using casein maltodextrin nanoparticles produced by the Maillard reaction. In this reaction, the ε-amine groups found on the protein's lysine residues are covalently bonded to the aldehyde of reducing sugars. Particles produced in this manner consist of an exterior composed by the bulky hydrophilic domains of casein. The result of this design is a particle with increased curvature, i.e., a smaller diameter, containing an outermost saccharide layer and a hydrophobic inner core. Once the optimal casein: maltodextrin ratios were determined for the formation of the conjugates, incorporation of oil- soluble vitamin D resulted in particles that were 30 nm in diameter and demonstrated significant protection of the vitamin at low pH values that simulated gastric juices [4].

Liposomes composed of polar lipids such as lecithin have been used as delivery systems for antimicrobials, colors, and antioxidants. However, best results have been reported incorporating an additional layer of material such as the cationic polysaccharide, chitosan. Liposomes composed of soy lecithin and prepared by homogenization, and combined with chitosan with stirring and sonication, were used to encapsulate grape seed extract [3]. In this instance, the particle size increased with the addition of the grape seed extract due to surface incorporation of the grape seed extract into the liposomes' layer. This was rectified by production of particles containing multiple polymer layers composed of chitosan and citrus pectin; grape seed polyphenols were no longer exposed to the matrix. Finally, microspheres with a mineral composition have also been developed for the encapsulation of nutrients. In this case, the encapsulation of water soluble polyphenols extracted from green tea has been accomplished using calcium carbonate salt solutions containing phosphate and carbonate[41].

2. THE GUT-ASSOCIATED LYMPHOID TISSUE

The proposed and anticipated uses of orally-delivered nanoparticles, the use of nanoparticles on food-contact surfaces, and the introduction of microencapsulated nutrients, necessitate an understanding of the events within the mucosal immune compartment known as the Gut-Associated Lymphoid Tissue (GALT), which is critically involved in the formation and maintenance of oral tolerance to introduced nutrient-derived antigens, and the generation of mucosal immune responsiveness to ingested pathogens and their toxins. The gastrointestinal tract is responsible for the digestion and absorption of ingested nutrients. This function is aided by the intestine's mucosal lining, whose absorptive surface is greatly increased by villi which project into the lumen and are composed of a single layer of epithelial cells and a rich network of capillaries and lymphatics. While the gastrointestinal tract is responsible for the absorption of nutrients, it is also the site of ongoing immune surveillance. The intestinal lumen normally contains dietary degraded products, commensal microbial flora, and any ingested contaminants including pathogenic bacteria and their products, viruses, fungi, or parasites. The resident gastrointestinal immune system must: 1) generate immunologic tolerance towards nutrients and the resident microflora, and 2) recognize and remove infectious agents and their toxins [42-44]. Oral tolerance is driven by prior administration of antigen by the oral route, generating suppressive regulatory T cells, but is also dependent upon the maintenance of an effective epithelial barrier. The role of the resident gastrointestinal CD4+ T cell population for the establishment and maintenance of the tolerant state is critical [45, 46]. Investigators have reported the formation of exosome-like structures, designated as "tolerosomes, " which are assembled in and released from small intestinal epithelial cells, that seem to play a crucial role for the induction of tolerance [47]. Breakdown of oral tolerance is thought to lead to the development of food allergy and some autoimmune diseases, including inflammatory bowel diseases (Crohn's disease and ulcerative colitis) and celiac disease.

The GALT of the gastrointestinal tract consists of Peyer's patches (PP) containing B cells, dendritics, and T cells (Figure 1), appendix, draining mesenteric lymph nodes, and lymphatic follicles distributed throughout the length of the intestinal tract. The first line of immunologic defense is the provided by antibodies of the secretory IgA type found in the mucosal secretions of the gut [48]. This is supported by the observation that individuals with IgA deficiencies demonstrate circulating immune complexes to bovine and milk proteins [49]. In this case, the lack of IgA permits the entrance of food-derived antigens into the peripheral circulation, resulting in immune complex formation. The production of IgA is now known to be induced by regulatory T cells that have been activated by CD11+ dendritic cells [50]. Additionally, lymphocytes are scattered within the columnar epithelial layer (Intraepithelial lymphocytes, or IEL) and throughout the lamina propria.

2.1 T cells of the Gut-associated lymphoid tissue

Beneath the epithelial layer of the mammalian gastrointestinal tract lies a rich source of immunocompetent cells within the submucosal lymphoid follicles known as the intraepithelial lymphocytes (IEL) that comprise a significant portion of the body's T cells. The peripheral immune system contains effector T lineage cells bearing the $\alpha\beta$ T cell receptor (TCR) which are either class II-restricted CD4+ T cells or class I-restricted CD8+T cells. Intraepithelial cells are distinguished by the predominant presence of homodimeric CD8$\alpha\alpha$+ T cells and T line age cells containing the $\gamma\delta$ TCR [51]; interestingly, the TCR$\gamma\delta$ lineage and TCR$\alpha\beta$+CD8$\alpha\alpha$ populations do not retain immunologic memory of infection. However, the $\gamma\delta$-T cell enriched IEL function as a surveillance system for damaged or infected epithelial cells, and may modulate local immune responses by controlling cellular traffic and limiting mucosal access of inflammatory cells [52]. The $\gamma\delta$ T cells are thought to play an important role in the pathophysiologic response to infections including Staphylococcal infection. In mice, 45% of the IEL present in the small intestine are estimated to be conventional thymus-derived lymphocytes that coexpress TCR-$\alpha\beta$ and classical CD8-$\alpha\beta$. These cells primarily exhibit a cytolytic function and are recognized residents of the lamina propria, yet retain the ability to dessiminate to various anatomical sites including the gut epithelium following an antigen priming [53]. However, there are also TCR bearing $\alpha\beta$ T cells in the lamina propria, the majority of which exhibit the activated/memory phenotype; the major histocompatibility (MHC) class II-restricted CD4+ T helper (Th) cells.

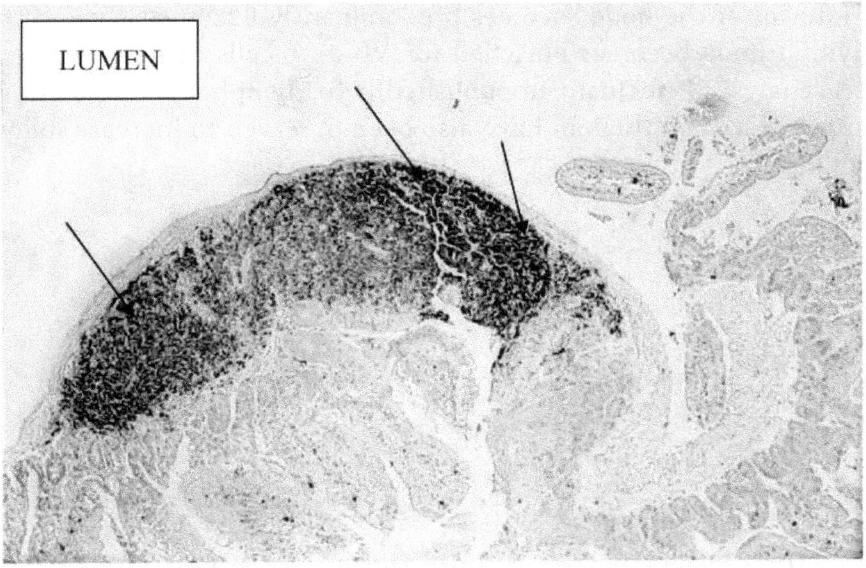

Figure 1. B220+ lymphocyte localization (indicated by arrows) in Peyer's Patch derived from normal C57Bl/10J mice. Formalin fixed tissue section [10X magnification) was stained using a monoclonal directed against B220 (RA3-6B2) and a horse-radish peroxidase conjugated antibody. B220 staining demonstrates a predominance of B cells in the unstimulated Peyer's Patch.

2.2 T cell immune activity in the GALT

During a gastrointestinal immune response, ingested antigens in the lumen enter the Peyer's Patches via the specialized epithelial cells known as M cells present in the epithelial layer overlaying the PP. The M cells can take up particulate antigen by endocytosis and transport the antigen into the interior of the PP where the dendritic cells process the antigen and present antigen to the T cell areas of the PP and MLN, initiating T cell activation and differentiation into effector cells, that will either mediate tolerance or immunologic responsiveness [54, 55]. In experiments using genetically-defined mice, ingestion of the superantigenic food toxin, Staphylococcal entertoxin B (SEB), has been demonstrated to increase the TCR- $\alpha\beta$ populations in PP (Figure 2) such that the predominant response is generated as a result of the binding between the toxin, target T cell receptor-bearing populations containing the defined Vβ-8 sequence, and antigen presenting cells [56, 57].The result of this interaction is the receptor-mediated induction of cytokine-driven T cell proliferation, resulting in a proliferation and expansion of the SEB-reactive Vβ-8+ T cells. As shown in Figure 2, normal PP contain an abundance of B220+ B cells. However, the distribution of B220+ B cells becomes dramatically altered following oral administration of SEB in quantities sufficient to induce illness in humans to genetically-defined C57Bl/10J mice. The B220+ populations become sequestered, and the interior of the node becomes predominantly B220 negative. In this case, the PP lymph node becomes enriched for Vβ-8+ T cells as determined by flow cytometric analysis (Principato, unpublished). γδ-T lymphocyte populations in PP, lamina propria, and epithelium have also been observed to increase following SE treatment [58].

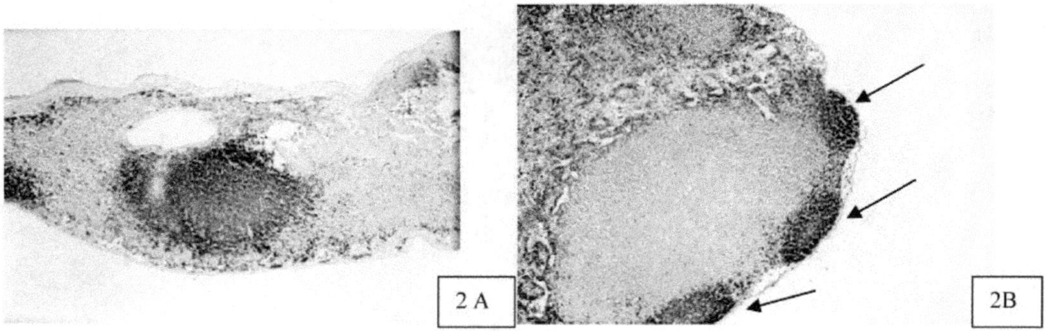

Figure 2. A. Distribution of B220+ B cells in normal C57Bl/10J mice Peyer's Patch. Formalin fixed tissue section was stained using a monoclonal antibody directed against murine B220 (RA3-6B2) and a horse-radish peroxidase conjugated antibody [10X magnification). B200 staining (brown areas) demonstrates a diffuse presence of B220+ B cells in the unstimulated Peyer's Patch. B. Expansion of non-B220+ (i.e., T cells) within Peyer's Patches 6 days following ingestion of Staphlyococcal enterotoxin B. A redistribution of the B220+ cells into aggregates forming below the PP epithelial capsule is indicated by arrows. [40X magnification).

2.3 T helper subsets of the GALT

The intestinal mucosa harbors all of the major T helper (Th) cell subsets (Th1, Th2, Treg (immunoregulatory), Th17) that are defined by their lineage-specific transcription factor expression, cytokine production, and immune function. The Th1 subset is critical for immune responses generated against intracellular pathogens, and provides cytokine-mediated "help" to the cytotoxic T lymphocytes. It is characterized by the production of interferon-gamma (IFN-γ) which is controlled by the transcription factor T-bet [59]. The Th2 subset provides help for B cells and is also implicated in allergic sensitization, including those attributable to foods [60]. The specific transcription factor for Th2 cells is GATA-3, which drives the synthesis of IL-4, IL-5, and IL-13 [61]. Tregs that have arisen from antigen-specific induction of CD4+Foxp3+ T cells are critical for the induction of oral tolerance[62]. The Th17 cells express retinoic acid-related orphan receptors (RORγt and RORα) that are needed for the transcription and synthesis of IL-17 [63, 64], and provide important protection of mucosal surfaces against extracellular bacteria.

2.4 Innate immunity in the GALT

The cells of the innate immune system include macrophages, dendritic cells, and Langerhan's cells, and are involved in critical activities pertaining to the initiation and support of T cell-mediated, antigen-specific immunity. Significantly, the distribution of these cell types includes the skin and epithelia that line the internal organs including the gastrointestinal tract. Macrophages and dendritic cells are situated below the single layer of epithelial cells that lines the Peyer's Patches and lamina propria [65]. Macrophages have long been identified as components of the reticuloendothelial system and are recognized for their ability to ingest extracellular matter including proteins, cellular fragments, and debris that is foreign to the body in the process known as phagocytosis. They are widely distributed within the tissues of the body, and are crucial components of immune responsiveness and inflammation. Initial binding of the target occurs on the surface of the cell, utilizing receptors with specific capabilities. Receptors identified include surface Fc receptors that bind the Fc portion of IgG immunoglobulin, complement C3b and C3d receptors, MHC Class I and Class II, Toll like receptors (TLR), cytokine receptors, and other membrane receptors such as the C- type lectins [66] that provide additional innate functionality which supports the binding and internalization of a wide variety of targets. Opsonization of target by plasma proteins is known to improve phagocytosis, and the endocytosing vesicles have been demonstrated to consist of clathrin structures [67, 68]. Interestingly, the endosome exhibits plasticity, and its shape has been demonstrated to change depending on the material that is engulfed [69]. In an early examination of macrophage activity, Unanue and coworkers demonstrated distinct differences in macrophage effector function based on the anatomical source of the macrophage. These investigators compared the ability of alveolar and peritoneal-derived macrophages to bind and present antigen, the intracellular pathogen Listeria monocytogenes, to previously sensitized T cells [70]. While alveolar and peritoneal macrophages both expressed class II Ia antigen, alveolar macrophages were less efficient with respect to the uptake and presentation of antigen to sensitized T cells as compared to the peritoneal macrophages. However, opsonizing Listeria using an anti-Listeria antiserum to coat the bacterium enhanced the alveolar macrophages' ability to engulf the bacterium and effectively present the antigen to sensitized T cells. Once internalized, the ingested antigen undergoes intracellular metabolic and proteolytic degradation, and modification. The resulting fragment [71, 72], is transported to the surface of the cell where it is presented in conjunction with the major histocompatibility (MHC) gene molecule. This structural relationship is critical for the activation of the appropriate responding T cell, which contains a great variability of gene sequences which must be rearranged to configure a mature, functional, TCR. This permits the specific recognition of the presented peptide sequence by the TCR of the responding T cell, and provides for the development of the adaptive immune response, which will also generate an immunologic memory of the peptide target.

Innate immunity through the Toll like receptors (TLR) is conferred with the task of recognizing a broad range of repetitive antigenic specificities that are found on a wide array of pathogens. With this type of recognition, pathogen detection is based on the ability to recognize pathogen-associated molecular patterns using evolutionarily conserved, germline encoded recognition receptors, the TLR [73-75]. Thus, while a strict sequence-dependent antigenic specificity is not required as with antigen-specific immune responsiveness, what is required is an ability to bind carbohydrate residues in a Ca++ dependent manner, and the recognition of conserved molecular patterns such as in bacterial cell wall components. Thus, LPS is the ligand for TLR 4, and targeted mutation of the TLR4 locus in mice results in LPS non-responsiveness [76]. TLR2 recognizes ligands found on yeast cell walls, bacterial lipoproteins [77], and lipoteichoic acid found in the cell walls of gram positive bacteria [78]; other TLR recognize bacterial DNA, or double stranded viral RNA. In humans and mice, there are now at least 10 such TLR identified. Unlike the sequence-specific receptors found on the antigen-binding T cells of the adaptive immune response, TLR are non-clonal and do not require gene rearrangement in order to become functionally mature.

Upon contact with a pathogen, the cells of the innate immune system become activated, the binding of their receptors initiating signaling cascades that turn on required transcription of target genes for the production of inflammatory cytokines, and the upregulation of costimulatory and MHC molecules necessary for the direct elimination of the infection or for the recruitment of adaptive immune responses. The binding of TLR with their target ligand induce costimulatory molecules that were first identified as the B7.1 and B7.2, or now referred to as CD80 and CD86. Thus, the responding T cell must recognize the modified target ligand which is expressed on the surface of the macrophage or dendritic cells in the context of both MHC and costimulator molecules with its sequence-specific TCR. It is clear that the cells of the innate immune system are critical to the establishment of an effective immune responsiveness against pathogens and for the recruitment of an efficient adaptive immune response. The extremely successful yellow fever vaccine, YF-17D, which induces both Th1/Th2 responses and generates powerful neutralizing antibodies in vaccine recipients, was shown to induce such a strong protective immunity as a result of its ability to stimulate multiple subsets of human dendritic cells and multiple TLRs [79]. Vaccine designs utilizing synthetic 300 nm PLGA nanoparticles containing antigen and ligands that bind TLR 4 and TLR 7 on the surface of dendritic cells, have successfully induced enhanced antigen-specific antibody responses against the immunizing antigen when injected into experimental mice[80]. The immunization protocol induced long-lived, high avidity antibody that was dependent upon the expression of the targeted TLR on both B cells and dendritics. The B cell response indicated the generation of memory-type B cells.

Macrophage and dendritic cells have been documented with respect to the striking specializations of the subsets. For instance, CD11b+ dendritic cells of the lamina propria can sample luminal microbes by extending their dendrites to interdigitate between neighboring intestinal epithelial cells [81], and have been reported to promote the differentiation of Th17+ regulatory T cells following activation of TLR 5 due to exposure to bacterial flagellin [82]. Interestingly, as previously observed by Unanue and coworkers [70], anatomic localization can denote distinctions in the functional effector function within subsets of cells. Thus, CD11b+CD103+ dendritic cells of the lamina propia are found preferentially in the duodenum and rarely in the colon during the steady state, but accumulate in the lamina propria of the colon along with Th17 cells during intestinal inflammation [83]. Macrophages of the lamina propria have been demonstrated be hyporesponsive to certain inflammatory stimuli, secrete IL-10, promote the differentiation of FoxP3+ regulatory T cells [84] and are able to dampen some immune responses and intestinal inflammation. For instance, a severe dextran sulfate-associated experimental colitis can be induced in a macrophage-depleted transgenic mouse or in clodronate-treated normal C57BL/6 or Babl/c mice [85]. Finally, CD103+ dendritic cells are known to assist in the antigen specific induction of FoxP3+ T regs necessary for tolerance induction[86]. Thus, the cells of the innate immune system maintain a balance between a normal state of tolerance, and inflammatory and autoimmune responses.

3. INGESTION OF NANOPARTICLES

The ingestion of nutrients with subsequent transit throughout the lumen of the gastrointestinal tract leads to the translocation of the material across the mucosa via the M cells of the epithelial layer. M cells are specialized cells that exhibit endocytic activity, and are known to transport antigens into the interior of the PP where the dendritic cells process the antigen and present antigen to the T cell areas of the PP and MLN, initiating T cell activation and differentiation into effector cells, that will either mediate tolerance or immunologic responsiveness [54, 55]. Multiple physiochemical properties, including size and surface charge, have been shown to influence nanoparticle uptake and absorption in the gut, and the extent and rate at which the particles are removed from the circulation and their ultimate biodistribution. Thus, orally-administered non-ionic nanoparticles of 100 nm or less have demonstrated preferential absorption in the Peyer's Patch and the small intestine. Focused, engineered targeting of particles to the GALT has reported success with respect to the induction of measurable antibody responses. However, the specific immunologic mechanisms inherent to nanoparticle intake and absorption within the gastrointestinal tract have not been adequately identified, and the effector pathways that generate the immune responses measured have not been characterized.

3.1 Influence of nanoparticle size and charge

Desai and coworkers demonstrated that 100 nm nanoparticles underwent a preferential uptake in the gastrointestinal tract [87] using an in situ rat ileal loop model. Polylactic polyglycolic acid (PLGA) nano- and microparticles were synthesized with averaged diameters of 100 nm, 500 nm, 1 µm, and 10 µm and infused into the tissue. Tissue uptake was quantified as weight of the nanoparticles (µg) (taking into account the density of the polymer and the

diameter of the microparticle) per square mm area of rat intestinal tissue. Infusion of the particles into gastrointestinal tissue demonstrated 100 nm particle uptake by both duodenal and ileal tissue. However, the ileum's Peyer's Patch and non-Peyer's Patch tissue demonstrated a higher uptake of 100 nm size particles. This observation was repeated using surrogate-loaded microparticles. Histologic examination of the tissue using fluorescent microscopy confirmed a greater retention of the 100 nm nanoparticles, with a concentration below the epithelial layer.

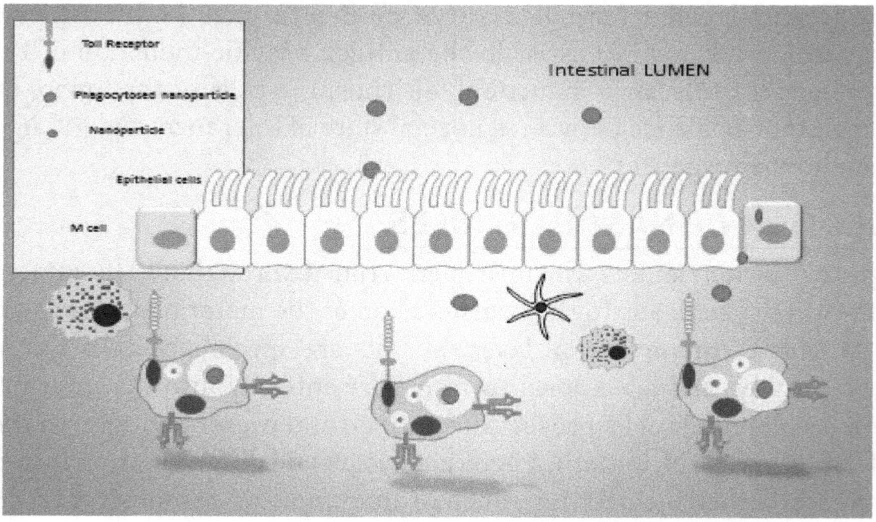

Figure 3. Schematic representation of the ingestion of nanoparticles. Macrophages and dendritic cells can be found beneath the eptihelial layer of the GALT. Ingested nonionic or targetted nanoparticles distribute preferentially below the intestinal epithelium, and can meet macrophages bearing class II and TLR molecules, and are phagocytosed by the macrophage. Actively phagocytosing macrophages are represented in the foreground; engorged macrophages are represented containing multiple particles.

The authors' observations recollect those of an earlier study utilizing latex particles [88]. In that study, Jani and coworkers conducted a 10-day feeding study in which non-ionic latex particles ranging in size from 100 nm, 500 nm, 1 micron, and 3 microns were fed to Sprague- Dawley rats. Their histologic and radiologic examination provided unequivocal evidence of a preferential tissue distribution of 100nm particles in which the Peyer's Patches, liver, and spleen demonstrated significant uptake. Significantly, their result confirmed the potential transport of particles from the gastrointestinal tract to the periphery via the lymphatics.

A separate 5 day feeding study in rats demonstrated the effects of a hydrophilic charge upon the tissue distribution of normally hydrophobic polystyrene particles [89]. In this study, commercial non-ionized polystyrene particles with a mean diameter of 60 nm were compared to similarly-sized particles coated with poloxamer 407. Their results confirmed the earlier observations by Jani and coworkers: a preferential uptake of uncharged polystyrene was noted in the small intestines and Peyer's Patches as measured using gel permeation chromatography to quantify polystyrene in the tissue, and by microscopy. Further, a smaller concentration of particles was observed to be in the mesenteric lymphatic tissue and liver. Collectively, these data indicate a movement of the particles from the lumen of the intestinal tract to the peripheral circulation with subsequent residence in other tissues. However, charged particles demonstrated a significant reduction in uptake, 1.5%- 2% of the total administered dose of particles were absorbed as opposed to 10% uptake using uncharged particles. Interestingly, the tissue distribution was altered as a result of the poloxamer coating: the particles were particularly concentrated within the tissues of the large intestine. Taken together, these results demonstrate the importance of particle size in determining the tissue range of ingested neutrally charged particles, and the critical role of charge as a particularly strong determinant of distribution within the body.

3.2 Biodistribution

The macrophage is most often implicated in the uptake of nanoparticles and opsonization will influence nanoparticle uptake into the cells. Nevertheless, final biodistribution and disposition is likely determined by the transport of particles by phagocytic and endocytotic cells. The intraperitoneal injection of 40 nm gold nanoparticles in mice has been demonstrated to result in the localization of particles in the Kupffer cells of the liver. In this research, commercially-produced colloidal gold nanoparticles containing a negative surface charge, in sizes of either 2 nm or 40 nm, were injected either intraperitoneally (ip) or intravenously (iv) into C57Bl/6 mice, and detected within cryostat sections of liver and other organs by autometallography, which amplifies the detection of gold. Interestingly, a preferential uptake of the 40 nm particles by Kupffer cells of the liver was observed 24 hours after ip injection. Very little uptake was observed 1 hour after injection. Animals who received the particles by ip injection also demonstrated particle uptake within the walls of the small intestine, mesenteric lymph node, and in the spleen illustrating that transit of the administered particles had occurred, most likely via the phagocytic cells [90]. Using a rabbit model, orally administered chitosan/PLGA spherical particles (averaged 283 nm) for the encapsulation of Vitamin A were found along the mucosal epithelium of the lumen, and within the intestinal epithelial cells, presumably due to endocytosis. Macrophages in the lamina propia showed evidence of particles as did the endothelial cells [38].

Variations to the particle, such as addition of a polymeric coating, will alter the biodistribution. Thus, coating polystyrene 60 nm and 5.25 μm particles with poloxamer polymers will decrease uptake by liver and spleen macrophages. Importantly, increasing the thickness of the coating will alter uptake as well; in this case it has resulted in a reduction of uptake by the peritoneal macrophage [91]. Experiments in which hydrophilic negatively-charged alginate-coated chitosan nanoparticles were passively absorbed into gastrointestinal tissue demonstrated localization beneath the follicle associated epithelium of the Peyer's Patches and agglomeration of the particles intracellularly, although the specific nature of the cell was not described [92].

Significant patterns in organ compartmentalization have also been described for metallic nanoparticles.

The fate of ingested silver salt and silver nanoparticle was examined in a feeding study in which separate groups of female Wistar rats were administered 9 mg silver acetate or 12.6 mg silver nanoparticle per kg of body weight daily for 28 days. It was estimated that 63% of the daily ingested dose was excreted in the feces. The overall accumulation of either silver ion or nanoparticle was similar, and appeared greatest in the small intestine while also detectable in liver, kidney, and stomach. Autometallographic staining (AMG) detects the presence of either silver acetate or nanoparticle; thus, silver was localized to the lamina propia and submucosa in the ileum. Interestingly, silver was concentrated around the veins and portal circulation of the liver and was not preferentially taken up by the Kupffer cells of the liver as was reported with injected gold [90]. Transmission electron microscopy displayed similar localizations for both the silver nanoparticles and silver acetate; the material was found within the lysosomes of the macrophages within the lamina propia of the ileum [93].

4. IMMUNE RESPONSES AND THE INGESTED NANOPARTICLE/MICROPARTICLE

Particulate antigens are known to induce stronger immune responsiveness to the antigen when compared to an immune response generated with the soluble form of antigen. Thus, vaccine design has recently emphasized the use of nanoparticles to maximize induction of the protective immune responsiveness. Mucosal immunizations have been viewed increasingly as an alternative to parenteral administration of vaccines, and features such as carbohydrate residue targeting by lectins has been examined by many groups [94-96]. Nevertheless, nanoparticle absorption within the GALT is still not well understood, and the effector cellular interactions involved in the generation of the induced immune response have not been fully defined.

Immunoglobulin production as a function of particle size, was measured by Gutierro and coworkers [97]. Bovine Serum Albumin (BSA) –loaded PLGA microspheres of 200 nm, 500 nm, and 1000 nm were constructed using a double emulsion technique; size was determined by laser diffractometry using a CoulterCounter® particle size analyzer. PLGA microspheres containing BSA target antigen was administered by each of three routes: subcutaneously, intranasally, or orally into 6-8 week old Balbc/J mice and the elicited immune response was measured by assaying IgG immunoglobulin production. Their results showed that IgG antibodies were elicited using each of the three sizes of microspheres when administered subcutaneously; one size did not elicit greater antibody production than the others. Further, all three sizes elicited antibody responses that were greater than that elicited using either soluble antigen or conventional adjuvant approaches. The oral immunization protocol consisted of orally feeding each of the three sizes of microspheres, each containing 500 ∝g BSA, on three successive days. Interestingly, oral administration of the loaded microspheres showed that the 200 nm and 500 nm sized particles elicited fewer antibodies than an administration of antigen with either alum or Freund's adjuvant. The greatest production of serum IgG was demonstrated using the 1000 um size particle and this group contained the higher percentage of individual responders. Analysis of serums at weeks 3 and 5 following immunization did not reveal differences in IgG2a/IgG1 isotype profiles, the latter being indicative of Th1/Th2 subset immunity and antigen-presenting differences. Ultimately, no differences were found among the various sized particles, suggesting that the method of antigen presentation was the same for all of the sizes tested. Again, the larger particles provided the higher immunoglobulin production, regardless of the mode of immunization. These results are interesting from the perspective of what has been reported [87, 88] regarding the effect of size and nanoparticle biodistribution, and what is known about the distribution of effector cells within the GALT. The present experiments used particles that were larger than those previously published; it is likely that the biodistribution affected the manner in which particulate antigen was presented for the induction of an immune response.

4.1 Targeting M cells in the GALT

Directed PLGA nanoparticles, using lectins to bind onto target sugar residues, has been shown to be a means to achieve organ targeting for the induction of a systemic immune response. In one study, PLGA nanoparticles were created by the double emulsion method and loaded with hepatitis B surface antigen (HBsAg) [95]. Lectin directed to α-L- fucose residues, Tetragonolobus purpureas, was bound to the nanoparticle using 1-ethyl-3-(3 dimethylaminopropyl) carbodiimide to produce TLA lectin-PLGA-HBsAg nanoparticles that were measured to be 270 + 23nm in size. Confocal microscopy confirmed binding of the particles to the M cells of the Peyer's Patches within immunized mice. Further, lectinized particles were stabilized by the addition of hydrophilic trehalose, which improves the release of antigen. Therefore, nanoparticles stabilized with trehalose demonstrated an increased antigen release of 43.2+2.7% after 35 days, as opposed to a release of 32.4 +2.3% by the non-stabilized equivalent. In this investigation, 10 mg of encapsulated antigen per dose was used for the oral immunization of 8 week old Balb/c mice, followed with a booster 2 weeks following the primary immunization. Thus, HBsAg -loaded PLGA nanoparticles, TLA lectin-PLGA-HBsAg nanoparticles, trehalose-stabilized HBsAg - loaded PLGA nanoparticles, and trehalose-stabilized TLA lectin-PLGA-HBsAg nanoparticles were compared for the induction of antibody. Significantly, this study demonstrated the successful induction of antigen-specific IgG antibody by each of the engineered nanoparticles as determined by ELISA assay of the immune seras, when compared to the levels of antibody produced by the animals immunized with an alum based antigen. Isotyping of the antibodies demonstrated induction of IgG1 antibody, indicative of a Th2 response, at levels that were twice those attained by IgG2a which is indicative of a Th1 response. While demonstrable levels of the Th1 cytokines, IL-2 and γ-IFN, were detected in the spleens of all nanoparticle-treated animals, greater levels of γ-IFN were obtained with TLA lectin-PLGA-HBsAg, with or without stabilization by trehalose. It is not known whether the engineered particle could have induced a greater γ-IFN response as PLGA nanoparticles without antigen were not used for comparison in this study.

A directed approach has been extremely successful using chitosan alginate microparticles [94]. In this instance, chitosan nanoparticles (CNP) prepared by the ionic gelation method were loaded with BSA test antigen, and coated with alginate. Thus, alginate was modified by using 1-ethyl-3-(3 dimethylaminopropyl) carbodiimide to form amide linkages between the carboxylate residues on alginate and the amino group of the lectin Ulex europaeus agglutinin (UEA-1). The lectin Ulex europaeus agglutinin (UEA-1) was used to direct the microparticles towards the α-L- fucose residues found on the surface of M cells. Confocal microscopy confirmed the targeting; punctate staining was visualized using the lectin-modified microspheres. The conjugation and loading resulted in a particle shift in size: the particle size of the original CNP particle is reported as 257+ 55.17 nm, while the lectin-modified antigen carrier CNP particle size increased to 1485 + 214.3 nm. Oral immunization of 6-8 week old Balb/c mice with each of the preparations and control antigen provided striking differences in the antibody responses against BSA antigen. The highest IgG titers were obtained using alum-absorbed BSA as the immunogen's positive control, and the lowest titers were obtained using BSA loaded CNP. In contrast, antigen encapsulated in lectin-modified alginate chitosan particles (LACNP) consistently generated IgG titers that were greater than those obtained with CNP or ACNP formulations. Demonstrable levels of antigen-specific IgG2a/IgG1 were detected with all three formulations. Significantly, the highest titers of antigen-specific IgG were obtained with lectin- modified microspheres, and the results seem to indicate that there was a greater IgG2a, or Th1 response, to antigen (BSA) with that particle. The original CNP particle and the alginate chitosan particles seemed to have induced a greater Th2 response.

Together, these studies demonstrate the induction of a Th1/Th2-induced immunity using engineered particles as do others [93]. However, it is not known whether α- fucose residues are found on macrophages and dendritic cells present at other body sites, possibly resulting in multiple pathways of immune responsiveness. As discussed earlier, Unanue and coworkers demonstrated distinct differences in macrophage effector function based on the anatomical source of the macrophage [69]. Further, while TLA lectin-PLGA-HBsAg nanoparticles induced the production of sIgA in saliva and gastrointestinal fluids, it was not reported whether the engineered particles in these reports ultimately interacted with CD11+ dendritic cells. IgA has been reported to be induced by regulatory T cells that have been activated by CD11+ dendritic cells [50]. Finally, the directed attachment of the particles to the endocytotic M cells of the epithelial layer presents the possibility that the particles were transcytosed by the M cells towards CD103+ dendritic cells, found beneath the epithelial layer. In that case, the possibility exists for the induction of tolerance [86]. Normal exposure to ingested, digested antigen results in the production of regulatory T cells that suppress an immune response in an antigen specific manner, resulting in tolerance and preventing food allergy. However, the targeted microparticles document the induction of Th1and Th2 responses.

5. FUTURE CONSIDERATION: INGESTED NANOPARTICLE AND IMMUNE ALLERGIC DYSFUNCTION TO FOODS

Proteins used in commercial food applications include casein, whey protein, collagen, egg white, and fish myofibrillar protein, and popular plant-based proteins including soybean protein and wheat gluten [7]. Compounds such as polysaccharides and proteins that are already in use within commercial food applications are attractive candidates for the production of new nanocomposite packaging and encapsulation material, as several are generally regarded as safe and are biodegradable. However, food allergy has emerged as a growing health problem throughout modern society, and current research efforts towards the identification and characterization of clinically relevant food allergens are critical to our understanding of their role in the immunopathogenic mechanisms involved in hypersensitivity reactions, and the safety of novel and proposed food-oriented nanotechnology. Thus, the characterization and identification of the proteins responsible for immune-mediated food allergies is critical.

In view of reported differences with respect to nanoparticle size and organ biodistribution, it is interesting to note that particles are often detected below the epithelium of the gastrointestinal tract. Since the Intraepithelial lymphocytes (IEL) reside below the epithelial layer of the mammalian gastrointestinal tract, an understanding of the interactions between particle and resident IEL is crucial. Following the ingestion of food, digested protein fragments, or antigens, cross the epithelium to be processed and presented on the surface of class II molecule-bearing antigen presenting cells for recognition by specific TCR-bearing T cells. Allergic sensitization in the presence of IL-4 results in the generation of Th2 cells that will assist the development of IgE+ B cells. A repeat encounter with the antigen will result in a food allergic response. This event generates a skewed Th2 response, and will occur when luminal antigen is introduced to IgE bound onto IgE Fc receptor on the surface of mast cells. Thus, crossing the epithelial barrier to reach the mast cells is a critical step. The binding of antigen to the receptor-bound complex will result in the release of histamine, serotonin and prostaglandins in anaphylactic reactions including those generated by food.

Recent studies suggest that intestinal epithelial cells play a central regulatory role in determining the rate and pattern of uptake of ingested antigens. This is particularly critical in food allergy within the antigen-sensitized gastrointestinal tract. Studies using rats sensitized to horseradish peroxidase (HRP) showed that intestinal antigen transport is keenly affected by antigen-specific sensitization and is composed of 2 phases. The first phase consists of the rapid transepithelial transport of specific antigen from the lumen, via endocytosis, into the lamina propia. This phase is antigen specific, implying the existence of an antigen -specific receptor on the surface of the epithelial cells, and occurs within 2 minutes in sensitized rats as compared to a transit time of 20 minutes in non-sensitized, normal control animals. This is followed by a flow of the antigen in tight junctions resulting in an increase of antigen across the tissue. The second phase of antigen transport is not antigen specific, but is markedly increased by antigen challenge in sensitized rats compared with non-sensitized controls [98], indicative of the paracellular penetration through the epithelium by antigen. These studies clearly demonstrate that the kinetics of transport of antigen during IgE-mediated reactions in the gastrointestinal tract is markedly increased across the epithelium. The result of this transport is the generation of a Th2 response.

Finally, a feeding study using mice orally sensitized to the known milk allergens, casein, β- lactalbumin, and β-lactoglobulin, provided compelling evidence regarding the importance of the form of the antigen (soluble vs. particulate) for the induction of anaphylaxis [99]. The soluble proteins, β-lactalbumin, and β-lactoglobulin, resulted in anaphylactic reactions when administered orally. Interestingly, the soluble proteins were detected in the lamina propria of the small intestine of sensitized mice indicating that these proteins were able to transcytose through the enterocytes in vivo. This observation was confirmed in vitro using Caco-2 cells. Further, the challenge with sensitizing antigen resulted in significant levels of serum IgG1, and low, but detectable levels, of serum IgE and IgG2a. Casein, normally present within micelles, demonstrated a significant difference in anaphylactic induction. Oral administration did not induce anaphylaxis. Instead, casein required a systemic administration (i.p. injection) in order to induce anaphylaxis; and it induced significantly higher serum IgE and IgG1 (Th2) allergic responses as compared to the soluble milk allergens. Further, transcytosis by casein through Caco2 monolayers was poor compared to the soluble milk allergens. When the tissue was examined by fluorescence microscopy, the casein was detectable in the Peyer's patches. Thus, these data indicated that the form of the sensitizing antigen was critical to the induction of an anaphylactic response. Next, the soluble allergens, β- lactoglobulin and soluble α-lactalbumin, were next converted into particulate aggregates by pasteurization; the process reportedly abolishes the monomeric form and supports the formation of aggregates of approximately 670 kDa. Pasteurization does not alter casein, and it exists in two predominant types as it would in its natural state: 180 kDa and 670 kDa. The conversion of soluble β- lactoglobulin and soluble α-lactalbumin into particulate aggregates by pasteurization altered the immunogenicity of the proteins such that they now required a systemic administration to induce anaphylaxis. Oral administration of either protein aggregate in sensitized mice did not induce anaphylaxis. The magnitude of the elicited serum IgG1 and IgE immunoglobulin production was much greater than that induced by their soluble forms. Further, the proteins were now detectable in association with the Peyer's Patches. Casein's induction was not altered by the process. Taken together, these results present the critical role of antigenic structure and its uptake across the epithelium as critical factors contributing to the allergic state.

The allergic state presents serious challenges to the incorporation of nanoparticles in food and food-associated products, particularly when considering the composition and ultimate biodistribution of the particles. The engineering of nanoparticle containing materials implicated, related, or identified as allergens raises concern for the initiation of alternate allergy- inducing pathways in the host. For instance, while casein is incorporated in a variety of foods and is generally regarded as safe, it is also known to elicit strong allergic responses in afflicted individuals with dairy intolerance. Disruption of the epithelial barrier is known to result in gastrointestinal illness [100]. Infection and inflammation are conditions associated with a disruption of the epithelial layer leading to the increased paracellular transport of luminal antigen. Cytokines such as IFN-γ and TNF-α directly affect barrier function of the epithelium, the latter being implicated in milk allergy [101-103]. Thus, the transit of nanoparticle through a sensitized gastrointestinal system might result in a more complicated scenario, depending upon the sensitizing antigen and the composition of the nanoparticle itself. Thus, casein nanoparticle constructs, with or without targeting lectins, might not be advisable for individuals with casein sensitivity. In this instance, the transit of the nanoparticle might be hastened across the layer, due to the pre-existing sensitivity, resulting increased transit through the layer, perhaps overwhelming the resident macrophage phagocytic activity (Figure 4), leading to exacerbation of the allergic state or the generation of alternative immunologic reactions.

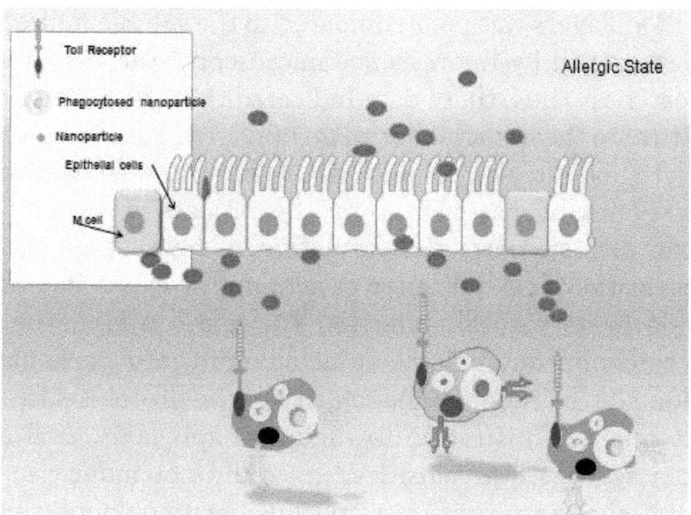

Figure 4. Schematic representation of nanoparticle transit through the epithelial layer and the allergic state. An increased rate of transit by the nanoparticles is theorized as a result of the induction or presence of an allergic state.

6. CONCISE SUMMARY

The selection of materials for formulating nanoparticles and spheres in encapsulated nutrients or supplements for ingestion is crucial to potential effects on mucosal immunity. An essential factor is whether the material will affect the induction of tolerance or active immunity to the injected nutrition due to its deposition in the gastrointestinal tract and potential interaction with resident effector cells.

The precise targeting of nanoparticles and spheres through unique ligand interactions offers a distinct advantage in this regard. Polymers incorporating natural biodegradable materials like chitosan, PLGA, whey, and casein provide significant benefits in this technology; however, they also pose additional challenges in comprehending the mechanisms that sustain gastrointestinal immune homeostasis and avert the onset or exacerbation of immune dysfunction.

REFERENCES

1. Anonymous. Nanotechnology 101. Nano.gov/nanotech-101/special. 2009. (accessed 3/22/12)
2. Cushen M, Kerry J, Morris M, Cruz-Romero M, Cummins E. Nanotechnologies in the food industry-Recent developments, risks and regulation. . Trends in Food Science and Technology. 2012;24:30-46.
3. Gibis M, Vogt E, Weiss J. Encapsulation of polyphenolic grape seed extract in polymer-coated liposomes. Food & function. 2012;3(3):246-54. Epub 2011/11/26.
4. Markman G and Yoav L. Maillard-conjugate based core shell co-assemblies for nanoencapsulation of hydrophobic nutraceuticals in clear beverges. Food & function. 2012;3:263-70.
5. Huang Q, Yu H, Ru Q. Bioavailability and delivery of nutraceuticals using nanotechnology. Journal of food science. 2010;75(1):R50-7. Epub 2010/05/25.
6. Anonymous FaDA. About the FCS Review Program 2010. http://www.fda.gov/Food/ FoodIngredientsPackaging/FoodContactSubstancesFCS/AbouttheFCSReviewProgram/default.htm (accessed 3/22/12)
7. Arora A, Padua GW. Review: nanocomposites in food packaging. Journal of food science. 2010;75(1):R43-9. Epub 2010/05/25.
8. Chaudhry Q, Scotter M, Blackburn J, Ross B, Boxall A, Castle L, Aitken R, Watkins R. Applications and implications of nanotechnologies for the food sector. Food additives & contaminants Part A, Chemistry, analysis, control, exposure & risk assessment. 2008;25(3):241-58. Epub 2008/03/04.
9. Augustin M, Sanguansri P . Nanostructured Materials in the Food Industry. Advancesin Food and Nutrition Research. Advances in Food and Nutrition Research. 2009;58:182-213.
10. Ray SS, Yamada K, Okamoto M, Ogami A, Ueda K. New polylactide/layered silicate nanocomposites. 3. High-performance biodegradable materials. Chem Mater. 2003;15(7):1456-65.

11. Ray SS, Yamada K, Okamoto M, Ueda K. Polylactide-layered silicate nanocomposite: A novel biodegradable material. Nano Letters. 2002;2(10):1093-6.

12. Ogata N, Jimenez G, Kawai H, Ogihara T. Structure and thermal/mechanical properties of poly(l-lactide)-clay blend. J Polym Sci Pol Phys. 1997;35(2):389-96.

13. Russel AD, and Hugo WB. Antimicrobial activity and action of silver. . Prog Med Chem. 1994;31: 351-70.

14. Brunne D. Metal release from dental biomaterials. Biomaterials. 1986;7:163-75.

15. Parsons D, Bowler PG, MylesV, Jones S. Silver Antimicrobial Dressings in Wound Management: Comparison of Antibacterial, Physical, and Chemical Charactristics. . Wound. 2005;7(8):222-32.

16. Illingworth B, Bianco RW, Weisberg S. In vivo efficacy of silver coated fabric against fungal burn wound pathogens. 2000;27:344-50.

17. Klasen HJ. Historical review of the useof silver in the treatment of burns. Burns. 2000;26:117-30.

18. Anonymous USEPA. EPA reregistration eligibility document for silver, case4082. 1993. http://www.epa.gov/oppsrrd1/REDs/old_reds/silver-pdf. (accessed 3/22/12)

19. Feng QL, Wu. J, Chen GQ, Cui FZ, Kim TN, Kim OJ. A mechanistic study of the antibacterial effect of silver ions in Escherichia coli and Staphylococcus aureus. Appl Environ Microbiol. 2008;74:2171-8

20. Sondi I, Salopek-Sondi B. Silver nanoparticles as antimicrobial agent: a case study on E-coli as a model for Gram-negative bacteria. J Colloid Interf Sci. 2004;275(1):177-82.

21. Bragg PD, Rainnie DJ. The effect of silver ions on the respiratory chain of Escherichia coli. Can J Microbiol 1973;20:883–9. .

22. Percival SL, Bowler PG, Russell D. Bacterial resistance to silver in wound care. The Journal of hospital infection. 2005;60(1):1-7. Epub 2005/04/13.

23. Raffi M, Hussain F, Bhatti TM, Akhter JI, Hameed A, Hasan MM. Antibacterial characterization of silver nanoparticles against E. coli ATCC-15224. J Mater Sci Technol. 2008;24(2):192-6.

24. Kim JS, Kuk E, Yu KN, Kim JH, Park SJ, Lee HJ, Kim SH, Park YK, Park YH, Hwang CY, Kim YK, Lee YS, Jeong DH, Cho MH. Antimicrobial effects of silver nanoparticles. Nanomed-Nanotechnol. 2007;3(1):95-101.

25. Pal S, Tak YK, Song JM. Does the antibacterial activity of silver nanoparticles depend on the shape of the nanoparticle? A study of the gram-negative bacterium Escherichia coli. Appl Environ Microb. 2007;73(6):1712-20.

26. Sarkar S, Jana AD, Samanta SK, Mostafa G. Facile synthesis of silver nano particles with highly efficient anti-microbial property. Polyhedron. 2007;26(15):4419-26.

27. Morones JR, Elechiguerra JL, Camacho A, Holt K, Kouri JB, Ramirez JT, Yacaman MJ. The bactericidal effect of silver nanoparticles. Nanotechnology. 2005;16(10):2346-53. Epub 2005/10/01.

28. Fernandez A, Soriano, E., Lopez-Carballo G., Picouet P, Lloret E, Gavara R, Hernandez-Munoz P. Preservation of aseptic conditions inabsorbent pads by using silver nanotechnology. Food Research International. 2009;42(8):1105-1112.

29. Shameli K, Ahmad MB, Zargar M, Yunus WM, Rustaiyan A, Ibrahim NA. Synthesis of silver nanoparticles in montmorillonite and their antibacterial behavior. International journal of nanomedicine. 2011;6:581-90. Epub 2011/06/16.

30. Cohen-Sela E, Chorny M, Koroukhov N, Danenberg HD, Golomb G. A new double emulsion solvent diffusion technique for encapsulating hydrophilic molecules in PLGA nanoparticles. Journal of controlled release : official journal of the Controlled Release Society. 2009;133(2):90-5. Epub 2008/10/14.

31. Jain R. The manufacturing techniques of various drug loaded biodegradable poly (Lactide-co-glycolide) (PLGA) devices. Biomaterials. 2001(21):2475-90.

32. Devrim B, Bozkir A, Canefe K. Preparation and evaluation of PLGA microparticles as carrier for the pulmonary delivery of rhIL-2:I. Effects of some formulation parameters on microparticle characteristics. J Microencapsul. 2011;28(6):582-94.

33. Kumar N, Ravikumar MN, Domb AJ. Biodegradable block copolymers. Adv Drug Deliv Rev. 2001;53(1):23-44. Epub 2001/12/06.

34. Patel MP, Patel RR, Patel JK. Chitosan mediated targeted drug delivery system: a review. Journal of pharmacy & pharmaceutical sciences : a publication of the Canadian Society for Pharmaceutical Sciences, Societe canadienne des sciences pharmaceutiques. 2010;13(4):536-57. Epub 2010/01/01.

35. Borchard G, LueBen H L, de Boer A G, Verhoef JC, Lehr CM, Juninger H E. The potential of mucoadhesive polymers in enhancing intestinal peptide drug absorption. III: Effects of chitosan-glutamate and carbomer on epithelial tight junctions in vitro. . J Controlled Release. 1996;39:131-8.

36. Artursson P, Lindmark T, Davis SS, Illum L. Effect of chitosan on the permeability of monolayers of intestinal epithelial cells (Caco-2). Pharmaceutical research. 1994;11(9): 1358-61. Epub 1994/09/01.

37. Albertini B, Di Sabatino M, Calogera G, Passerini N, Rodriguez L. Encapsulation of Vitamin A palmitate for animal supplementation:Formulation , manufacturing, and stability implications. J Microencapsulation. 2010;27(150-161.).

38. Murugeshu A, Astete C, Leonardi C, Morgan T, Sabliov CM. Chitosan/PLGA particles for controlled release of alpha-tocopherol in the GI tract via oral administration. Nanomedicine (Lond). 2011;6(9):1513-28. Epub 2011/06/29.

39. Chen L, Subirade M. Effect of preparation conditions on the nutrient release properties of alginate-whey protein granular microspheres. European journal of pharmaceutics and biopharmaceutics : official journal of Arbeitsgemeinschaft fur Pharmazeutische Verfahrenstechnik eV. 2007;65(3):354-62. Epub 2006/12/08.

40. Hebrard G, Hoffart V, Beyssac E, Cardot JM, Alric M, Subirade M. Coated whey protein/alginate microparticles as oral controlled delivery systems for probiotic yeast. J Microencapsul. 2010;27(4):292-302.

41. Elabbadi A, Jeckelmann N, Haefliger OP, Ouali L. Complexation/encapsulation of green tea polyphenols in mixed calcium carbonate and phosphate micro-particles. J Microencapsul. 2011;28(1):1-9. Epub 2010/10/16.

42. Luongo D, D'Arienzo R, Bergamo P, Maurano F, Rossi M. Immunomodulation of gut-associated lymphoid tissue: current perspectives. Int Rev Immunol. 2009;28(6): 446-64. Epub 2009/12/04.

43. Strober W, Kelsall B, Marth T. Oral tolerance. J Clin Immunol. 1998;18(1):1-30. Epub 1998/02/25.

44. van Wijk F, Cheroutre H. Intestinal T cells: facing the mucosal immune dilemma with synergy and diversity. Seminars in immunology. 2009;21(3):130-8. Epub 2009/04/24.

45. Khoo UY, Proctor IE, Macpherson AJ. CD4+ T cell down-regulation in human intestinal mucosa: evidence for intestinal tolerance to luminal bacterial antigens. J Immunol. 1997;158(8):3626-34. Epub 1997/04/15.

46. Chen Y, Kuchroo VK, Inobe J, Hafler DA, Weiner HL. Regulatory T cell clones induced by oral tolerance: suppression of autoimmune encephalomyelitis. Science. 1994;265(5176):1237-40. Epub 1994/08/26.

47. Karlsson M, Lundin S, Dahlgren U, Kahu H, Pettersson I, Telemo E. "Tolerosomes" are produced by intestinal epithelial cells. European journal of immunology. 2001;31(10):2892-900. Epub 2001/10/10.

48. Mestecky J, Lue C, Russell MW. Selective transport of IgA. Cellular and molecular aspects. Gastroenterology clinics of North America. 1991;20(3):441-71. Epub 1991/09/01.

49. Cunningham-Rundles C, Brandeis, W., Good, Robert A., Day, N.K. . Milk precipitin circulating immune complexes and IgA deficiency. Proc Natl Acad Sci USA 1978;75:3387-9.

50. Johansson C, Kelsall BL. Phenotype and function of intestinal dendritic cells. Seminars in immunology. 2005;17(4):284-94. Epub 2005/06/28.

51. LeFrancois L and Puddington L. Basic Aspects of Intraepithelial lymphocytes immunobiology. In: L Ogra, J Mestecky, ME Lamm, W Strober, J Bienstock, JR McGhee (eds) Mucosal Immunology. SanDiego : Academic Press; 1999.

52. D'Souza CD, Cooper AM, Frank AA, Mazzaccaro RJ, Bloom BR, Orme IM. An anti- inflammatory role for gamma delta T lymphocytes in acquired immunity to Mycobacterium tuberculosis. J Immunol. 1997;158(3):1217-21. Epub 1997/02/01.

53. Masopust D, Vezys V, Marzo AL, Lefrancois L. Preferential localization of effector memory cells in nonlymphoid tissue. Science. 2001;291(5512):2413-7. Epub 2001/03/27.

54. Macpherson AJ, Uhr, T. Induction of protective IgA by intestinal dendritic cells carying commensal bacteria. Science. 2004;303:1662-5.

55. Weiner HL. Oral tolerance: immune mechanisms and the generation of Th3-type TGF-beta-secreting regulatory cells. Microbes Infect. 2001;3(11):947-54. Epub 2001/09/21.

56. Spiekermann GM, Nagler-Anderson C. Oral administration of the bacterial superantigen staphylococcal enterotoxin B induces activation and cytokine production by T cells in murine gut-associated lymphoid tissue. J Immunol. 1998;161(11):5825-31. Epub 1998/12/02.

57. Principato MA, Raybourne, RB, Hakim FT. Analysis of TCR Vβ8 expression in Peyers Patch lymphocytes. FASEB conference proceedings 1990; 4(7) 1132. Washington,DC.

58. Perez-Bosque A, Miro L, Polo J, Russell L, Campbell J, Weaver E, Crenshaw J, Moreto M. Dietary plasma proteins modulate the immune response of diffuse gut-associated lymphoid tissue in rats challenged with Staphylococcus aureus enterotoxin B. The Journal of nutrition. 2008;138(3):533-7. Epub 2008/02/22.

59. Szabo SJ, Sullivan BM, Stemmann C, Satoskar AR, Sleckman BP, Glimcher LH. Distinct effects of T-bet in TH1 lineage commitment and IFN-gamma production in CD4 and CD8 T cells. Science. 2002;295(5553):338-42. Epub 2002/01/12.

60. Berin MC, Mayer L. Immunophysiology of experimental food allergy. Mucosal Immunology. 2009;2(1):24-32.

61. Zheng W, Flavell RA. The transcription factor GATA-3 is necessary and sufficient for Th2 cytokine gene expression in CD4 T cells. Cell. 1997;89(4):587-96. Epub 1997/05/16.

62. Hadis U, Wahl, B., Scholz, O., Hardtke-Wolenski, M., Schippers, A., Wagner, N., et. al. Intestinal tolerance requires gut homing and expansion of FoxP3+ regulatory T cells in the lamina propia. Immunity. 2011;34(2):237-46.

63. Ivanov, II, McKenzie BS, Zhou L, Tadokoro CE, Lepelley A, Lafaille JJ, Cua DJ, Littman DR. The orphan nuclear receptor RORgamma directs the differentiation program of proinflammatory IL-17+ T helper cells. Cell. 2006;126(6):1121-33. Epub 2006/09/23.

64. Yang XO, Pappu BP, Nurieva R, Akimzhanov A, Kang HS, Chung Y, Ma L, Shah B, Panopoulos AD, Schluns KS, Watowich SS, Tian Q Jetten AM, Dong C. T helper 17 lineage differentiation is programmed by orphan nuclear receptors ROR alpha and ROR gamma. Immunity. 2008;28(1):29-39. Epub 2008/01/01.

65. Stumbles PA, McWilliam AS, Holt P G, Dendritic Cells and Macrophages. In: L Ogra, J Mestecky, ME Lamm, W Strober, J Bienstock, JR McGhee (eds) Mucosal Immunology. SanDiego : Academic Press; 1999.

66. Weis WI, Taylor ME, Drickamer K. The C-type lectin superfamily in the immune system. Immunological reviews. 1998;163:19-34. Epub 1998/08/13.

67. Aggeler J, and Werb, Z. . Initial events during phagocytosis by macrophages viewed from the outside and inside the cell: membrane-particle interactions and clathrin. J Cell Biol. 1982;94:613-23.

68. Unanue ER, Ungewickell E, Branton D. The binding of clathrin triskelions to membranes from coated vesicles. Cell. 1981;26(3 Pt 1):439-46. Epub 1981/11/01.

69. Knapp PE, Swanson JA. Plasticity of the tubular lysosomal compartment in macrophages. Journal of cell science. 1990;95 (Pt 3):433-9. Epub 1990/03/01.

70. Weinberg DS, Unanue, E.R. . Antigen –presenting function of alveolar macrophages:uptake and presentation of Listeria moncytogenes. J Immunol. 1981;126:794-9.

71. Unanue ER, Allen PM. The basis for the immunoregulatory role of macrophages and other accessory cells. Science. 1987;236(4801):551-7. Epub 1987/05/01.

72. Zeigler K, Unanue, ER. Identification of a macrophage antigen-processing event required for I-region-restricted antigen presentation to T lymphocytes. J Immunol. 1981;127:1869-75.

73. Kawai T, Akira S. The role of pattern-recogntion receptors in innate immunity: update on Toll-like receptors. Nature Immuno. 2010;11:373–84

74. Kawai T, Akira S. The roles of TLRs, RLRs and NLRs in pathogen recognition. Inter national immunology. 2009;21(4):317-37. Epub 2009/02/28.

75. Janeway CA, Jr., Medzhitov R. Innate immune recognition. Annual review of immunology. 2002;20:197-216. Epub 2002/02/28.

76. Hoshino K, Takeuchi O, Kawai T, Sanjo H, Ogawa T, Takeda Y, Takeda K, Akira S. Cutting edge: Toll-like receptor 4 (TLR4)-deficient mice are hyporesponsive to lipopolysaccharide: evidence for TLR4 as the LPS gene product. J Immunol. 1999;162:3749–52.

77. Aliprantis AO, Yang RB, Mark MR, Suggett S, Devaux B, Radolf JD, Klimpel GR, Godowski P, Zychlinsky A. Cell activation and apoptosis by bacterial lipoproteins through toll-like receptor-2. Science. 1999;285(5428):736-9.

78. Takeuchi O HK, Kawai T, Sanjo, H, Takada H, OgawaT, Takeda K, Akira S. Differential roles of TLR2 and TLR 4 in recognition of gram negative and gram positive bacterial cell wall components. Immunity. 1999;11:443-51.

79. Querec T, Bennouna S, Alkan S, Laouar Y, Gorden K, Flavell R, Akira S, Ahmed R, Pulendran B. Yellow fever vaccine YF-17D activates multiple dendritic cell subsets via TLR2, 7, 8, and 9 to stimulate polyvalent immunity. The Journal of experimental medicine. 2006;203(2):413-24. Epub 2006/02/08.

80. Kasturi SP, Skountzou I, Albrecht RA, Koutsonanos D, Hua T, Nakaya HI, Ravindran R, Stewart S, Alam M, Kwissa M, Villinger F, Murthy N, Steel J, Jacob J, Hogan RJ, Garcia-Sastre A, Compans R, Pulendran B. Programming the magnitude and persistence of antibody responses with innate immunity. Nature. 2011;470(7335):543-7. Epub 2011/02/26.

81. Rescigno M, Urbano M, Valzasina B, Francolini M, Rotta G, Bonasio R, Granucci F, Kraehenbuhl JP, Ricciardi-Castagnoli P. Dendritic cells express tight junction proteins and penetrate gut epithelial monolayers to sample bacteria. Nature immunology. 2001;2(4):361-7. Epub 2001/03/29.

82. Uematsu S, Fujimoto K, Jang MH, Yang BG, Jung YJ, Nishiyama M, Sato S, Tsujimura T, Yamamoto M, Yokota Y, Kiyonon H, Miyasaka M, Ishii KJ, Akira S. Regulation of humoral and cellular gut immunity by lamina propria dendritic cells expressing Toll-like receptor 5. Nature immunology. 2008;9(7):769-76. Epub 2008/06/03.

83. Denning TL, Norris BA, Medina-Contreras O, Manicassamy S, Geem D, Madan R, Karp CL, Pulendran, B. Functional Specializations of Intestinal Dendritic Cell and Macrophage Subsets That Control Th17 and Regulatory T Cell Responses

Are Dependent on the T Cell/APC Ratio, Source of Mouse Strain, and Regional Localization. J Immunol. 2011;187(2):733-47.

84. Denning TL, Wang YC, Patel SR, Williams IR, Pulendran B. Lamina propria macrophages and dendritic cells differentially induce regulatory and interleukin 17-producing T cell responses. Nature immunology. 2007;8(10):1086-94. Epub 2007/09/18.

85. Qualls JE, Kaplan AM, van Rooijen N, Cohen DA. Suppression of experimental colitis by intestinal mononuclear phagocytes. Journal of leukocyte biology. 2006;80(4): 802-15. Epub 2006/08/05.

86. Coombes JL, Siddiqui KR, Arancibia-Carcamo CV, Hall J, Sun CM, Belkaid Y, Powrie, F. A functionally specialized population of mucosal CD103+ DCs induces Foxp3+ regulatory T cells via a TGF-beta and retinoic acid-dependent mechanism. The Journal of experimental medicine. 2007;204(8):1757-64. Epub 2007/07/11.

87. Desai MP, Labhasetwa r V, Amidon GL, Levy RJ. Gastrointestinal uptake of biodegradable microparticles: effect of particle size. Pharmaceutical research. 1996;1 3(12): 1838-45. Epub 1996/12/01.

88. Jani P, Halbert GW, Langridge J, Florence AT. The uptake and translocation of latex nanospheres and microspheres after oral administration to rats. The Journal of pharmacy and pharmacology. 1989;41(12):809-12. Epub 1989/12/01.

89. Florence AT, Hillery AM, Hussain N, Jani PU. Factors a ffecting the oral uptake and translocation of polystyrene nanoparticles: histological and analytical evidence. Journal of drug targeting. 1995;3(1):65-70. Epub 1995/01/01.

90. Sadauskas E, Wallin H, Stoltenberg M, Vogel U, Doering P, Larsen A, Danscher, G. Kupffer cells are central in the removal of nanoparticles from the organism. Part Fibre Toxicol. 2007;4:10. Epub 2007/10/24.

91. Illum L, Jacobsen LO, Muller RH, Mak E, Davis SS. Surface characteristics and the interaction of colloidal particles with mouse peritoneal macrophages. Biomaterials. 1987;8(2):113-7. Epub 1987/03/01.

92. Borges O, Cordeiro-da-Silva A, Romeijn SG, Amidi M, de Sousa A, Borchard G, Junginger, HE. Uptake studies in rat Pe yer's patches, cytotoxicity and release studies of alginate coated chitosan nanoparticles for mucosal vaccination. Journal of controlled release : official journal of the Controlled Release Society. 2006;114(3):348-58. Epub 2006/08/15.

93. Loeschner K, Hadrup N, Qvortrup K, Larsen A, G ao X, Vogel U, Mortensen A, Lam HR, Laresen EH. Distribution of silver in rats following 28 days of repeated oral exposure to silver nanoparticles or silver acetate. Part Fibre Toxicol. 2011;8:18. Epub 2011/06/03.

94. Malik B, Goyal AK, Markandeywar TS, Rath G, Zakir F, Vyas SP. Microfold-cell targeted surface engineered polymeric nanoparticles for oral immunization. Journal of drug targeting. 2012;20(1):76-84. Epub 2011/09/29.

95. Mishra N, Tiwari S, Vaidya B, Agrawal GP, Vyas SP. Lectin anchored PLGA nanopart icles for oral mucosal immunization against hepatitis B. Journal of drug targeting. 2011;19(1):67-78. Epub 2010/03/26.

96. Sarti F, Perera G, Hintzen F, Kotti K , Karageorgiou V, Kammona O, Kiparissides,C, Bernkop-Schnurch A. In vivo evidence of oral vaccination with PLGA nanoparticles containing the immunostimulant monophosphoryl lipid A. Biomaterials. 2011;32(16):4052-7. Epub 2011/03/08.

97. Gutierro I, Hernandez RM, Igartua M, Gascon AR, Pedraz JL. Size dependent immune reponse after subcutaneous, oral, and intrana sal administration of BSA loaded nanospheres. Vaccine. 2002;21:67-77.

98. Berin MC, Killiaan AJ, Yang PC, Groot J, Taminiau JAJM, Perdue MH. RapidTransepithelial Antigen Transport in Rat Jejunum:Impact of Sensitization and the Hypersensitivity Raction. Gastroenterology. 1997;113:856-64.

99. Roth-Walker F, Berin MC, Arnaboldi P, Escalante CR, Dahan S, Rauch J, Jensen-Jarolim E, Mayer L. Pasteurization of milk proteins promotes allergic sensitization b y enhancing uptake through Peyer's Patches. Allergy. 2008;63:882-90.

100. Walker WA, Sanderson IR. Epithelial barrier function to antigens. An overview. Annals of the New York Academy of Sciences. 1992;664:10-7. Ep ub 1992/01/01.

101. Madara J, Stafford J. Interferon-gamma directly affects barrier function of cultured intestinal eptihelial monolayers. J Clin Invest. 1989;83:724-7.

102. Rodriguez P, Heyman M., Candalh C, Blaton MA, Bouchard C. Tumor necrosis factor-alpha induces morphollgical and functional laterations of intestinal HT-29 cl.19A cell monolayers. Cytokine. 1995;7:441-8.

103. Heyman M, Darmon N, Dupont C, Dugas B , Hirribaren A, Blaton MA, Desjeux,JF. Mononuclear cells from infants allergic to cow's milk secrete tumor necrosis factor alpha, altering intestinal function. Gastroenterology. 1994;106(6):1514-23. Epub 1994/06/01.

23 Chapter

YEAST: THE PINNACLE OF CULINARY EXPERTISE

1. INTRODUCTION: AN OVERVIEW

Yeast is the most rudimentary eukaryotic creature currently known. They are unicellular microbes categorized within the kingdom Fungi. Nonetheless, yeasts were likely the first microorganisms to be tamed and have been utilized daily in bread production and alcoholic beverages from the early stages of human history. Currently, yeast has emerged as a crucial microorganism in various industrial and food processing sectors, including the production of beer, wine, cheese, and bread. Its application in the baking business is particularly significant due to the fundamental function of bread as a dietary staple globally. Furthermore, yeasts are considered of considerable interest as sources of nutrients and health benefits for both people and animals. We boldly designate yeast as one of the world's premier chefs.

Yeasts inhabit a variety of natural settings, ranging from terrestrial to aerial and aquatic ecosystems. They inhabit rotting fruit, dirt, serve as opportunistic diseases in humans, reside in fish intestines, and exist freely in marine environments. Generally, they facilitate the decomposition of organic matter, however their effective colonization is closely linked to their ability to physiologically adapt to various environments. To date, around 1500 species have been described [1].

This chapter seeks to provide a clear study of the significance of yeasts in contemporary feed practices, particularly with the yeast Saccharomyces cerevisiae. This yeast is recognized by various designations: "Baker's yeast" in the baking and confectionery sectors, "Brewer's yeast" among industrial and artisanal beer manufacturers, and the less well known "Wine yeast" by producers of wine-like alcoholic beverages. We will initially examine various physiological features of yeast metabolism, particularly related to glucose catabolism in anaerobic environments (fermentation) and aerobic situations. The majority of our focus is directed on the glycolysis pathway and alcoholic fermentation to adequately prepare the reader for subsequent discussions. Considerable notice will be paid to the intervention of yeast in alcoholic beverages, in particular beer and wine, important economical industries of our times. The particular role of S. cerevisiae in baking industry, interactions with lactic acid bacteria (LAB) in sourdoughs, and also the scientific approaches/advances for sustainability were exhaustively reviewed, recently by us in [2] but also by [3], therefore we will focus preferentially on the production of commercial yeast for baking. Ultimately, we will succinctly examine the alternative applications of S. cerevisiae in lesser-known items, encompassing animal and fish feed formulations as well as biotic supplements.

1.1 Yeast metabolism

Yeasts, resembling other heterotrophic organisms, have the energy and carbon metabolism operating in concert, i.e., anabolism is coupled with catabolism. Their chemical energy, in the form of ATP, results from the oxidation of organic molecules and is used as energy resource by the cell. On the other hand, those organic molecules can also operate as carbon sources for biosynthesis. Yeast environmental diversity leads to a vast metabolic complexity, due to the multiplicity of carbon and the energy sources available in nature. This includes polyols, alcohols, organic acids and amino acids yet, yeasts preferentially metabolize sugars.

1.1.1 Greedy yeasts – Sugar metabolism

The yeast metabolize diverse sugars, hexoses such as glucose, fructose, galactose or mannose, some can use pentoses like xylose or arabinose, disaccharides as maltose or sucrose; yet, glucose and fructose are the preferred substrates. The metabolic routes for the dissimulation of hexoses and disaccharides share the same pathways, with the great majority of the metabolic elements arising from intermediaries of glycolysis, the tricarboxylic acid cycle (TCA) and the pentose phosphate pathway, and differ only in the initial basic steps of metabolism.

The sugar dissimilation may occur in anaerobic or in aerobic environment. In the first case is called fermentation and in the presence of oxygen is named respiration. The most common process is the glucose dissimilation, generally known as alcoholic fermentation, which occurs anaerobically and yields as final products: ethanol and CO2.

For the sugar utilization, yeast has primarily to sense the presence of glucose in the environment and then to transport it across the plasma membrane [4, 5]. The presence and levels of glucose sensed by the yeast can influence the enzyme levels through several processes, alteration of mRNA translation rates; mRNA stability or protein degradation, but also the concentration of intracellular metabolites (for a review see [6]). Yet, the major outcome is the extensive transcriptional regulation of a large number of genes leading to the adaptation to fermentative metabolism (alcoholic fermentation). These encompasses the induction of genes required for the utilization of glucose, such as genes encoding glycolytic pathway enzymes (discussed below), whereas genes required for the metabolism of alternative substrates, and those encoding proteins in the gluconeogenic and respiratory pathways are repressed by glucose (for reviews see [6] and [7]).

The gene family of hexose transporters in S. cerevisiae consists of more than 20 members: i) 18 genes encoding transporters (HXT1-HXT17, GAL2), being the most relevant Hxt1p and Hxt3p, with a low affinity for glucose and high transport capacity, and Hxt2p, Hxt4p and Hxt7p, with a high affinity and low transport capacity and; ii) at least two genes encoding sensors (SNF3, RGT2), although several points of evidence suggest that GPR1 and HXK2 also sense and signal glucose levels [6, 8]. All these, sensors and transporters are therefore the primary interveners on sugar metabolism. After glucose uptake, it enters in the glycolytic pathway (Figure 1 – Steps from glucose to pyruvate) in order to be metabolized to pyruvate, whereby production of energy in form of ATP is coupled to the generation of intermediates and reducing power in form of NADH for biosynthetic pathways (for reviews see [5, 6, 9]) [10].

The first step of the glycolytic pathway consists on the phosphorylation of glucose to glucose 6-phosphate by the action of the hexokinases (Hxkp) and the glucokinase (Glkp); which are linked to high-affinity glucose uptake. Then glucose-6-phosphate is isomerized by the phosphoglucose isomerase, encoded by PGI gene, to fructose-6-phosphate. The next step, done by the phosphofructokinase (Pfkp) requires energy, in the form of ATP, to convert fructose-6-phosphate into fructose 1,6-biphosphate.

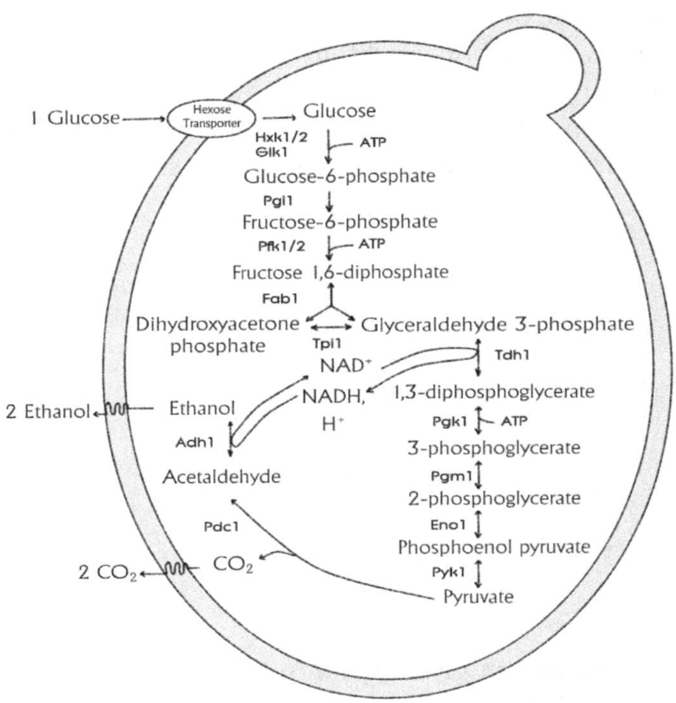

Figure 1. Alcoholic fermentation - enzymatic steps on S. cerevisiae (adapted from [11]).

Yeast phosphofructokinase, Pfkp, is a heterooctameric enzyme subject to a complex allosteric regulation. Aldolase (fructose 1,6-bisphosphate aldolase- Fbap) in turn, catalyses the reversible cleavage of fructose 1,6-bisphosphate to glyceraldehyde 3-phosphate and dihydroxyacetone phosphate. These two compounds can be converted one into another, again in a reversible way, by the triosephosphate isomerase (Tpip). Subsequently, glyceraldehyde 3-phosphate yields the pyruvate by the action of a series of acting enzymes, whereas some of the dihydroxyacetone phosphate follows gluconeogenesis. Glyceraldehyde 3-phosphate is firstly oxidised by NAD+ (with the production of a reducing equivalent, which will take part in the latter steps of glycolysis when acetaldehyde gives ethanol (Figure 1)) and then phosphorylated, under the catalysis of the 3-phosphate dehydrogenase (Tdhp). The resulting 1,3-diphosphoglycerate, by the action of phosphoglycerate kinase (Pgkp), donates a phosphate group to an ADP molecule originating the 3 phosphoglycerate and releasing 1 molecule of energy (ATP). The next step is just a relocation of the phosphate group on position 2, done by the phosphoglycerate mutase (Pgmp); preparing this way the following reaction, the dehydration by the enolase (Enop) and from which results the phosphoenol pyruvate, a high energetic molecule. This is then phosphorylated by the pyruvate kinase (Pykp) giving the pyruvate and also releasing another molecule of ATP.

At this point, pyruvate can follow distinguished metabolic routes (Figure 2) depending on the environmental conditions, which in turn regulate the enzymes involved as well as their kinetics properties, but also of the yeast species [12]. Conversely, the carbon flux gets to a branching point in which may be divided among the respiratory and the fermentative pathways.

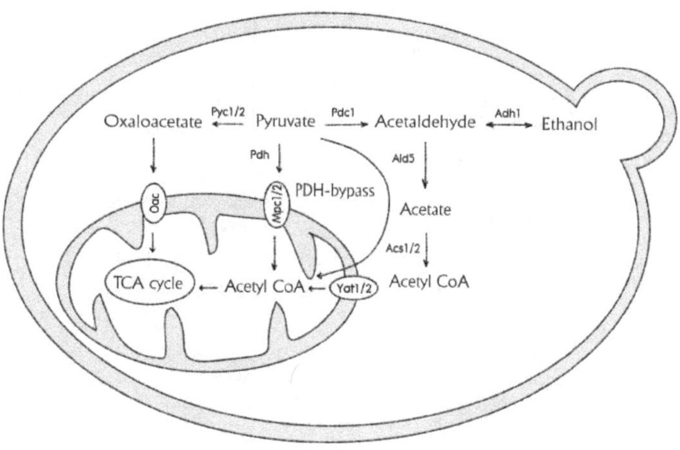

Figure 2. Pyruvate formed in glycolysis alternative metabolic routes. Pyruvate can be converted into 2 intermediates of TCA cycle: acetyl–CoA by the pyruvate dehydrogenase complex (Pdhp) and transported to the mitochondria by mitochondrial oxaloacetate carrier (Oacp); and/or oxaloacetate by pyruvate

carboxylase (Pyc1p/2p) whose mitochondrial carrier is (Mpc1p/2p). Pyruvate can also be decarboxylated to give acetaldehyde by the pyruvate decarboxylase (Pdc1p). Adh1p - alcohol dehydrogenase; Ald5p - acetaldehyde dehydrogenase; Acs1p/2p - acetyl–CoA synthase; Yat1p/2p - carnitine acetyltransferase (adapted from [9]).

In alcoholic fermentation, pyruvate is decarboxylated to give acetaldehyde and CO2, by the pyruvate decarboxylase (Pdc1p). In the final reaction, catalysed by the alcohol dehydrogenase (Adhp), acetaldehyde is reduced yielding the ethanol and promoting the re-oxidation of NADH to NAD+. At the same time, and in addition to the 2 molecules of CO2 and of ethanol, formed per molecule of glucose, the sugar is incorporated into other by-products such as yeast biomass, acids (pyruvic, acetaldehyde, ketoglutaric, lactic) and also importantly glycerol. This is generated from dihydroxyacetone-phosphate and is, to a certain extent, very desired by the wine producers in order to get fuller bodied wines (discussed below). Furthermore, alcoholic fermentation is a redox-neutral process; given that the NADH produced during the oxidation of glyceraldehyde 3-phosphate is afterwards reoxidized in the reduction of acetaldehyde to ethanol [13]. Yet, one must keep in mind that with fermentation is associated culture growth and, biomass composition is more oxidized than glucose, consequently an excess of reducing equivalents may be attained. The way yeast circumvent this problem, under anaerobic conditions, consists on the production of glycerol by reduction of the glycolytic intermediate dihydroxyacetone phosphate to glycerol 3-phosphate catalysed by NAD+-dependent glycerol 3-phosphate dehydrogenase (encoded by the two isogenes GPD1 and GPD2), and its subsequent dephosphorylation due to the action of glycerol 3-phosphatase (encoded by GPP1 and GPP2) [14-16].

Although fermentation usually happens in the absence of oxygen, this is not a strict rule. Even in the presence of high levels of oxygen, if the sugars are fully accessible to be metabolized, yeasts choose to ferment instead of respire. This phenomenon is called the Crabtree effect [17], defined as the inhibition of aerobic metabolism when glucose is available, which occurs both in the presence or absence of oxygen. For instance, S. cerevisiae is known as Crabtree negative yeast, since is able to produce ethanol aerobically in the presence of high external glucose concentrations. These high concentrations promote the acceleration of glycolysis, producing appreciable amounts of ATP through substrate-level phosphorylation. Simultaneously, it reduces the need of oxidative phosphorylation done by the TCA cycle via the electron transport chain, inhibits respiration and ATP synthesis, and therefore decreases oxygen consumption. Conversely, Crabtree negative yeasts produce biomass via TCA cycle, but these should not be mystified with obligate aerobes: Crabtree-negative yeasts are able to ferment, yet usually only ferment under anaerobic conditions, since there is no inhibition of aerobic respiration in the presence of glucose and this is a more efficient form of energy metabolism. Obligate aerobe yeasts, on the other hand, cannot ferment and only respire aerobically, providing another category of metabolic diversity. Moreover, Crabtree effect is not specific to yeasts: many mammalian tumour cells display a Crabtree effect as well [18-20].

In aerobic respiration (Figure 3), the pyruvate is converted to Acetyl-CoA due to an oxidative decarboxylation, catalysed by the pyruvate dehydrogenase multi enzyme complex. In this way starts TCA cycle, which major issue is to supply the respiratory chain with reducing equivalents (in form of NADH and FADH2) obtained from the oxidative decarboxylation of Acetyl-CoA, which is then used to generate energy through the highly conserved electron transport chain. Moreover TCA cycle also has anabolic functions, almost all intermediates are utilized in other metabolic reactions; exception is made to isocitrate, including the synthesis of amino acids and nucleotides (for reviews see [9, 21]).

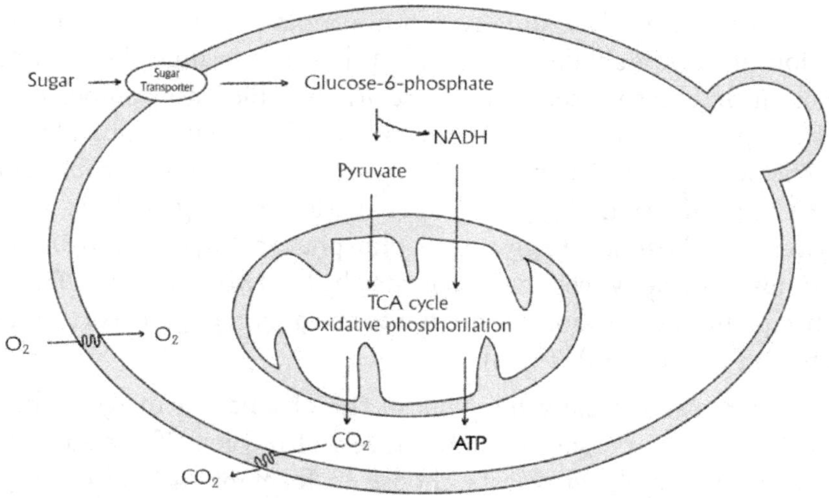

Figure 3. Aerobic respiration in S. cerevisiae (adapted from [22]).

2. S. CEREVISIAE, THE PARTY STARTER

Beverages with an alcoholic content are largely consumed by mankind since ancient times. Such beverages made from fermentation of sugar-rich goods, namely cereals and fruits, are present in oldest records [23]. Beer, made from germinated barley, and wine, produced from grapes, are among the most popular and their worldwide consumption is second only to non-alcoholic drinks as water, tea and coffee [24].

Wine and beer history is hand to hand with human civilization history, as most likely only the agriculture advent and the establishment of permanent settlements provided the conditions for its production. Nevertheless, wine and beer are most probably result of an “accident”, as some harvested grapes were not consumed rapidly enough or some cereal wet pulp was left aside, and S. cerevisiae took advantage of the sugary free meal. The result should have been pleasant enough, especially the mild psychotropic effect, therefore the early farmers must have tried to repeat such "accident". The high ethyl alcohol content of this beverage and its analgesic, disinfectant and conservative properties contributed to a widespread utilization as a drug. Hence, our successful partnership with this yeast began.

In fact, our relationship with S. cerevisiae can be traced back as far as 7,000 years ago in China, with the first fermented beverages similar to beer, and Mesopotamia, with the first wines as well as domesticated vines [11, 25]. Conversely, the oldest known written account on the production of beer was found in Sumeria, modern day Iraq, in a stone tablet dating from 2,000 BC. This tablet, called "Hymn to Ninkasi", describes the production of beer and its especially relevant role in religious ceremonies to Ninkasi, the Sumerian Goddess of beer [26]. Babylonians succeeded the Sumerians and kept producing the fermented beverage. They became so skilful at this art that at least 20 different brews were produced and exported, to as far as Egypt, at the zenith of Babylonian empire. Egyptians were so adept of this "imported beer" that they started producing their own from unbaked dough and even created a special hieroglyph for this new craft. Furthermore, records show that pyramid workers were paid in beer, a readily storable merchandise, and Pharaohs were entombed with model breweries to ensure an afterlife beer supply [26, 27]. Beer popularity grew and this beverage spread for the entire Europe, especially in the Mediterranean region [28].

Wine was also very popular in the ancient cultures, with references to this beverage in religious ceremonies of Egypt and Phoenicia. Pharaohs tombs were frequently adorned with vintage scenes and jars filled with wine accompanied the Kings afterlife [11, 23]. Wine consumption spread with the rise of the Greek and Roman Empires. Under the Greek and Roman influence, wine earned the status of "Civilized" drink, becoming very popular with the Empire upper classes, and beer was labelled as a "Barbarian" drink. Wine production and vine cultivation spread across Europe and replaced beer as the main drink in many countries. Some of these are still nowadays associated with wine production like Portugal, Spain and France. The production and consumption of beer continued mainly in northern borders of the Roman Empire, where Germanic tribes ruled and Roman influence was weaker [26, 28].

In Middle Ages, wine and beer production gained a new impetus with the shift from the familiar production to a more centralized production in monasteries [28]. Such happened because, at the time, water was frequently polluted, so alcoholic beverages were safer than water for the monks' consumption. Additionally, during the long fasting periods that the monks subjected themselves, the drinking of these highly nutritious beverages became common to satisfy hunger. This happened because wine and beer were considered similar to water and didn't constitute a breach of fast. In fact, in some monasteries monks were allowed to drink up to 5 litres of beer per day [28]. Southern monasteries produced mainly wine, as the weather was warmer and suitable for vines, but in the north the colder weather was more fitting barley and wheat growth and therefore Northern monasteries were more devoted to beer production. Each monastery developed its own methodologies for wine and beer making, leading to new wines and new brews and to a great technical improvement. Later these products become a source of income for the monasteries.

From the sixteenth century, with the discovery of the New World by Portuguese and the Spanish explorers, wine and beer spread to new territories. Vines were introduced in Brazil by the Portuguese around 1500 [29], and in Africa by the Dutch around 1650. In the Australian continent and North America this happened later, around 1800.

In the nineteenth century, wine and beer making suffered probably the major scientific advances. Around 1860 Louis Pasteur, a name forever associated with wine and beer production, developed studies on the conservation of wine through a heating-cooling process later known as "pasteurization", showing that wine could be stored for longer periods after such treatment. Moreover, in 1870 Pasteur made known to the world the role of S. cerevisiae in the fermentation process. Later, in 1876 Pasteur conducted similar studies in beer in his work "Études sur la bière". Such works were based in the first observations of yeast and bacteria by Antonie Van Leeuwenhoek in late seventeenth century [11, 26, 27]. Only almost a decade after it was isolated a pure yeast culture resulting from a single cell by Emil Christian Hansen at the Carlsberg brewery, Denmark and to that followed the name of several yeast species [30]. A few years later, based on Hansen's work, Hermann Müller-Thurgau introduced the notion of inoculating wine fermentations with pure yeast starter cultures [11].

The intensive study of this amazing microorganism and its role in fermentation showed the specificities of each yeast species and strains. The necessity of consistent properties and quality in different fermentations, both in brewing and winemaking, paved the way for the selection of the right yeast for the job. The quest for stable and improved yeast began.

2.1 The "Right" yeast for the job

S. cerevisiae, known as "Wine's Yeast" or "Brewer's Yeast" (and also as "Baker Yeast", see below), is the main responsible for some of the world's most important fermented beverages. However, brewing and winemaking have inherent differences: i) the culture medium, ii) the bioreactor and iii) the yeast starter culture.

The final product, either wine or beer, is greatly influenced by the sugar-rich fermentable broth, grape juice or malted cereals, with different composition in fermentable sugars and nitrogen sources. The progression of the fermentation is another very important aspect of winemaking and brewing, e.g., the oxygen available, the temperature and pH variations during substrate consumption and ethanol and CO2 production. But the most significant difference is the yeast starter culture, its physiological state, whether it is dried yeast or a fresh inoculum, how well it ferments the available sugars and resists to fermentation by-products and its ability to flocculate at the right moment.

All S. cerevisiae strains described so far are capable of fermenting sugars to ethanol, but centuries of partnership with mankind directed the yeast evolution. Such evolutionary pressure resulted in a selection of distinct yeast strains for different applications, to produce wine and beer you need "Wine's Yeast" and "Brewer's Yeast", respectively.

2.1.1 Yeast physiology

Beer is the denomination commonly attributed to a carbonated alcoholic beverage produced by fermentation of malted barley, while wine is made of the fermented juice of any of several types of grapes. However, there are as many different wines and beers as there are different producers, all with their unique character and flavour influenced by the selected ingredients, kind of fermentation and yeast selected.

As said, the choice of the ingredients greatly impacts the fermentation final product; usually beer is the product of malt, hops, water and yeast. Malt is the result of germinating and drying (kilning) barley, yet other cereals besides barley can be used to produce beer, as wheat and rye. Malt extract will provide the entirety of the carbohydrates and nitrogen to the fermentation process and as such, it will influence the final ethanol concentration as well as colour and flavour development. Conversely, another important aspect is the intervention of hop, the female flower cluster of Humulus lupulus, which acts as bacteriostatic agent against Gram-positive bacteria, helping to control unwanted microorganism during brewing. It also functions as bittering agent, disguising beer natural sweet taste [31, 32]. In winemaking, the maceration of grapes is the starting point to wine production. The variety of grapevines, as well as the weather and cultivation/soil conditions, greatly influences the wine final properties. In fact, the environment has such influence that some type of wines can only be produced in certain regions, like the Porto wine in Douro region, Portugal, and Champagne wine in Champagne region, France.

Brewer's yeast can be distinguished in top fermenting and bottom fermenting yeasts, based in the position at which the fermentation occurs. This division accounts with the yeast flocculation behaviour, and it is such an important element of brewing that defines the two main classes: ale beers (top fermenting) and lager beers (bottom fermenting). Such categories were devised as soon as the first pure yeast culture was isolated. Hansen was able to purify two different species, a top fermenting appropriate for ale brewing, S. cerevisiae, and a bottom fermenting, S. carlsbergensis, suitable for lager beer [26]. Such taxonomic classification was reviewed several times [1], and the top fermenting yeasts are now included in the S. cerevisiae and S. bayanus species and the bottom fermenting yeasts fit to the S. Pastorianus species, all belonging to the Saccharomyces sensu stricto genus [27, 33].

These brewer's yeasts present several differences in their genomes. Lager yeast strains present complex polyploidy genomes, with evidences of contribution from distinct Saccharomyces species [33]. Usually these complex genomes are tetraploid, which may result from the fusion of diploid parental strains or from duplication of the genetic information after the original cell fusion. Analysis on lager yeast genomes revealed other changes such as chromosome loss and/or duplications, likely due to human selection of relevant phenotypes [33, 34]. In reference [34] lager yeast genomes were analysed and classified in two groups. In group I, cells present one S. cerevisiae genome equivalent and, in group II, cells present two S. cerevisiae genome equivalents. Both groups exhibit one S. bayanus genome equivalent and the remaining genome was mostly hybrid chromosomes from both species. These different yeast must be related with the conditions they are exposed, meaning for instance to lager beer fermentation, yeast has to react to conditions inherent to beer production, as cropping and pitching, and to bottom fermentation specificities, e.g., temperature of reaction [35].

As for ale yeasts, studies revealed that these strains are closely to S. cerevisiae [34, 36]. In fact, a genotype analysis of 651 S. cerevisiae strains revealed that ale strains were more closely related to wine and bread strains (above referred as Baker's yeast), than to lager brewer's yeast strains [36]. Reports of hybrids in ale yeasts showed that strains traditionally classified as S. cerevisiae may indeed be the result of hybridization events [37]. Ale beer has less representation in worldwide markets, and as a consequence less studies and information are available on the corresponding yeasts. As such, the considerations on beer yeast physiology will be focused on lager strains.

In winemaking, most wine yeasts belong to S. cerevisiae species, but S. bayanus has also been detected. Yet, wine fermentations also present yeast, derived from vineyard environment, belonging to the genera Candida, Debaryomyces and Brettanomyces. But, yeast is mainly selected for its resistance to ethanol, favouring S. cerevisiae. There is also selection for capacity to float or to flocculate, important for some specific wines. While in most wines the ability to flocculate is important to improve the filtration, in some wines such as sherry wine, the formation of a floating film is vital. This vellum is formed at the surface of the wine and promotes oxidative metabolism. Sherry wine is characterized by high ethanol content and low aldehyde. Its featured nutty flavour can be ascribed to partial oxidation of ethanol to acetaldehyde [11].

The utilization of dried yeast as a starter culture is very common in the wine industry. Cells are dehydrated through a cycle of filtrations and centrifugation to remove external water and then submitted to streams of dehumidified hot air. Such procedure can reduce yeast cells' content in water to as low as 6%. However, even though yeast cells can survive such treatment, it causes cellular damage. Damages to cell wall and plasma membranes caused by the changes on cell size and shape, as well as damages to proteins produced by free radicals were reported [35].

One of beer brewing specificities is the utilization of a freshly grown starter culture. In one hand, it ensures a healthy population fully adapted to growth medium. Cells are usually collected at the late exponential phase, preventing a large percentage of aged cells, and at the same time ensuring metabolic fitness. On the other hand, it meets the requirement for flavour consistency of the final product, even though it is more expensive than the alternatives. The pattern of metabolic products of yeast is highly dependent on its growth conditions, and cells fully adapted to wort produce a more consistent flavour. The first batches inoculated with dried yeast are often of inferior quality, with by-products of fermentation conferring off-flavours, which compromises the regularity of a brand product [30]. An alternative that meets the requirements of fresh grown cells and flavour consistency, but at the same time reduces the process duration and costs, is the fed-batch technology. Cells are kept in late exponential phase by leaving a certain amount of yeast in the reactor and adding fresh wort, shortening the fermentation time and maintaining the beers properties [38].

During fermentation, yeast is constantly facing new pressures. The high osmotic stress due to the sugar high content of wort and must is just the beginning. Wort is a rich and complex medium, composed of carbohydrates (90%), nitrogen sources (5%) and small amounts of inorganic ions, lipids and polyphenols. Wort composition, being highly dependable of the quality of the cereal and the process used to malt it, is usually enriched in fermentable sucrose (5%), monosaccharides (10%), maltotriose (15%) and maltose (50%). About 20-30% of total carbohydrates are non-fermentable dextrins, polysaccharides result from starch degradation [35]. On the other hand, grape juice is rich in fructose and glucose, presenting small amounts of sucrose. Grape variety influence the ratio glucose/fructose, Chardonnay is a high fructose variety, whereas Zinfandel is regarded as high glucose variety. Such high-gravity worts, 12-18 g of extract per 100 mL, subject cells to high osmotic pressure. The production of compatible solutes, as glycerol and trehalose, and a "robust" plasma membrane composition seem to be the main adaptations to withstand such stress. The cells fully adapted to wort used to inoculate (pitch) fermentations are important to avoid extended lag phases where cells are adapting its physiology. In wine the use of active dried yeast (ADY) is common and no effect on fermentation time was reported [39].

Nitrogen assimilation is especially important in flavour development. The main sources of nitrogen are free amino acids and ammonium ions, which are used by the cell for protein formation [35, 40]. Such amino acids are also relevant for the production of alcohols and esters, important in these beverages flavour. During fermentation, amino acids are always used following a certain order, independent from the fermentation conditions. Group A, including arginine, asparagine, aspartate, glutamate, glutamine, lysine, serine and threonine, are used first. Group B amino acids are utilized slowly and include histidine, isoleucine, leucine, methionine and valine. Group C is composed by alanine, glycine, phenylalanine, tyrosine, tryptophan, and are only absorbed after the complete exhaustion of group A. Group D is composed of proline, which require an aerobic metabolism for its uptake and it is poorly used during fermentations [40].

In winemaking, fermentations are usually developed under anaerobic conditions, but it is common in brewing to oxygenate the wort. So another important stressor is the dissolved oxygen, which may lead to the formation of reactive oxygen species (ROS). ROS, as hydrogen peroxide or superoxide radical, can promote damages in cell main constituents, DNA, proteins and lipids. However, oxygen is very important for the synthesis of sterols and unsaturated fatty acids ensuring the physiological fitness for cell replication. The control of dissolved oxygen is vital to ensure a healthy population. An important problem is the excessive growth of yeast cells when exposed to high amounts of dissolved oxygen, at the expenses of ethanol production [41].

The inorganic ions are necessary, but at nanomolar concentrations. These trace elements, as calcium, zinc or copper, are mainly required as cofactors of enzymes or in the flocculation process. For instance, the response to oxidative stress is dependent on enzymes such as the different superoxide dismutase isoforms that require manganese, zinc or copper [32]. On the other hand, calcium is vital for the flocculation advance [42]. Conversely, insufficient amounts of such elements can lead to cellular damage and stress, and consequent stuck fermentations.

The use of antimicrobials in vineyards is common to control fungi that spoil grapes. But, when grapes are macerated these compounds are incorporated into the juice. Even though they may help to prevent the wine oxidation and microbial spoilage, a concentration to high may lead to off-flavours and in worst case, yeast death. So antimicrobials, especially sulphur dioxide, are an important stress to yeast during fermentation. Commercially available yeast also has to deal with toxins produced by wild yeasts derived from the vineyards. These toxins are produced to give those wild yeast advantages over others species in accessing to the nutrients. Isolation of strains resistant to both antimicrobials and natural toxins is an important research field [11, 43].

Certain compounds are extremely important for brewing and wine making not as substrates but as by-products. Such metabolites greatly influence the final product's colour and flavour, as well as its stability. In fact, the importance of these compounds is such that lager beers are usually stored from several days to weeks, lagering, solely to remove diacetyl, an off-flavour causing metabolite. This time consuming maturation phase consists in a second fermentation at low temperature to eliminate the butter-like flavour caused by this vicinal diketone. Studies are being conducted in order to minimize this metabolite formation and reduce the maturation time [30]. Sulphur containing compounds are other family of by- products receiving great attention. Such group comprises sulphite, sulphide and dimethyl sulphide, and while sulphite is a beneficial and flavour stabilizing metabolite, the remaining compounds are responsible for off-flavours. The equilibrium of such compounds formation could lead to better wine and beer and shorter fermentations [44].

Ethanol is one of the most important by-products of beer fermentation. Nevertheless, it represents an important stressor for yeast cells due to its high toxicity. Ethanol concentration can reach 10% in higher gravity fermentations, and acts especially upon biological membranes [35]. Reports showed ethanol effects in growth inhibition [45], lipid modification and loss of proton motive force across the membrane and increased membrane permeability/ fluidity [46]. Yet, cells exposed to oxygen, with high levels of sterols in membranes, and adequate levels of nutrients, amino acids and trace elements in the fermentation broth are able to respond efficiently to such effects [35].

Nutritional stress occurs at the end of fermentation and cells enter stationary phase. This occurs because fermentable carbon sources tend to be depleted, and cells have to change their metabolism from fermentative to respiratory (explained in section 1), entering in a quiescent state [30]. Such phenomenon induces flocculation, a cell-cell interaction process dependent of lectins and calcium that promotes sedimentation. Flocculation in turn is influenced by several other factors besides nutrient depletion. Reports showed the influence of ethanol content, calcium concentration, pH changes, oxygen concentration and temperature [47]. The onset of flocculation is an important area of interest in brewing. If flocculation occurs to soon, stuck fermentation may occur, which results in a high sugar and low ethanol content. If, on the other hand, happens in a later stage, it has a high impact in beer filtration as most cells tend to be kept in suspension.

Fermentation is the most yeast-dependent phase of these alcoholic beverages production, but yeast also interferes with others proceedings. The metabolic fitness of the starter culture, the storage and maintenance of both dried and fresh yeasts, and the storage of the products submit yeast to different conditions to which they have to respond/adapt. To obtain detailed information on such processes please see reviews [35] and [43].

2.2 Old beverages, new solutions

Wine and beer industrial production led to a demand for better and more efficient yeast. Yeasts with improved utilization of substrates, carbohydrates and nitrogen, and consistent flocculent behaviour, as well as high fermentative capacity and high ethanol production are the industry goal. Enhancing beer and wine flavour through modification of by-products formation is another field of intensive research [30, 39]. As it is the improvement of the fermentation process, through encapsulation / immobilization of yeast [48].

Large collections of yeast were assembled, as the Centraalbureau voor Schimmelcultures (CBS) collection, in The Netherlands, and the Carlsberg collection, in Denmark. Manipulation of these strains to improve wine and beer properties has been performed in several ways, from spores manoeuvring and natural mutants' survey to genetic engineering (GE). A rather recent and extensive review in strategies for the improvement of S. cerevisiae industrial strains can be found in [2]. As referred above, an efficient utilization of substrates by yeast during fermentation is extremely important for wine and beer industries. Such efficiency will yield higher amounts of by-products, as ethanol, and reduce the fermentation time. Furthermore, glucose repression on other sugars and consumption of unusual nitrogen sources are vital research areas [30].

The presence of glucose, even in small amounts, represses the simultaneous uptake and consumption of several sugars, namely maltose and galactose. Maltose (50%), maltotriose (15%) and sucrose (5%) are the main sugars in wort under glucose repression. Such repression leads to late fermentation of these sugars and slower fermentations, as their utilization is dependent on glucose depletion. This process is controlled at the transcriptional level, through the action of the proteins Mig1p, Ssn6p and Tup1p. Mig1p, a zinc finger protein, binds to specific sequences in the promoter region of the glucose-repressed genes and recruits the SSN6-TUP1 complex, the responsible for the actual repression [49]. Mig1p binding sites were found in genes associated with the utilization of sucrose (SUC2), maltose (MALR,

MALS and MALT) and galactose (GAL1-5) [50]. Therefore, MIG1 presents itself as a potential target to improve yeast sugar consumption.

Conversely, in sucrose metabolism, glucose repression addresses the sucrose conversion in fructose and glucose under the action of Suc2p. Studies showed that the disruption of MIG1 lessen the glucose repression on the transcription of this excreted invertase in both lab and industrial strains. Therefore, the lag in sucrose utilization was greatly diminished. Besides Mig1p, another zinc finger protein, Mig2p, was associated with glucose repression of SUC2 [51]. The interruption of both MIG1 and MIG2, in S. cerevisiae strains led a high sucrose metabolism in the presence of high glucose concentrations [52].

Maltose metabolism is more complex than sucrose, as it responds to both glucose repression and maltose induction. Maltose induction is under the influence of the locus MAL, a closely integrated group of genes. This sugar presence induces MALR, a transcription factor, which in turn will induce MALT, coding for a maltose permease, and MALS, coding for a maltase [30]. Up to 5 different loci have been detected in S. cerevisiae industrial strains; still haploid lab strains present a single locus. In both situations these loci are under repression of Mig1p [53]. However, disruption of MIG1 in industrial strains only alleviated the sucrose metabolism [50], but didn't cause any effect regarding maltose metabolism. Even though, in haploid lab strains MIG1 disruption lifted the glucose repression; the same has not happen in the polyploid strains, presenting multiple loci. Complex regulation between the different genes must be under way and most probably not solely controlled by Mig1p [53].

Finally, maltotriose, a glucose tri-saccharide, is the second most abundant sugar in wort (15%). This is under similar regulation by the presence of glucose and maltose, therefore most studies have focused in an efficient uptake of this carbohydrate [54]. Those works showed that overexpression of maltotriose transporters lead to positive effects on its metabolism [54, 55].

In winemaking and brewing, where flavour has such importance, amino acids metabolism has a notorious place. As said, amino acids are involved in formation of higher alcohols and esters that significantly contribute to beer and wine flavour. Since yeast cannot hydrolyse must and wort proteins, it depends on the available ammonium and amino acids in solution [11, 30]. However, the predominant amino acid in both must and wort, proline, is the less assimilated [40]. As such, improvements in yeast ability to uptake this amino acid has been attempted. Efficient proline uptake was reported in lager beer yeast expressing a mutagenized proline permease, Put4p. The site-directed mutagenesis stabilized the permease and enhanced amino acid utilization without affecting the beer quality [56]. The study of the same problematic in winemaking led to the disruption of URE1, a repressor of permease encoding PUT1 and pyrroline-5-carboxylate dehydrogenase PUT2, with significant improvements in fermentation rate and vigour described [57].

Flocculation is a phenotype of industrial interest. It facilitates the filtration process in the end of fermentation, saving both time and money. In the case of brewing, it also serves the cropping (recover of part of the yeast population of the fermentation to pitch the next). Flocculation is a reversible aggregation of cells, where lectins recognize sugar residues in neighbour cells. Two industrially relevant flocculation phenotypes are well-known, Flo1 and NewFlo. Both are under the control of FLO genes, Flo1 phenotype is repressed by mannose and NewFlo by mannose, glucose and sucrose. Almost every industrial strain is NewFlo, associated with FLO10 [30]. The main approach to improve these phenotypes is to put FLO genes under a promoter active only in stationary phase. Promoters of HSP26 and HSP30 were proven as the most suitable for induction at this late growth phase in lab strains [30].

The control of by-products production in order to improve wine and beer organoleptic properties is an expanding research area. The production of glycerol, to improve wine and beer's fullness, as well as sulphite, to improve stability, and the reduction in diacetyl and sulphide content are the main targets. Glycerol, as the second fermentation metabolite, is rather important to wine and beer; the increase of its concentration to improve these beverages sensory character is an active field. Overexpression of GPD1, encoding glycerol-3-phophate dehydrogenase, is the main approach. However, this change resulted in a redox imbalance with increased production of unwanted metabolites [39]. This point has been fully discussed in [2].

The presence of sulphite, an antioxidant and flavour stabilizer, and reduction of off-flavour producing sulphide is another important problem addressed by the industry. Both these goals can be achieved at the same time with the directed mutagenesis of NADPH-dependent sulphite reductase, an important enzyme in sulphur–containing amino acids synthesis. This strategy lowered this enzyme activity and increased the amount of sulphite in wine while reducing the sulphide presence in wine [58].

The reduction of diacetyl has special importance, as the maturation time (lagering) is directly dependent on this compound concentration. The expression of the bacterial enzyme acetolactate decarboxylase (ALDC) in yeast is the main approach to reduce the amounts of this compound. ALDC catalyses the reaction of α-acetolactate to acetoin, preventing the formation of diacetyl. However, after heterologous expression of ALDC, the yeast became auxotrophic for some amino acids and the growth rate was very low in wort. An alternative approach was the interruption of ILV2, encoding acetolactate synthase. Such strategy also resulted in an auxotrophic strain with slow growth. The search for natural mutants in ILV2 with an appropriated growth rate is now the major strategy [30].

Improvement of yeast to render fermentations faster and cheaper is an industry goal, but the enhancing of the fermentation process itself is another alternative. The fed-batch technology has already proved its benefits [38], and improvements of such process with yeast immobilization/encapsulation are now under the spotlight. This results in much faster fermentation rates as compared to the existing free cell fermentations. However, it has some disadvantages, such as: i) complexity of production process including the choice of the suitable carrier materials, ii) bioreactors design, iii) fine-tuning of the flavour formation during fermentation processes, and iv) cost constraints [59].

3. BAKER'S YEAST – MAGIC ON BREAD MAKING

The process of bread making relies on the fermentation carried out by a mixture of yeast and bacteria. Even when all this was unknown and the flour leavening seen as "magic", bread was already produced and extensively consumed. On those ancient times, the leavening resulted presumably due to the action (fermentation) of the natural microbial contaminants of flour or dough ingredients. This was obviously not a controlled process, yet with the practice of maintaining a fresh inoculum from one preparation to the next, promoted the selection of yeast and bacteria biodiversity. Nowadays, some types of bread are still prepared in this fashion, sourdoughs are one example (for a review see [2]), but the baking industry moved for the use of commercially baker's yeast, typically the strain S. cerevisiae, for the bread production. And, while the flour types, geographical origin and mixtures introduce organoleptic differences in bread, the globalization of commercial baker's yeast market decreased worldwide bread diversity (for a review see [2]).

3.1 Commercial baker's yeast production – The break of spell

Commercial baker's yeast is produced in several forms in order to meet specific requirements of climate, technology, methodology, transportation, storage and final product. As with all biotechnology processes, this is in constant development/ undergoing research not only to optimize the process technology and its components, but as to produce faster growing strains with the characteristics to deliver better quality end products.

Molasses (beet and cane molasses), the common carbon and energy source used in the production of baker's yeast, is a by-product of sugar refining industries, therefore cheaper than the formerly used cereals grain. Furthermore the sugars present on those molasses (around 50%), consisting on a mixture of sucrose, fructose and glucose, are ready to be fermented by the yeast. In order to obtain the proper broth for the optimum yeast biomass yield; the mixture of molasses has to be supplemented with nitrogen sources, minerals, salts and vitamins [60, 61].

After the preparation and sterilization of the broth, the production of baker yeast can take place. It begins by the inoculation of a small closed test flask containing the prepared sterilized broth with a pure yeast culture. The growth is allowed and careful screened, and when the culture reaches an elevated density, it is transferred to larger vessels and supplied with more broth, fed-batch reactors. This scale-up process continues until a desirable biomass quantity is attained, the so-called commercial starter, able to inoculate industrial fermenters/ reactors, which production ranges from 40,000 to 200,000L [62].

The entire fermentation process of baker's yeast has to be directed towards maximum biomass production; by-products such as ethanol are not desired. As we saw in section 1.1.1, in anaerobic dissimilation of sugars (alcoholic fermentation) the ATP yield is quite low comparing with respiratory dissimilation, affecting drastically the biomass yield. The way to avoid anaerobic ethanol production is the use of the mentioned fed-batch reactors, in which is possible to control the specific growth rate and sugar concentration by controlling the fed of reactors with fresh broth [63].

Nowadays, during the industrial large reactors the addition of nutrients and regulation of pH, temperature and airflow are carefully monitored and controlled by computer systems during the entire production process. In this way, the tones of baker's yeast obtained in the end of the fermentation have the same quality/ characteristics/properties as the original pure yeast culture that started the process. These tones of yeast are suspended in a large amount of water, resulting in a creamy suspension of active yeast, being necessary the so-called downstream processes to obtain the concentrated yeast [64, 65]. At the end of the fermentation, the yeast culture is concentrated using a series of combined centrifugation and washing steps, into a yeast cream with concentration of approximately 20%. The yeast is then cooled to approximately 4°C, and stored. It can be sold in this form – Cream Yeast- however is quite expensive due to the manipulations required because of high content in water. Cream yeast can be further processed, compressed or dried originating the Granular Yeast or Instant Dried Yeast, if then converted in small granules, or Cake Yeast or Active Dry Yeast, if as an alternative the dried yeast is extruded or cut into blocks/cakes. All these yeast types are then packaged, typically vacuum packed to reduce the risk of contamination, and distributed to wholesalers or traders. The shelf life of Active Dry/Cake Yeast and Instant Dry/ Granular Yeast at ambient temperature is 1 to 2 years.

3.2 Idol baker's yeast

Yeast has a significant role on bread making, greatly influencing the final product properties. The most important contribution is in the leavening phase; after the dough has been kneaded and the gluten network start to develop, yeast starts to consume the available fermentable sugars and to produce ethanol and CO2, as mentioned in section 1. As fermentation occurs, the dough is gradually depleted of O2 present in the air bubbles trapped in the dough, leaving small bubble nuclei full of N2. As the metabolism is more and more stimulated, with incubation at optimal temperatures, the CO2 starts to saturate the liquid phase of the dough and starts to accumulate in the bubble nuclei [66]. This leads to dough rising provided that a mature gluten network, capable of ensure the dough foam-like structure, is formed. The amount of time such process occurs will influence dough gumminess and rheology, as well as crust colour, crumb texture, and firmness of the bread (reviewed by [2, 3]). The amount of sugars present in the dough that are actually fermented, as well as the efficient secretion of enzymes as invertase, responsible for the conversion of sucrose to glucose and fructose, has a great impact in the flavour characteristics of the bread. The action of several enzymes, namely proteases, lecithinases, lipases α-glucosidase and β-fructosidase, leads to different utilization of the dough substrates. In bread making, the flavour is also greatly influenced by the metabolic by-products of the yeast fermentation, and while the most important by-product of yeast metabolism is certainly the CO2, the production of metabolites such as alcohols, esters, and carbonyl compounds have also a deep impact. More than 300 volatile compounds associated with bread's flavour and aroma, are produced by yeast. Although, some are dependent more on the substrates, the vast majority of such compounds are yeast dependent, introducing an important variability to bread [67].

As mentioned, the common procedure of bread making today, at least in developed countries, consists of using this commercial baker's yeast. Its quality/ individuality depends on storage stability, osmotolerance and freeze-thaw resistance. Considerable efforts have been made to obtain the Idol Yeast, including evolutionary engineering, genetic engineering (mainly to provide yeast with high capacity to tolerate freeze-thaw treatments). Yet, there is still considerable space for improvement. Those several strategies to achieve the Idol Yeast has been thoroughly revised and discussed in a previous work from the beginning of this year [2] as well as on [3, 68].

4. YEAST À LA CARTE

Fresh S. cerevisiae consists of approximately 30–33% of dry materials, 6.5–9.3% of nitrogen, 40.6–58.0% of proteins, 35.0–45.0% of carbohydrates, 4.0–6.0% of lipids, 5.0–7.5% of minerals and various amounts of vitamins, depending on its growth conditions [69]. So, today yeasts are acquiring increasingly more attention for other uses, besides the production of alcoholic beverages and the baling industry. The products of modern yeast biotechnology form the backbone of many commercially important sectors, including functional foods (for animals, fish and humans), health food supplements, including additives, conditioners and flavouring agents, as pharmaceutical products, for the production of microbiology media and extracts, as well as livestock feed or even as agents of detoxifying effluents containing heavy metals [70, 71].

4.1 Yeast treats for animals

The pioneering research conducted almost a century ago by Max Delbrück and his collea gues was the first to highlight the value of surplus brewer's yeast as a feeding supplement for animals [72]. Yeasts have been fed to animals for more than a hundred years, either in the form of yeast fermented mash produced on the farm, yeast by-products from breweries or distilleries, or commercial yeast products specifically produced for animal feeding. In animals, including pets, this practice is used to compensate for the amino acid and vitamin deficiencies of cereals [73, 74], and in fish as a substitute for other ingredients [71].

Brewer's yeast biomass, as described above, which results from the cultivation of S. cerevisiae on malted barley, separated after the wort fermentation, debittered and then dried, is the second major by-product from brewing industry, just after the brewer spent grain [73]. This biomass is an excellent source of proteins, peptides and amino acids, vitamins (especially of B-group: B1, B2, PP, B5, B6, B8, B12), minerals and trace elements (calcium, phosphorus, potassium, magnesium, copper, iron, zinc, manganese, chromium, selenium), carbohydrates (glucans and mannans), as well as phospholipids [75]. The winemaking industry also generates a huge amount of microbial biomass - leeds. Yet, this incorporates the yeasts (that die due to nutrient depletion) but also other microorganisms, suspended solids, colloids, and organic matter, and have been shown to display quite low nutritional value to be considered for use as a supplement in animal feed [76].

Those yeast used for monogastrics food or feeding rations is generally inactivated because feeding of live yeast might cause avitaminosis due to the depletion of B-vitamins in the intestine [77]. They can also cause adverse fermentation in the digestive tract of swine leading to diarrhoea and bloating [75]. Yeasts can be killed through application of heat or using chemicals. High temperature destroys the yeast membrane, but does not necessarily inactivate all yeast enzymes, unless quite elevated temperatures are applied. Alternatively, there are the chemical treatments with propionic acid or formic acid, which also act as a preservatives for yeast, and contribute to the feed value of the yeast [73].

Yeasts have been used in diets of numerous species with varying levels of success. Yeast for swines is sold for feed applications as wet slurry, as dried brewer's yeast, or in mixtures with other brewery by-products [73]. It is ideal for their feed as a good protein source, it contains most of the essential amino acids in adequate quantities, and numerous vitamins, selenium, copper, and phosphorus. Selenium concentrations are much higher in yeast than in soybean meal, and deficiency of this compound in the swine's alimentation has been the cause of higher swine mortality [78, 79]. Additionally, dried brewer's yeast contains mannan oligosaccharides, which have been reported to increase the growth performance and intestinal health of pigs [80]. Benefits have been described as well for nursing and weanling pigs [81].

Either live or inactivated brewer's yeast have been used as well in ruminants diets, consequently it was observed an increase in productivity of animal meat or milk [73, 82], but also live yeast cultures have been used. These are prepared by inoculating wet cereal grains or grain by-products with live yeast, partially fermenting the mash, and then drying the entire medium without killing yeast or destroying vitamins and enzymes [73]. Live yeast is reported to stimulate fermentation in the rumen through its ability to stimulate the development of anaerobic, cellulolytic and acid lactic bacteria fermentations. In addition, the ingestion of yeast offers continuous supply of vitamins, dicarboxylic acids, removal of oxygen, buffering effect, and reduction in the number of protozoa. As a result, there is an improvement of the digestion of the fibrous and cellulolytic portion of the diet, which leads to a greater intake of food and better performance [83, 84].

Dried yeast was used traditionally in poultry diets in the past as a source of aminoacids and micronutrients, and though the broiler growth was improving this practice was largely discontinued for economic reasons [77]. On the other hand, brewer's yeast appears to be especially beneficial for breeder turkeys and laying hens [85]. Reproductive improvements are attributed to the high level of dietary biotin and selenium in yeast, which is more beneficial than inorganic selenium added to poultry diets [86] and it contributes for the prevention of biotin deficiency in poultry diets, which may result in reduced feed conversion, low egg production, and poor hatchability [87]. Brewer's yeast is also very rich in folic acid, an important vitamin for turkeys [73].

Brewer's yeast has been recognized to have potential as well as a substitute for live food in the production of certain fish or as a potential replacement for fishmeal [88-90]. In addition, it has low content in phosphorous, meaning less water and environmental contamination than common fish meal and other plant-based alternate protein sources [91]. Multiple studies have demonstrated the immunostimulant properties of yeasts, such as their ability to enhance non-specific immune activity [92]. That reaction can be related to β-glucans, nucleic acids as well as mannan oligosaccharides [93]. Brewer's yeast may serve as an excellent health promoter for fish culture as even when administered for relatively long periods is able to enhance immune responses as well as growth of various fish species, without causing immunosuppression [94, 95]. Furthermore, the relative high levels of nucleic acid nitrogen present (mostly in the form of RNA) that in humans and most monogastric animals can became toxic if taken in excess, as the capacity of excretion of the uric acid formed is limited [96], in fish does not happen due to their very active liver uricase [97].

4.2 Human little treats, big benefits

In a world of rapidly increasing population and low agricultural production, yeasts are relatively cheap and easily produced on an industrial scale representing a sustainable alternate protein source to cover the population nutritional demands. The first time that yeast was cultivated in large scale for human nutritional use was in Germany during both World Wars [72]. The yeasts S. cerevisiae as baker yeast, Candida utilis as torula yeast, and Kluyveromyces fragilis as whey yeast, when produced on suitable, food grade substrates (e.g., sugars, ethanol, and lactose), are permitted in many foods around the world [70]. Besides the alcoholic beverages and baking products referred above, yeast are used in the health food industry; as additives, conditioners, and flavouring agents; as sources of high nutritional value proteins, enzymes, nucleic acids, nucleotides, and cell wall polysaccharides [98] as well as for the production of food-grade yeast extracts and autolysates [69, 99]. Yeast extract from dried brewer's yeast cells can be used by enzymatic treatment in a wide variety of foods (e.g., meat products, sauces and gravies, soups, chips and crackers) as flavours enhancers or potentiators [73, 100]. β-Glucan obtained from brewer's yeast can be used in food products as a thickening, water-holding, or oil-binding agent and emulsifying stabilizer [101]. The probiotic activity is an additional role of some yeast that is attracting increasing interest [102].

S. cerevisiae has been studied extensively for its medicinal properties and several beneficial/ probiotic effects on human health and well-being have been reported, including prevention and treatment of intestinal diseases and immunomodulatory actions, are the most well- known. Probiotics are viable microorganisms that are beneficial to the host when consumed in appropriate quantities [103]. The probiotic properties of yeasts reported refer the ability to survive through the gastrointestinal tract and interact antagonistically with gastrointestinal pathogens. As described above, S. cerevisiae have been used as supplements to animal and fish feeds with reported improvements on the growth and health of the hosts [104]. Regarding humans, S. cerevisiae var. boulardii has been successfully used as an oral biotherapeutic agent to treat patients with severe cases of diarrhoea (e.g., antibiotic-associated diarrhoea and traveller's diarrhoea) and other gastrointestinal disorders (e.g., irritable bowel syndrome and Crohn's disease) [105]. Several studies have shown that S. cerevisiae var. boulardii confer beneficial effects against various enteric pathogens, involving different mechanisms as: i) prevention of bacterial adherence and translocation in the intestinal epithelial cells, ii) production of factors that neutralize bacterial toxins and iii) modulation of the host cell signalling pathway associated with pro-inflammatory response during bacterial infection [106]. As reviewed in detail in [106] prevention of bacterial adherence and translocation in the intestinal epithelial cells is due to the fact that the cell wall of S. cerevisiae var. boulardii has the ability to bind enteropathogens, which results in a decrease of their adherence to host epithelial cells. This yeast also produces proteins that are responsible for degradation, neutralisation or dephosphorylation of bacterial toxins. Moreover, the mechanism by which S. cerevisiae var. boulardii modifies host cell signalling pathways associated with pro-inflammatory response is based on blocking activation of nuclear factor-kappa B (NF-κB) and mitogen activated protein kinase (MAPK) which decreases the expression of inflammation-associated cytokines such as interleukin 8 (IL-8), tumor necrosis factor alpha (TNF-α) and interferon gamma (IFN- γ). Conversely, S. cerevisiae var. boulardii also stimulates the peroxisome proliferator-activated receptor gamma (PPAR-γ) expression in human colonocytes and reduces the response of human colon cells to pro-inflammatory cytokines. There are several studies indicating the stimulation of the host cell immunity, both innate and adaptive immunity, by yeast in response to pathogen infections. Furthermore, it has been shown that S. cerevisiae var. boulardii also has a role in the maintenance of epithelial barrier integrity; during bacterial infection the tight junctions are disrupted and this yeast enhances the ability of intestinal epithelial cells to restore the tight-junction structure and the barrier permeability [106].

The benefits from ingesting yeasts do not stop here, many other have been reported as: selectively stimulating the growth and/or activity of one or a limited number of bacteria in the colon [107], e.g., fructooligosaccharides [108]; decreasing the serum cholesterol levels [109], in the treatment of diabetes (regulation of insulin levels) and chronic acne, reducing the appetite, for healthy hair and nails [70, 110, 111] and also for promoting the bioavailability of minerals through the hydrolysis of phytate, folate biofortification and detoxification of mycotoxins due to surface binding to the yeast cell wall [106]. Yet, still are some concerns about public health safety as cause of crescent reports associating the intake of yeast with cases of fungaemia [112, 113].

5. CONCISE SUMMARY

Given that S. cerevisiae var. boulardii is acknowledged as a variant of the species S. cerevisiae, it is probable that additional strains within S. cerevisiae may also exhibit probiotic qualities. To date, significant efforts have focused on harnessing the probiotic properties of lactic acid bacteria, but very less attention has been directed towards the advantageous effects provided by yeast. Nevertheless, yeasts present numerous advantages in comparison to lactic acid bacteria (LAB). They possess a more varied enzymatic profile and seem to exert a more adaptable influence on the immune system. They offer protection against pathogenic microorganisms and harmful substances through surface binding and seem to be more effective for nutritional enhancement and the transport of bioactive chemicals. Furthermore, yeast has greater resilience than lactic acid bacteria (LAB), making it simpler to produce and disseminate, particularly in underdeveloped regions [106]. Moreover, despite the potential for enhancement, the encapsulation technology utilized for probiotics has demonstrated advantages; for instance, S. cerevisiae var. boulardii in microspheres safeguards the yeast from degradation in the gastrointestinal tract, thereby augmenting the intestinal delivery of the viable probiotic. [114].

REFERENCES

1. Kurtzman CP, Fell JW, Boekhout T. The Yeasts: A Taxonomic Study. Amsterdam: Elsevier; 2011.
2. Tulha J, Carvalho J, Armada R, Faria-Oliveira F, Lucas C, Pais C, Almeida J, Ferreira C. Yeast, the Man's Best Friend. In: Valdez B. (ed.) Scientific, Health and Social Aspects of the Food Industry. Rijeka: InTech; 2012. p255-278.
3. Salim ur R, Paterson A, Piggott JR. Flavour in sourdough breads: a review. Trends in Food Science & Technology 2006;17(10) 557-566.
4. Spencer-Martins I. Transport of sugars in yeasts: implications in the fermentation of lignocellulosic materials. Bioresource technology 1994;50(1) 51-57.
5. Kruckeberg AL, Dickinson JR. Carbon metabolism. In: Dickinson JR and Schweizer M. (ed.) The metabolism and molecular physiology of Saccharomyces cerevisiae. London: CRC; 2004. p42-103.
6. Gancedo JM. The early steps of glucose signalling in yeast. FEMS Microbiology Reviews 2008;32(4) 673-704.
7. Galdieri L, Mehrotra S, Yu S, Vancura A. Transcriptional regulation in yeast during diauxic shift and stationary phase. Omics: a journal of integrative biology 2010;14(6) 629-638.
8. Rintala E, Wiebe M, Tamminen A, Ruohonen L, Penttilä M. Transcription of hexose transporters of Saccharomyces cerevisiae is affected by change in oxygen provision. BMC microbiology 2008;8(1) 53.
9. Rodrigues F, Ludovico P, Leão C. Sugar metabolism in yeasts: an overview of aerobic and anaerobic glucose catabolism. Biodiversity and ecophysiology of yeasts 2006;101-121.
10. Nelson DL, Cox MM. Lehninger Principles of Biochemistry. New York: W. H. Freeman; 2008.

11. Pretorius IS. Tailoring wine yeast for the new millennium: novel approaches to the ancient art of winemaking. Yeast 2000;16(8) 675-729.

12. Pronk JT, Steensma HY, Van Dijken JP. Pyruvate metabolism in Saccharomyces cerevisiae. Yeast 1996;12(16) 1607-1633.

13. van Dijken JP, Scheffers WA. Redox balances in the metabolism of sugars by yeasts. FEMS microbiology letters 1986;32(3-4) 199-224.

14. Eriksson P, André L, Ansell R, Blomberg A, Adler L. Cloning and characterization of GPD2, a second gene encoding a DL-glycerol 3-phosphate dehydrogenase (NAD+) in Saccharomyces cerevisiae, and its comparison with GPD1. Molecular Microbiology 1995;17(1) 95-107.

15. Larsson K, Ansell R, Eriksson P, Adler L. A gene encoding a glycerol 3-phosphate dehydrogenase (NAD+) complements an osmosensitive mutant of Saccharomyces cerevisiae. Molecular Microbiology 1993;10(5) 1101-1111.

16. Norbeck J, Påhlman AK, Akhtar N, Blomberg A, Adler L. Purification and characterization of two isoenzymes of DL-glycerol-3-phosphatase from Saccharomyces cerevisiae. Journal of Biological Chemistry 1996;271(23) 13875-13881.

17. Crabtree HG. Observations on the carbohydrate metabolism of tumours. Biochemical journal 1929;23(3) 536.

18. De Deken R. The Crabtree effect: a regulatory system in yeast. Journal of General Microbiology 1966;44(2) 149.

19. Golshani-Hebroni SG, Bessman SP. Hexokinase binding to mitochondria: a basis for proliferative energy metabolism. Journal of bioenergetics and biomembranes 1997;29(4) 331-338.

20. Skinner C, Lin SJ. Effects of calorie restriction on life span of microorganisms. Applied Microbiology and Biotechnology 2010;88(4) 817-828.

21. Murray DB, Haynes K, Tomita M. Redox regulation in respiring Saccharomyces cerevisiae. Biochimica et Biophysica Acta 2011;1810(10) 945-958.

22. Feldmann H. Yeast Metabolism. In: Feldmann H. (ed.) Yeast Molecular Biology - A Short Compendium on Basic Features and Novel Aspects. Munich: Adolf-Butenandt- Institute - University of Munich; 2005. http://biochemie.web.med.uni-muenchen.de/ Yeast_Biol/index.htm

23. Michel RH, McGovern PE, Badler VR. The first wine and beer: chemical detection of ancient fermented beverages. Analytical chemistry 1993;65(8) 408A-413A.

24. Fillaudeau L, Blanpain-Avet P, Daufin G. Water, wastewater and waste management in brewing industries. Journal of cleaner production 2006;14(5) 463-471.

25. McGovern PE, Zhang J, Tang J, Zhang Z, Hall GR, Moreau RA, Nuñez A, Butrym ED, Richards MP, Wang C. Fermented beverages of pre-and proto-historic China. Proceedings of the National Academy of Sciences of the United States of America 2004;101(51) 17593.

26. Feasey P. A brief history of beer. Food Australia 2007;59(5) 191.

27. Sicard D, Legras JL. Bread, beer and wine: Yeast domestication in the Saccharomyces sensu stricto complex. Comptes Rendus Biologies 2011;

28. Poelmans E, Swinnen JFM. From Monasteries to Multinationals (and Back): A Historical Review of the Beer Economy. Journal of Wine Economics 2011;6(02) 196-216.

29. Dickenson J. Viticulture in pre-independence Brazil. Journal of Wine Research 1995;6(3) 195-200.

30. Donalies UE, Nguyen HT, Stahl U, Nevoigt E. Improvement of Saccharomyces yeast strains used in brewing, wine making and baking. Advances in Biochemical Engineering/ Biotechnology 2008;111(1) 67-98.

31. Poelmans E, Swinnen J. From Monasteries to Multinationals (and Back): A Historical Review of the Beer Economy. 2012; Lodolo EJ, Kock JL, Axcell BC, Brooks M. The yeast Saccharomyces cerevisiae — the main character in beer brewing. FEMS Yeast Research 2008;8(7) 1018-1036.

32. Querol A, Bond U. The complex and dynamic genomes of industrial yeasts. FEMS microbiology letters 2009;293(1) 1-10.

33. Dunn B, Sherlock G. Reconstruction of the genome origins and evolution of the hybrid lager yeast Saccharomyces pastorianus. Genome Research 2008;18(10) 1610-1623.

34. Gibson BR, Lawrence SJ, Leclaire JP, Powell CD, Smart KA. Yeast responses to stresses associated with industrial brewery handling. FEMS Microbiology Reviews 2007;31(5) 535-569.

35. Legras JL, Merdinoglu D, Cornuet JM, Karst F. Bread, beer and wine: Saccharomyces cerevisiae diversity reflects human history. Molecular Ecology 2007;16(10) 2091-2102.

36. Pope GA, MacKenzie DA, Defernez M, Aroso MAMM, Fuller LJ, Mellon FA, Dunn WB, Brown M, Goodacre R, Kell DB. Metabolic footprinting as a tool for discriminating between brewing yeasts. Yeast 2007;24(8) 667-679.

37. Thiele F, Back W. The assimilation technique as a basis for optimal yeast management. Brauwelt 2005;145(50) 1594-1598.

38. Schuller D. Better Yeast for Better Wine — Genetic Improvement of Saccharomyces cerevisiae Wine Strains. In: Rai M and Kövics G. (ed.) Progress in Mycology. Heidelberg: Springer 2010. p1-49.

39. Pierce J. The role of nitrogen in brewing. Journal of the Institute of Brewing 1987;93(5) 378-381.

40. Briggs DE, Boulton CA, Brookes PA. Brewing: science and practice. Cambridge: CRC; 2004.

41. Jin Y-L, Alex Speers R. Flocculation of Saccharomyces cerevisiae. Food Research International 1998;31(6–7) 421-440.

42. Bauer FF, Pretorius IS. Yeast stress response and fermentation efficiency: how to survive the making of wine — a review. South African Journal Enology Viticulture 2000;21(Special Issue) 27-51.

43. Bamforth C, Stewart G. Brewing — The transformation of a craft into a technology Biologist 2010;57(3) 139.

44. Canetta E, Adya AK, Walker GM. Atomic force microscopic study of the effects of ethanol on yeast cell surface morphology. FEMS microbiology letters 2006;255(2) 308-315.

45. Marza E, Camougrand N, Manon S. Bax expression protects yeast plasma membrane against ethanol-induced permeabilization. FEBS Letters 2002;521(1-3) 47-52.

46. Miki B, Poon NH, James AP, Seligy VL. Possible mechanism for flocculation interactions governed by gene FLO1 in Saccharomyces cerevisiae. Journal of Bacteriology 1982;150(2) 878-889.

47. Brányik T, Vicente AA, Dostálek P, Teixeira JA. Continuous Beer Fermentation Using Immobilized Yeast Cell Bioreactor Systems. Biotechnology progress 2005;21(3) 653-663.

48. Carlson M. Glucose repression in yeast. Current Opinion in Microbiology 1999;2(2) 202-207.

49. Klein C, Olsson L, Rønnow B, Mikkelsen JD, Nielsen J. Alleviation of glucose repression of maltose metabolism by MIG1 disruption in Saccharomyces cerevisiae. Applied and Environmental Microbiology 1996;62(12) 4441-4449.

50. Lutfiyya LL, Johnston M. Two zinc-finger-containing repressors are responsible for glucose repression of SUC2 expression. Molecular and cellular biology 1996;16(9) 4790-4797.

51. Klein CJL, Rasmussen JJ, Rønnow B, Olsson L, Nielsen J. Investigation of the impact of MIG1 and MIG2 on the physiology of Saccharomyces cerevisiae. Journal of biotechnology 1999;68(2) 197-212.

52. Federoff H, Eccleshall T, Marmur J. Carbon catabolite repression of maltase synthesis in Saccharomyces carlsbergensis. Journal of Bacteriology 1983;156(1) 301-307.

53. Day RE, Rogers PJ, Dawes IW, Higgins VJ. Molecular analysis of maltotriose transport and utilization by Saccharomyces cerevisiae. Applied and Environmental Microbiology 2002;68(11) 5326-5335.

54. Dietvorst J, Londesborough J, Steensma H. Maltotriose utilization in lager yeast strains: MTT1 encodes a maltotriose transporter. Yeast 2005;22(10) 775-788.

55. Omura F, Fujita A, Miyajima K, Fukui N. Engineering of yeast Put4 permease and its application to lager yeast for efficient proline assimilation. Bioscience, biotechnology, and biochemistry 2005;69(6) 1162-1171.

56. Salmon JM, Barre P. Improvement of nitrogen assimilation and fermentation kinetics under enological conditions by derepression of alternative nitrogen-assimilatory pathways in an industrial Saccharomyces cerevisiae strain. Applied and Environmental Microbiology 1998;64(10) 3831-3837.

57. Sutherland CM, Henschke PA, Langridge P, Lopes MB. Subunit and cofactor binding of Saccharomyces cerevisiae sulfite reductase — towards developing wine yeast with lowered ability to produce hydrogen sulfide. Australian journal of grape and wine research 2003;9(3) 186-193.

58. Zamora-Vega R, Montañez-Soto JL, Martínez-Flores HE, Flores-Magallón R, Muñoz- Ruiz CV, Venegas-González J, Ariza Ortega TJ. Effect of incorporating prebiotics in coating materials for the microencapsulation of Saccharomyces boulardii. International Journal of Food Sciences and Nutrition 2012;doi: 10.3109/09637486. 09632012.09687364.

59. Rehm HJ, Reed G, Nagodawithana TW. Biotechnology: Enzymes, biomass, food and feed. Darmstadt: VCH; 1995.

60. Chen SL, Chiger M. Production of bakers' yeast. In: Moo-Young M. (ed.) Comprehensive Biotechnology & Bioengineering. Oxford: Pergamon Press; 1985. p429-455.

61. van Hoek P, de Hulster E, van Dijken JP, Pronk JT. Fermentative capacity in high-cell density fed-batch cultures of baker's yeast. Biotechnology and bioengineering 2000;68(5) 517-523.

62. Beudeker FR, van Dam HW, van der Plaat JB, Vellenga K. Developments in baker' s yeast production. In: Verachtert H. (ed.) Yeast: Biotechnology and Biocatalysis. New York: Taylor & Francis; 1989.

63. Reinhart P, Manville R. The Bread Baker's Apprentice: Mastering the Art of Extraordinary Bread. Berkeley: Ten Speed Press; 2001.

64. Gisslen W. Professional Baking. Darmstadt: Wiley; 2008.

65. Cauvain SP. Bread Making: Improving Quality. Cambridge: CRC Press; 2003.

66. Heenan SP, Dufour J-P, Hamid N, Harvey W, Delahunty CM. Characterisation of fresh bread flavour: Relationships between sensory characteristics and volatile composition. Food Chemistry 2009;116(1) 249-257.

67. Chambers PJ, Bellon JR, Schmidt SA, Varela C, Pretorius IS. Non-Genetic Engineering Approaches for Isolating and Generating Novel Yeasts for Industrial Applications. In: Satyanarayana T and Kunze G. (ed.) Yeast Biotechnology: Diversity and Applications. Dordrecht: Springer Netherlands; 2009. p433-457.

68. Pandey A. Concise encyclopedia of bioresource technology. New York Food Products Press; 2004.

69. Bekatorou A, Psarianos C, Koutinas AA. Production of Food Grade Yeasts. Food Technol. Biotechnol. 2006;44(3) 407-415.

70. Ferreira IM, Pinho O, Vieira E, Tavarela JG. Brewer's Saccharomyces yeast biomass: characteristics and potential applications. Trends in Food Science & Technology 2010;21(2) 77-84.

71. Castrillo JI, Ugalde U. Mycoprotein And Related Microbial Protein Products. In: (ed.) Fungal Biotechnology in Agricultural, Food, and Environmental Applications. Cambridge: CRC Press; 2003.

72. Huige N. Brewery By-Products and Effluents. In: Priest FG and Stewart GG. (ed.) Handbook of Brewing, Second Edition. Cambridge: CRC Press; 2006. p655-713.

73. Sure B. Biological value of food yeast proteins and their role as supplements to the proteins of the cereal grains. Journal of the American Dietetic Association 1946;22(1) 114-116.

74. Bhattacharjee JK. Microorganisms as Potential Sources of Food. In: Perlman D. (ed.) Advances in Applied Microbiology. Academic Press; 1970. p139-161.

75. Rivas B, Torrado A, Moldes AB, Domínguez JM. Tartaric Acid Recovery from Distilled Lees and Use of the Residual Solid as an Economic Nutrient for Lactobacillus. Journal of agricultural and food chemistry 2006;54(20) 7904-7911.

76. Blair R. Nutrition and feeding of organic poultry. Wallingford CABI; 2008.

77. Mahan DC, Parrett NA. Evaluating the efficacy of selenium-enriched yeast and sodium selenite on tissue selenium retention and serum glutathione peroxidase activity in grower and finisher swine. Journal of Animal Science 1996;74(12) 2967-2974.

78. Oldfield JE. Some recollections of early swine research with selenium and vitamin E. Journal of Animal Science 2003;81(14 suppl 2) E145-E148.

79. Kogan G, Kocher A. Role of yeast cell wall polysaccharides in pig nutrition and health protection. Livestock Science 2007;109(1–3) 161-165.

80. LeMieux FM, Naranjo VD, Bidner TD, Southern LL. Effect of Dried Brewers Yeast on Growth Performance of Nursing and Weanling Pigs. The Professional Animal Scientist 2010;26(1) 70-75.

81. Newbold CJ, Wallace RJ, Mcintosh FM. Mode of action of the yeast Saccharomyces cerevisiae as a feed additive for ruminants. British Journal of Nutrition 1996;76(1) 249-261.

82. Denev SA, Peeva TZ, Radulova P, Stancheva N, Staykova G, Beev G, Todorova P, Tchobanova S. Yeast Cultures in Ruminant Nutrition. Bulgarian Journal of Agricultural Science 2007;13(1) 357-374.

83. Patra AK. The Use of Live Yeast Products as Microbial Feed Addictives in Ruminant Nutrition. Asian Journal of Animal and Veterinary Advances 2012;7(5) 366-375.

84. Alabi OJ, Shiwoya EL, Ayanwale BA, Mbajiorgu CA, Ng'ambi JW, Egena SSA. Effects of dried baker's yeast inclusion in rice husk-based diets on performance and egg quality parameters in laying hens. Indian Journal Animal Research 2012;46(1) 56-60.

85. Watanabe N, Inaida O, Yamada K, Karakawa T. New approaches to using spent brewer's yeast. Journal of the American Society of Brewing Chemists. 1980;38(1) 5-8.

86. Scott ML. Importance of biotin for chickens and turkeys. Feedstuffs 1981;53(8) 59-60.

87. Oliva-Teles A, Gonçalves P. Partial replacement of fishmeal by brewers yeast (Saccharomyces cerevisae) in diets for sea bass (Dicentrarchus labrax) juveniles. Aquaculture 2001;202(3–4) 269-278.

88. Ozório ROA, Portz L, Borghesi R, Cyrino JEP. Effects of Dietary Yeast (Saccharomyces cerevisiae) Supplementation in Practical Diets of Tilapia (Oreochromis niloticus). Animals 2012;2(1) 16-24.

89. Coutteau P, Lavens P, Sorgeloos P. Baker's Yeast as a Potencial Substitute for live Algae in Aquaculture Diets: Artemia as a Case Study. Journal of the World Aquaculture Society 1990;21(1) 1-9.

90. Cheng ZJ, Hardy RW, Huige NJ. Apparent digestibility coefficients of nutrients in brewer's and rendered animal by-products for rainbow trout (Oncorhynchus mykiss (Walbaum)). Aquaculture Research 2004;35(1) 1-9.

91. Bagni M, Romano N, Finoia MG, Abelli L, Scapigliati G, Tiscar PG, Sarti M, Marino

92. G. Short- and long-term effects of a dietary yeast β-glucan (Macrogard) and alginic acid (Ergosan) preparation on immune response in sea bass (Dicentrarchus labrax). Fish Shellfish Immunology 2005;18(4) 311-325.

93. White LA, Newman MC, Cromwell GL, Lindemann MD. Brewers dried yeast as a source of mannan oligosaccharides for weanling pigs. Journal of Animal Science 2002;80(10) 2619-2628.

94. Lara-Flores M, Olvera-Novoa MA, Guzmán-Méndez BzE, López-Madrid W. Use of the bacteria Streptococcus faecium and Lactobacillus acidophilus, and the yeast Saccharomyces cerevisiae as growth promoters in Nile tilapia (Oreochromis niloticus). Aquaculture 2003;216(1–4) 193-201.

95. Ortuno J, Cuesta A, Rodriguez A, Esteban MA, Meseguer J. Oral administration of yeast, Saccharomyces cerevisiae, enhances the cellular innate immune response of gilthead seabream (Sparus aurata L.). Veterinary Immunology and Immunopathology 2002;85(1-2) 41-50.

96. Schulz E, Oslage HJ. Composition and nutritive value of single-cell protein (SCP). Animal Feed Science and Technology 1976;1(1) 9-24.

97. Rumsey GL, Kinsella JE, Shetty KJ, Hughes SG. Effect of high dietary concentrations of brewer's dried yeast on growth performance and liver uricase in rainbow trout (Oncorhynchus mykiss). Animal Feed Science and Technology 1991;33(3–4) 177-183.

98. Ferreira IM, Pinho O, Vieira E, Tavarela JG. Brewer's Saccharomyces yeast biomass: characteristics and potential applications. Trends in Food Science & Technology 2010;21(2) 77-84.

99. Fleet G. The Commercial and Community Significance of Yeasts in Food and Beverage Production Yeasts in Food and Beverages. In: Querol A and Fleet G. (ed.) Berlin: Springer 2006. p1-12.

100. Chae HJ, Joo H, In MJ. Utilization of brewer's yeast cells for the production of food- grade yeast extract. Part 1: Effects of different enzymatic treatments on solid and protein recovery and flavor characteristics. Bioresource technology 2001;76(3) 253-258.

101. Thammakiti S, Suphantharika M, Phaesuwan T, Verduyn C. Preparation of spent brewer's yeast β-glucans for potential applications in the food industry. International Journal of Food Science & Technology 2004;39(1) 21-29.

102. Czerucka D, Piche T, Rampal P. Review article: yeast as probiotics - Saccharomyces boulardii. Alimentary Pharmacology and Therapeutics 2007;26(6) 767-778.

103. Fuller R. Probiotics in man and animals. Journal of Applied Bacteriology 1989;66(5) 365-378.

104. Patterson J, Burkholder K. Application of prebiotics and probiotics in poultry production. Poultry Science 2003;82(4) 627-631.

105. McFarland LV. Systematic review and meta-analysis of Saccharomyces boulardii in adult patients. World Journal of Gastroenterology 2010;16(18) 2202-2222.

106. Moslehi-Jenabian S, Lindegaard L, Jespersen L. Beneficial Effects of Probiotic and Food Borne Yeasts on Human Health. Nutrients 2010;2(4) 449-473.

107. Gibson GR, Roberfroid MB. Dietary modulation of the human colonic microbiota: introducing the concept of prebiotics. Journal of Nutrition 1995;125(6) 1401-1412.

108. Maugeri F, Hernalsteens S. Screening of yeast strains for transfructosylating activity. Journal of Molecular Catalysis B: Enzymatic 2007;49(1–4) 43-49.

109. Psomas EI, Fletouris DJ, Litopoulou-Tzanetaki E, Tzanetakis N. Assimilation of cholesterol by yeast strains isolated from infant feces and Feta cheese. Journal of Dairy Science 2003;86(11) 3416-3422.

110. McCarty M. High-chromium yeast for acne? Medical Hypotheses 1984;14(3) 307-310.

111. Bahijiri SM, Mira SA, Mufti AM, Ajabnoor MA. The effects of inorganic chromium and brewer's yeast supplementation on glucose tolerance, serum lipids and drug dosage in individuals with type 2 diabetes. Saudi Medical Journal 2000;21(9) 831-837.

112. Muñoz P, Bouza E, Cuenca-Estrella M, Eiros JM, Pérez MJ, Sánchez-Somolinos M, Rincón C, Hortal J, Peláez T. Saccharomyces cerevisiae Fungemia: An Emerging Infectious Disease. Clinical Infectious Diseases 2005;40(11) 1625-1634.

113. Baldo BA, Baker RS. Inhalant Allergies to Fungi: Reactions to Bakers' Yeast (Saccharomyces cerevisiae) and Identification of Bakers' Yeast Enolase as an Important Allergen. International Archeology Allergy Immunology 1988;86(2) 201-208.

114. Graff S, Hussain S, Chaumeil J-C, Charrueau C. Increased Intestinal Delivery of Viable Saccharomyces boulardii by Encapsulation in Microspheres. Pharmaceutical Research 2008;25(6) 1290-1296

24 Chapter

UTILIZATION OF CHEESE WHEY, A BY-PRODUCT OF THE DAIRY SECTOR

1. INTRODUCTION: AN OVERVIEW

Whey is a by-product of cheese or casein manufacturing, holding significant relevance in the dairy industry due to its substantial production quantities and nutritious composition. Global whey output is projected to be approximately 180 to 190 million tons per year, with only 50% of this quantity being processed. Approximately 50% of global cheese-whey (CW) production is processed and converted into diverse food and feed products. Approximately 50% of this quantity is utilized in liquid form, 30% as powdered cheese whey, 15% as lactose and its derivatives, and the remainder as cheese whey protein concentrates [2].

The European Union produces 40×10^6 tons of whey annually, with a surplus of 13×10^6 tons that includes approximately 619,250 tons of lactose. Currently, this surplus is not employed for additional lactose production; so, whey disposal poses a significant issue from both economic and environmental perspectives. Conversely, the recovery of whey components and/or the utilization of whey as a fermentation medium may be beneficial not only for the environment but also for a sustainable economy [4,5].

Whey comprises over fifty percent of the solids found in whole milk, including whey proteins (20% of total protein) along with the majority of lactose, water-soluble vitamins, and minerals. Thus, whey can be regarded as a significant by-product having multiple applications in the culinary and medicinal sectors.

From a valorization perspective, two distinct approaches to CW management can be contemplated: the first involves the utilization of technologies to extract valuable molecules, like proteins and lactose. At present, valorization methods for CW represent the primary method for managing this by-product, surpassed only by the manufacturing of powdered CW.The second option relies on the application of fermentation processes to obtain value added products [6] such as: organic acids (e.g. lactic, succinic and propionic), single cell proteins and oils, biopolymers (enzymes, polyhydroxyalkanoates, exopolysaccharides) and bacteriocins. Sometimes whey permeate, obtained from ultrafiltration step, has been used as fermentation medium; in this case, both the management options are applied.

The ultrafiltration technique generates a whey permeate abundant in lactose, comprising around 80% of the original lactose content in milk. Innovative methods, such as nanofiltration and reverse osmosis, have been created for the concentration of lactose, applicable in the confectionery industry or pharmaceutical fermentation processes [7]. Whey permeate, in addition to lactose, contains other nutrients vital for microbial growth; thus, its potential as a fermentation medium for producing high-value products is a significant opportunity that should not be overlooked[4]. Furthermore, whey permeate serves as a promising source of oligosaccharides for prospective use in human nutrition [8].

Among the different possibility of whey valorization, reported in Figure 1, individual whey protein purification and application of fermentation technology on whey permeate will be discussed.

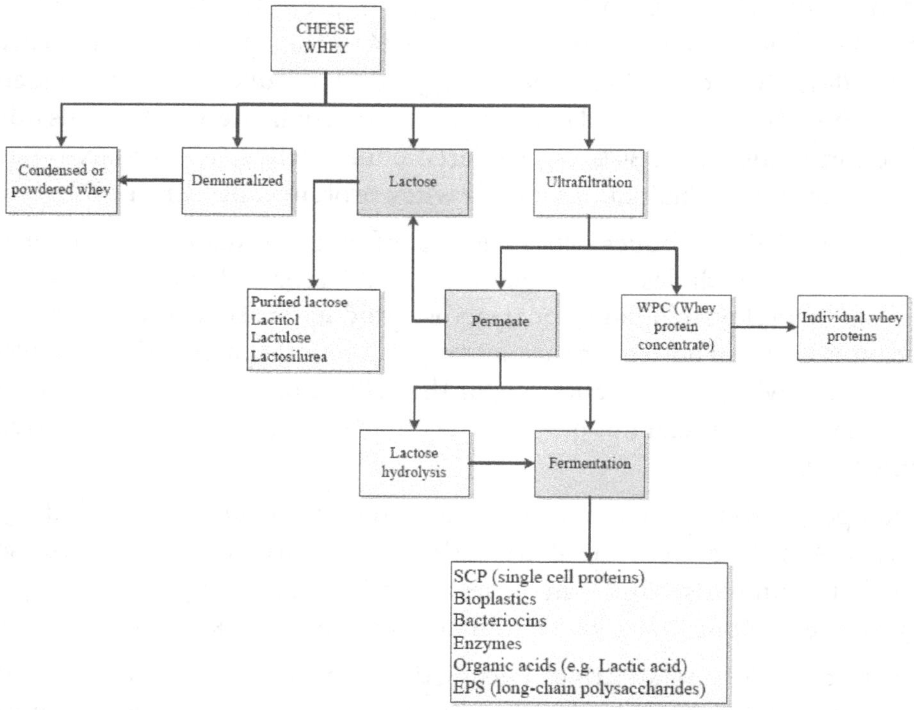

Figure 1. Scheme of current possibility of whey valorization.

2. WHEY PROTEINS

Today, whey is a popular dietary protein supplement purported to provide antimicrobial activity, immune modulation, and to prevent cardiovascular disease and osteoporosis. In addition, whey has the ability to act as an antioxidant, antihypertensive, antitumoral, hypolipidemic, antiviral, antibacterial, and chelating agent [9].

Advances in processing technology, including ultrafiltration, microfiltration, reverse osmosis, and ion-exchange, have resulted in development of several different finished whey products: whey protein concentrates (WPC) containing between 50 - 85% protein on a dry basis, whey protein isolate (WPI) containing between 90-98% protein and very small amounts of lactose and fat, reduced lactose whey, demineralized whey and hydrolyzed whey [10]. Each whey product varies in the amount of protein, carbohydrates, immunoglobulins, lactose, minerals, and fat in the finished product [9].

Nowadays whey ultrafiltration (UF) and diafiltration (DF) are standard operations in the dairy industry that allow protein recovery without significant loss of their functional properties and with a low salt content, making it suitable for human consumption [11,12].

The recovery of proteins by UF and DF represents the first step in whey valorisation. UF has been used in the dairy industry to produce WPC, because this technology allows the selective concentration of the proteins in relation to the retention of protein and selective permeation of lactose, minerals, water and compounds of low molecular weight [10]. DF is used for the production of WPC with a high protein content and purification grade. WPC, which are obtained by whey UF and DF are available in great variety according to protein content and functional properties [7,9].

Whey proteins have a high nutritional value, due to the high content of essential amino acids, especially sulfur-containing ones [13].They are high quality proteins with a protein efficiency ratio (PER) of 3.4, higher than casein (2.8) and similar to egg albumin [14].

Moreover whey proteins have functional properties (e.g. high solubility, water absorption, gelatinization and emulsifying capacities) essential in food application [15].

Thanks to the excellent nutritional and functional properties, commercial value of WPC is from 3 to 40 times greater than that of whey powder [1].

Moreover, the possibility of a different use of whey proteins is taken into account to obtain the so-called "functional foods". Individual whey proteins have their own unique nutritional, functional and biological characteristics that are unrealised in whey protein concentrates. WPC micro-, submicro- and nanocapsules have been applied in the encapsulation of bioactives of interest in the development of novel functional foods (e.g. the antioxidant β-carotene) [16].

The major components among whey proteins are β-lactoglobulin (β-LG), α-lactalbumin (α- LA), bovine serum albumin (BSA) and immunoglobulin (IG), representing 50%, 20%, 10% and 10% of the whey fraction, respectively. All of these major proteins, except for BSA and IG, are synthesized by epithelial cells in the mammary gland. Besides these, whey contains also numerous minor proteins, called low abundance proteins, such as lactoferrin (LF), lactoperoxidase (LP), proteose peptone (PP), osteopontin (OPN), lisozyme (LZ), among others; LF and LP are the most abundant minor proteins [17,18].

The concentration of whey proteins depends on the type of whey (acid or sweet), the source of milk (bovine, caprine or ovine), the time of the year, the type of feed, the stage of lactation, and the quality of processing. Whey proteins are globular molecules with a substantial content of α-helix motifs, in which the acidic/basic and hydrophobic/hydrophilic amino acids are distributed in a fairly balanced way along their polypeptide chains [19]. Major characteristics of whey proteins are summarized in Table 1.

Table 1. Major characteristics of whey proteins [20,18,21].

Protein	Molecular mass (Kg/mol)	Isoelectric point	Concentration (g/l)	Number of amino acids
β-LG	18	5.4	3.2	162
α-LA	14	4.4	1.2	123
Immunoglobulin G	150	5 ÷ 8	0.7	*
BSA	66	5.1	0.4	582
LF	77	7.9	0.1	700
LP	78	9.6	0.03	612
*variable values				

The three major forms in which whey protein products are available, such as concentrates (WPC), isolates (WPI), and hydrolysates (WPH), have limited acceptance by the food processing industry because of the lack of consistency in the gross composition and functionality. Whereas each whey protein has unique attributes for nutritional, biological and food ingredient applications; otherwise individual milk proteins exhibit better functionality than in their native protein mixtures [22,9].

As a matter of fact whey represents a rich mixture of proteins with wide-ranging chemical, physical and functional properties. These proteins play an important role in nutrition and, in a number of instances, also appear to have specific physiological actions, such as: ability to bind metals, functions related to the immune or digestive systems, source of essential amino acids also branched chain amino acids (leucine, isoleucine, and valine) which are thought to play a role as metabolic regulators in protein and glucose homoeostasis, involvement in lipid metabolism, etc. [23,24].

By this way these bioactive compounds are able to reduce disease risk and/or to prevent disease development and have been reported to have utility in many different applications ranging from effects on bone, muscle, blood, brain, pancreas, immune, cancer, infection, metabolism, wound healing, learning, and aging [25]. Moreover, these proteins, once partially digested, serve as a source of bioactive peptides with further physiological activities.

All of these biological and physiological activities offer to the food industry several opportunities; in particular they provide the basis for development of valuable whey protein-based food ingredients targeted to the functional food sector [26].

The term functional food was first introduced in Japan in 1980's: it refers to healthy food similar in appearance to conventional food, consumed as part of a usual diet, and claimed to have physiological benefits like health-promoting or disease-preventing properties beyond the basic function of supplying nutrients. Various definitions of functional food, proposed by authorities, academic bodies and industries, exist worldwide; the difficulty to give a unique definition depends on the fact that foods consumed perform some functions in one way or the other, in particular depending on the state of health of the consumer. Moreover various different terms, listed in Table 2, are sometimes linked or interchanged with the term functional foods [13,27,28]. Considering on one hand the ambiguity among these definitions and on the other the wide set of functions of whey proteins and related peptides, it is quite difficult to establish whether they fall within a definition or into another.

Table 2. Different terms linked or interchanged with functional foods [29,28].

Terms	Definition
Bioactive compounds:	they are chemical compounds derived from a plant, animal, or marine source, that exert the desired health/ wellness benefit.
Dietary supplements:	they act as a supplement to the diet in which the active ingredient is added to the food or it can be consumed in the form of pills, powders, or in liquid forms; they do not replace the complete food or meal.
Functional ingredients:	they are preparations, fractions or extracts containing bioactive compounds of varying purity, that are used as ingredients by manufacturers in the food.
Medical foods:	they are formulated to be administered under the supervision of a physician, for the specific dietary management of a disease or condition for which distinctive nutritional requirements are established.
Natural health products (NHP):	they include homeopathic preparations, substances used in traditional medicines, minerals or trace elements, vitamins, amino acid, essential fatty acids, or other botanical, animal, or microorganism derived substances. They are generally sold in medicinal to diagnose, treat, or prevent disease, restore or correct function, or to maintain or promote health. NHP also include nutraceuticals.

Nutraceuticals:	they are substances, either a food or part of a food, that provide medical or health benefits, including the prevention and treatment of disease. They are derived from foods and can be used in the form of pills, capsules, potions, and liquids.

2.1 Whey proteins separation

Protein functions are related to their native structure, which depends on pH, temperature, pressure and solvent effects. Changes in native structure affect functional properties, by this way in the last years there has been a great interest in developing efficient separation and purification processes that prevent denaturation and loss of biological activity. On the other hand fractionation can emphasize the functional and nutritional properties of the individual proteins [30,31].

Separation processes tend to exploit to the maximum extent different molecular masses, concentrations, and isoelectric points of whey proteins. The main available processes for commercial-scale production of whey protein fractions belong to four categories: selective precipitation, membrane filtration, selective adsorption, and selective elution [20].

A brief description and some examples for each separation category are after-presented.

- Selective precipitation involves adjusting the solution physical properties to promote insolubility. Proteins are typically least soluble at a pH near the isoelectric point (pI) and in low ionic strength solutions, and most likely to aggregate under these conditions. While β-LG precipitates rapidly and selectively at high temperature (70-120 °C) and pH near neutral (pH 8), α-LA precipitates and aggregates better at acidic pH (3.5-5.5) and moderate temperature (50-65 °C) with long reaction times, usually accompanied by the precipitation of bovine serum albumin, immunoglobulins and lactoferrin, while β-LG remain soluble [32,33].
- Membrane filtration is commonly used to make whey protein concentrate, a mixture of all the proteins in whey, with the aid of membranes with a 5,000 to 10,000 g/mol molecular mass rating. As regard fractionation, traditionally it has been based solely on differences in molecular mass; by this way in the past fractionation was possible only for proteins with great differences in molecular mass (e.g. α-lactalbumin vs. IGG) or for proteins with a great combined difference in charge and size. Nowadays it is possible to achieve separation also when proteins have little or no difference in molecular mass together thanks to a careful adjustment of the solution pH

and ionic strength [34,35]. The electrostatic rejection by the membrane, due to their slight residual charge can be enhanced or reduced by adjusting the solution pH. Furthermore separation is enhanced by operating near the pI of the smaller protein and far from the pI of the larger protein to maximize the difference in effective hydrodynamic size: this because the effective diameter of a protein increases with decreasing ionic strength. Low salt concentrations (1–20 mmol/L) increase electrostatic and steric rejection by the membrane: by this way a multi-step adjustment of the pH and ionic strength of the whey may allow fractionation of proteins using a sequence of membrane separation processes [20]. As an example reference [36] is reported; in this experimental study, the effect of working pH was evaluated, employing a 300 kDa tubular ceramic membrane in a continuous diafiltration mode, by measuring the flux-time profiles and the retentate and permeate yields of α-LA, β-LG, BSA, IgG and LF. It was found that a 300 kDa membrane could be employed to fractionate the original array of whey proteins in two parts: α-LA and β-LG in the permeate and BSA, IGG and LF in the retentate. As a matter of fact important protein yields for α-LA and β-LG were obtained in the permeate (except for pH 4 and 5), while for the rest of the proteins studied, there was no significant diffusion through the membrane.

- In selective adsorption, a single purified protein is produced in conjunction with a treated whey solution depleted in that protein; the cost of manufacture must be borne by the income generated from that single purified protein product and the depleted whey solution [20,36]. There are many examples of selective adsorption processes for whey proteins: the immobilized hexapeptide ligand affinity resin or immobilized phenyl groups were used for α-LA, while for β-LG immobilized retinal or ceramic hydroxyapatite chromatography with sodium fluoride as a displacer were applied [37-40].
- In selective elution all the proteins in a mixture are trapped simultaneously onto the adsorbent, rinsed free of contaminants, and then eluted one by-one to obtain different purified proteins. The process uses an adsorbent and buffers that are inexpensive and food- grade, and it is operated at a high flow rate. Further on the cost of manufacture is spread among many purified protein products with the possibility to manufacture different products simply by using different elution buffers; by this way it is considered an attractive alternative to selective adsorption [41]. Differently from precipitation and membrane separation processes, which are volume-dependent separation methods, selective adsorption and selective elution

processes are less volume dependent because adsorbent capacity depends mostly on the mass of protein recovered, not the volume of liquid processed [42]. Various studies about the development of ion exchange or affinity chromatography processes to separate whey proteins exist; referring as a case in point [43] various superparamagnetic anion-exchangers and their use together with cation-exchangers in the fractionation of bovine whey proteins (LF, LP, IG, and β-LG) were studied. While in reference [41] all the positively charged proteins in whey were bound simultaneously to a cation exchange column, rinsed free of contaminants and then eluted selectively to produce different fractions: with a single piece of equipment they were able to manufacture WPI, or α-LA and WPI depleted in α-LA or LF and LP only.

2.2 Biological properties of individual whey proteins

2.2.1 β-lactoglobulin (β-LG)

β-lactoglobulin, a member of the lipocalins family, is the major whey protein of ruminant species, 58% (w/w). It is present in many mammalian species but absent in human milk. Genetic variants of a single gene, of which β-LG A is the most common, have been widely reported particularly in the cow [44,21].

The lipocalins family presents various kind of features, most of which involve some ligand- binding functions; these last must be the physiological reason for the significant quantities of β-LG found in milk (the domestic cow produces 2 to 3g L–1) [45].

In isolated form, despite its globular nature, it exhibits a low solubility and a low ionic strength. It contains normally 162 aminoacidic residues and has a molecular weight of 18,362 g/mol. The secondary structure of this protein comprises nine strands of β structure, a short α helix segment and three helicoidal turns [46].

Its quaternary structure depends on the medium pH: at pH values between 7 and 5.2 it is a stable dimer (molecular weight of 36,700 kDa); at pH values between 5.2 and 3.5, an octamer (molecular weight of ca. 140,000 kDa); and below pH 3.0 and above 8.0, a monomer, with two-cysteine residues per monomer [47].

2.2.1.1 β-LG biological functions

The endogenous function of β-LG is not clear, but it is known that it is a source of amino acids. Indeed β-LG, over-expressed in the lactating mammary gland of many species, is primarily an important source of amino acids for the offspring of those animals that produce it [45]. The amino acid content fuels muscle growth and it is a source rich in cysteine, which is important for the synthesis of glutathione [13].

It participates in the digestion of milk lipids in the neonate: β-LG binds to free fatty acids as they are released by pregastric lipases, by this way the digestion of milk fat is facilitated [48].

β-LG is the major allergen in cow's milk, responsible for causing milk allergy [25]. It was evidenced that milk's major allergen can be rendered non-allergenic and it can also be modified or administered inhibiting rather than stimulating the allergic process. β-LG has been conjugated with acidic oligosaccharides to reduce its antigenicity. Immunization of mice with the conjugates leads to a reduced T cell response, predominantly Th1-mediated, suggesting that the conjugates may have utility in preventing Th2-mediated allergy [49]. Oral administration of recombinant Lactococcus lactis expressing bovine β-LG induces a specific Th1 response, suggesting that probiotics expressing β-LG could be useful in the management of food allergy [50]. Intranasal co-administration of live lactococci expressing IL-12 and β-LG produces a protective Th1 response that inhibites allergic airway disease in mice [51]. Acidic β-LG-derived peptides hydrolyzed with Lactobacillus paracasei peptidases represse lymphocyte stimulation, upregulate IL-10 production, and downregulate IFN-γ and IL-4 secretion [52].

β-LG structure comprise a ligand-binding site. The ligands that bind to β-LG tend to be hydrophobic and include long-chain fatty acids, triglyceride, retinoids, cholesterol and, more weakly, hydrocarbon molecules [53].

β-LG can also be a source of peptides with different functions:

- Lactokinins, Tyr-Leu (f102–103) and Ala-Leu-Pro-Met-His-Ile-Arg (f142–148), are inhibitors of angiotensin-I-converting enzyme (ACE) and represent potential nutraceuticals/functional food ingredients for the prevention and/or treatment of high blood pressure [54].
- β-lactorphin (f102–105) has ACE-inhibitory activity, improved vascular relaxation in spontaneously hypertensive rats, and it is an opioid receptor agonist suggesting that it can modulate absorption processes in the intestinal tract [55].
- Beta-lactotensin, His-Ile-Arg-Leu (f146 –149), is an ileum-contracting peptide, which exhibits hypertensive activity. It is a natural ligand for neurotensin NT2 receptors, has an anti-stress effect, promotes the abolition of fear memory, reduces sensitivity to painful stimuli, and consolidates memory. It is able to reduce blood cholesterol levels, but only when given parenterally, and not orally [56-58].

2.2.2 α-lactalbumin (α-LA)

α-lactalbumin is one of the most studied proteins: it is an important component of milk and it contributes significantly to its physical, biological and nutritional characteristics [59]. It is biosynthesized in the mammary gland during lactogenesis, and participates actively in the synthesis of lactose [60].

In human whey, α-LA is a major protein (1.7 mg/mL); however, in bovine whey it is ranked second with respect to protein content after β-LG. The bovine protein is characterized by its relatively low molecular mass (14.2 kDa) and its acidity (isoelectric point of 4.8); α-LA possesses a high mineral content and a balanced amino acid composition indeed its sequence is composed of 123 amino acids, that includes four residues of tryptophan and a high proportion of essential amino acids, Cys, Ile, Leu, Lys [61,62].

The structure of bovine α-LA is highly stabilized by four disulphide bonds and by the association of Ca2+ at the binding loop that joins together two domains of the protein: a large α- helical and a small β-sheet domains [63,64].

In addition to Ca2+ ion, α-LA is able to bind several other cations such as Mg2+, Mn2+, Na+ and K+ and also contains various Zn2+ binding sites [65]. Nevertheless, only the binding of Ca2+ ion is fundamental for the maintenance of the native conformation of the protein [66,67].

2.2.2.1 α-LA biological functions

α-LA and its biological and functional peptides are characterized by potential health benefits; as regard of that there is a growing scientific interest in the context of health-promoting functional foods. In particular, α-LA and α-LA-peptides can be used as supplements of essential amino acids in food to improve/maintain the immune system, to reduce the stress, for opioid activity, antihypertensive action, regulation of cell growth, immunomodulation etc. [68].

α-LA is cytotoxic and this property can be exploited for therapeutic uses. α-LA has protec tive properties against mucosal injury[69,70].

α-LA is particularly rich in essential amino acids. It has a high content of lysine and cysteine and a particularly high content of tryptophan (5.9% of the total amino acid content) [61].

The high content in tryptophane makes α-LA a nutraceutical; in particular it may help improve mood, sleep, and cognitive performance [71,72]. On the other hand the content of cysteine is valuable in boosting the immune system and promoting wound healing. As a general consideration, thanks to the high content in essential amino acids, α-LA is an invaluable supplement for infant formulas [68].

The protein may possess bactericidal or antitumor activity. The active form of the protein is called "human α-LA made lethal to tumor cells" (HAMLET), a complex formed by which induces apoptosis in tumor cells but spares mature cells and has received much attention due to its potential use as a new therapeutic agent against tumor cells [73].

The digestion of α-LA with trypsin and chymotrypsin creates some polypeptides with bactericidal properties, mostly against Gram-positive bacteria and a weak bactericidal activity against Gram-negative strains. The peptide α-lactorphin behaves like opioid receptor agonists [74,75].

α-LA has a marked suppressive effect against the increased release of proinflammatory cytokines and tumor necrosis factor-α, from the D-galactosamine induced liver injury rat model or ischemia/reperfusion induced intestinal injury rat model [76]. Two synthetic peptides corresponding to the sequences f50-51 (Tyr-Gly) and f18-20 (Tyr-Gly-Gly) of α-LA also enhance both the in vitro proliferation and protein synthesis of concanavalin A-stimulated human peripheral blood lymphocytes [77].

2.2.3 Immunoglobulins (IG)

Immunoglobulins constitute a complex group of elements produced by B-lymphocytes and their concentration in whey is 0.7 g/l. IG are divided into three basic classes: IGG, IGA and IGM. IGG is often sub-divided into two subclasses, IGG1 and IGG2. IGG represents up to 80% (w/w) of all IG in milk or whey. Qualitatively the family of IG found in bovine whey and colostrum include IGA, secretory IGA, IGG1, IGG2, IGG fragments, IGM, IGE, J-chain or component, and free secretor component [78,21].

IG make a significant contribution to the whey protein content and they exert an important immunological function (especially in colostrum). IG are subject to postnatal transfer via colostrum because the placenta does not permit passage of macromolecules. These proteins are present in the serum and physiological fluids of all mammals; some of them attach to surfaces, where they behave as receptors, whereas others function as antibodies, which are released in the blood and lymph [79].

In terms of quaternary structure, IG are either monomers or polymers of a four-chain molecule, consisting of two light polypeptide chains (with a molecular weight in the range 25,000 kDa) and two heavy chains (with molecular weight of 50,000–70,000 kDa) [80].

2.2.3.1 IG biological functions

Milk immunoglobulins normally provide passive immunity for the neonate, but they are also potentially powerful agents that could be incorporated into diets to remove toxic, or undesirable dietary factors. As an example, naturally occurring antibody in milk can be extracted which binds to cholesterol in the human digestive tract and prevents its absorption into the bloodstream. This anti-cholesterol antibody can be a useful food supplement for the functional food market [25].

Concerning IG antimicrobial and antiviral properties it is known that concentration of colostrum whey antibodies against a particular pathogen can be raised by immunising cows with the pathogen or its antigens. Antibody concentrates derived from immune milk collected from cows immunised with inactivated human rotavirus possess preventive/treatment features in enteric disease caused by said viruses in therapeutics of child infections [81]. The hyperimmune whey that results can potentially provide prophylactic protection against various infectious gut microbes including rotavirus and Helicobacter pylori [82]. Infant gastritis originated by H. pylori is well fought via a diet including immune milk containing specific anti-H. pylori antibodies [83]. There is also evidence of protection via bovine antibodies against dental caries caused by cariogenic streptococci [84].

IG can also act in the immune system modulation; they are recognised to provide protection against diseases in the newborn through passive immunity. Systemic immunisation of pregnant cows increases the levels of antibodies against immunising bacteria, and also reduces susceptibility to disease [85]. Vaccination of pregnant cows originates colostrum characterized by high concentrations of specific antibodies against the antigens of the vaccine used [86].

Other methabolic features of IG are also known. Immune milk is suggested to lower blood pressure. In reference [87] a clinical-trial study regard the effects on reduction of cholesterol and blood pressure of immune milk, produced by dairy cows previously hyper-immunised with a multivalent bacterial vaccine, was described. The involved human hypercholesterolemic subjects consumed 90 g of immune milk daily: this was a useful adjunct in the dietary management of hypercholesterolemia.

2.2.4 Bovine Serum Albumin (BSA)

Bovine serum albumin represents the 5% of the whey proteins and its concentration in whey is about 0.4 g/l. It appears in milk following passive leakage from the blood stream indeed it is not synthesized in the mammary gland [21,25]. The most outstanding property of BSA is its ability to bind reversibly various ligands; in particular it is the principal carrier of fatty acids [88].

It contains 582 amino acid residues, which lead to a molecular weight of 66,267 kDa; it also possesses 17 intermolecular disulphide bridges and one thiol group at residue 34 [89]. BSA molecule is heart-shaped; it consists of three homologous α-helical domains and each domain contains two sub-domains that share common structural motifs [88].

Thanks to its size and higher levels of structure, BSA can bind to free fatty acids and other lipids, as well as flavor compounds, this feature is severely injured upon denaturation [90].

Its heat-induced gelation at pH 6.5 is initiated by an intermolecular thiol-disulphide interchange, similar to what happens with β-LG [78].

2.2.4.1 BSA biological functions

BSA has the ability to inhibit tumor growth thanks to the modulation of activities of the autocrine growth regulatory factors; this was evidenced by in vitro incubation with human breast cancer cell line MCF-7 [91]. In relation to this activity [92] the delivery system MTO- BSA-NS (mitoxantrone nanoparticles loaded with bovine serum albumin) was studied; human MCF-7 breast cancer in nude mice and animal model of P388 lymphnode metastases in Kunming mice were applied to investigate the therapeutic efficiency. The inhibition rate of the nanospheres against breast cancer was high and lymphnode metastases were efficiently inhibited.

BSA is able to bind fatty acids stored in the human body as fat; this allows BSA to participate in synthesis of lipids [93]. BSA has also antioxidant activities; in vitro it is able to protect lipids against phenolic-induced oxidation [94]. BSA is also a source of essential amino acids, whose therapeutic potential is largely unexplored.

Biological functions of some BSA-derived peptides have also been examined. The peptide serorphin (Tyr-Gly-Phe-Gln-Asn-Ala) (f399–404) has opioid agonist activity, while the peptide albutensin A (Ala-Leu-Lys-Ala-Trp-Ser-Val-Ala-Arg) (f208 –216) is an ACE inhibitor and is reported to have ileum contracting and relaxing activities [54].

2.2.5 Lactoferrin (LF)

Lactoferrin together with lactoperoxidase is one of the minor whey proteins. It belongs to the transferrin family which is composed by proteins capable of binding and transferring $Fe3+$ ions. LF is a glycoprotein characterized by a molecular weight of 80,000 kDa. It is synthesized by glandular epithelial cells and mature neutrophils, and can be found in milk, saliva, tears, nasal and intestinal secretions, pancreatic juice, seminal fluid, and in secondary granules of neutrophils. Bovine milk contains between 0.02 and 0.35 mg/ml of LF, depending on the period of lactation [25,95].

LF is a bilobal protein that contains two homologous metal-binding sites with high affinities for ferric iron. The iron-binding sites are situated in the inter-domain clefts. The requirement for an anion, bound synergistically with the Fe ion is a unique feature of LF. Iron release can occur at low pH and is associated with a large-scale conformational change in which the two domains that enclose each iron-binding site move wide apart [96].

LF possess a wide range of biological functions, many of which are not connected with its iron binding ability; it is the most valuable biomedical protein present in whey due to the various therapeutic properties it exhibits [97,25].

2.2.5.1 LF biological functions

LF has quite important role in iron metabolism. It may contribute to local iron accumulation at sites of inflammation and it has been known to be responsible for hypoferraemia through binding free iron and shuttling it back to macrophages [98,97]. LF might also have a control function in situations when increased amounts of iron are released from its depots [99]. Further on LF seems to affect intestinal iron absorption in infants depending on the organisms need for iron [100].

LF is a part of the innate immune system and it also takes part in specific immune reactions; it represents one of the first defense systems against microbial agents [101,95].

LF can have a bacteriostatic effect thanks to its ability to bind free iron, essential for the growth of bacteria (e.g. E. coli). LF, as an iron donor, supports the growth of Lactobacillus sp. or Bifidobacterium sp., characterized by a lower iron demands and generally considered as beneficial [102,103]. LF owns also a bactericidal effect, which is not iron-dependent, against Gram-negative and Gram-positive bacteria. In the first case there is the disruption in the cell wall because the microorganisms have the receptors for the N-terminal region of LF; while for Gram-positives, changes in the permeability of the membrane occurs mediated by electrostatic interactions between the lipid layer and LF surface [104,105]. LF may contribute to defense against the invasion of facultative intracellular bacteria such as: E. coli HB101, Yersinia enterocolica, Y. pseudotuberculosis, Listeria monocytogenes, Streptococcus pyogenes, and Staphylococcus aureus. LF binds both target cell membrane glycoaminoglycans and bacterial invasins. The proteolytic activity of LF inhibits the growth of Shigella flexneri and E.coli through degrading proteins necessary for colonization [105,106].

The main contribution to antiviral defense of LF consists in its binding to glycosaminoglycans of cell membrane: viruses (Herpes simplex virus, cytomegaloviruses, human immunodeficiency virus) cannot enter cells and infection is stopped at an early stage [106]. LF is also capable of binding certain DNA and RNA viruses [107].

LF may support the proliferation, differentiation, and activation of immune system cells and strengthen the immune response; LF also acts as an anti-inflammatory factor. Thanks to its antimicrobial activity and capability of binding components of bacterial cell walls or their receptors, LF may prevent the development of inflammation and subsequent tissue damage caused by the release of pro-inflammatory cytokines and reactive oxygen species [101].

As regard the influence of LF on tumor growth, it is known that it is able to halt the growth of human mammary gland carcinoma cells between the G1 and S stage; this effect may be ascribed to the altered expression or activity of regulatory proteins

[108]. The lactoferrin-dependent, cytokine-mediated stimulation of activity of NK cells and lymphocytes CD4+ and CD8+, represents an important factor in defense against tumor growth: after the oral administration of lactoferrin the number of these cells increases [109]. Even if the exact mechanism has to be discovered completely, it seems that LF-mediated inhibition of tumor growth might be related to apoptosis of these cells induced by the activation of the Fas signaling pathway [110].

LF is also able to act as a growth factor activator; it can have effect on small intestine epithelial cells and endometrium stroma cells [111]. It has also been identified as a transcription factor [112].

LF contributes to the stabilization of the osseous tissue; it may affect bone cells through the inhibition of osteolytic cytokines whose levels rise during inflammation. LF is a potent anabolic factor affecting osteocytes; it stimulates osteoblast proliferation, enhances thymidine incorporation into osteocytes, and reduces apoptosis of osteoblasts. By this way LF might be potentially useful in the treatment of diseases such as osteoporosis [113].

2.2.6 Lactoperoxidase (LP)

Lactoperoxidase is member of the family of mammalian peroxidases. It is present in a variety of animal secretions (e.g. tears, saliva and milk), it is one of the most abundant enzymes in plain milk and represents 1% (w/w) of the total protein pool in whey; its concentration in whey corresponds to 0.03 g/l [114].

LP consists of a single polypeptide chain containing 612 amino acid residues and its molecular mass is about 80 kDa. It contains 15 half-cystine residues and carbohydrate moieties comprise about 10% of the weight of the molecule [115].

The complete LP system (i.e. enzyme plus substrate) was originally characterized in milk [116]; its activity depends on many factors (e.g. animal species, breed and lactation cycle). Other members of that group of oxidoreductases include myeloperoxidase (present in neutrophils and monocytes), eosinophil peroxidase and thyroid peroxidase; they are characterized by a close evolutionary relationship [21].

2.2.6.1 LP biological functions

LP is characterized by the antimicrobial activity related to the LP system, formed by LP, thiocyanate anion, and hydrogen peroxide, which is active only in the presence of all these three components: LP catalyses the oxidation of thiocyanate by H2O2 and generates intermediate products with a broad spectrum of antimicrobial effects against bacteria, fungi and viruses. The LP system is naturally occurring in milk and saliva and has been used in foods, cosmetics and in clinical applications because of its safety and broad spectrum of action; in particular it is fundamental in the dairy industry for the preservation of raw and pasteurized cheese, and yogurt [117,118].

The thiocyanate anion is significantly present in saliva, milk and airway secretions. The amount of the anion in cow's milk ranges from 0.1 to 15 mg/kg and its concentration varies according to animal species, breed, lactation cycle, season and composition of feed [114,21].

Hydrogen peroxide is not normally detected in raw milk. It may be generated endogenously by polymorphonuclear leucocytes or under aerobic conditions by many lactobacilli, lactococci, and streptococci. H2O2 is normally present to very small levels because its content is rapidly reduced by catalases and peroxidases that are adventitious in milk [119-121].

The major intermediary oxidation product, at physiological pH, is hypothiocyanate; it mediates bacterial killing, as it is cell-permeable, and can inhibit glycolysis, as well as (NADH)/ (NADPH)-dependent reactions in bacteria. It is bactericidal for enteric pathogens including multiple antibiotic resistant strains of E. coli. Other reaction products of the LP system, such as cyanosulphurous acid and cyanosulphuric acid, are able to oxidise sulphidrile groups of bacterial proteins [114,21].

The LP system inhibit Gram-negative, catalase positive organisms, such as pseudomonas, coliforms, salmonellae and shigellae; these microorganisms can also be killed if H2O2 is supplied exogenously. Gram-positive, catalase negative bacteria, such as streptococci and lactobacilli are generally inhibited but not killed by the LP system [122,121].

The LP system exerts both bacteriostatic and bactericidal activities against strains of Salmonella typhimurium, S. aureus, and L. monocytogenes; the system is bactericidal against Campylobacter jejuni, Brucella melitensis. As regard Bacillus cereus, Streptococcus mutans, Streptococcus sanguis, Streptococcus mitis, and Streptococcus salivarius the LP system has an inhibitory effect [123-127,121].

The antifungal activity of the LP system with glucose oxidase as H2O2 source has been reported against Rhodutorula rubra, Saccharomyces cerevisiae, Mucor rouxii, Aspergillus niger, and Byssochlamys fulva in salt solution and in apple juice. The LP-thiocyanate-H2O2 system was found to inhibit the growth and proliferation of many fungal species (e.g. Aspergillus flavus, Trichoderma spp.); the same system showed also antifungal activity against Alternaria spp., Penicillium chrysogenum and Claviceps spp. [128,121].

Finally, as regard the antiviral properties of the LP system, the ability to kill both poliovirus and vaccina virus with halides (I-, Br-) as electron donors has been reported [129]. In references [130,131] the LP- H2O2-halide system virucidal activity against HIV-1 was reported.

3. WHEY FERMENTATION PRODUCTS

During last 50 years, cheese whey was used in different bioconversions; for examples the microbial biomass production for animal feed supplement [132], biogas production using anaerobic methanogenic bacteria [14], bioethanol production by Kluyveromyces marxianus [133,134] or recombinant Saccharomyces cells [135,136], hydrolized lactose solution in sweeteners and dietary supplements production [14].

Therefore, at present, it can be very interesting and promising to consider again the possibility to use cheese whey and, particularly, glycidic component because of new demands, as bioplastic synthesis (poly-hydroxyalkanoates -PHA- and polylactate acid -PLA-), antimicrobial peptides (bacteriocins), enzymes and esopolisaccharides (EPS).

Just below, there are the information regarding the fermentative production of the mole cules and macromolecules previously mentioned.

3.1 Polyhydroxyalcanoate

Polyhydroxyalkanoates (PHA) are aerobic bacteria synthesized macromolecules (polyesters); they are carbon and energy reserve, accumulated as intracellular granules. These microbial compounds play the same role as glycogen and starch in animal and plant cells, respectively [137].

Among biodegradable plastics, PHA are very interesting polymers because of their chemical, physical and mechanical properties comparable to petroleum-derived plastics (polyethylene and polypropylene); otherwise, PHA are different for their complete biodegradability, UV-resistence, oxygen-impermeability (fundamental property for food packaging) and biocompatibility (essential property for medical and surgery applications) [138,139].

Under unbalanced growth conditions (limitation of essential nutrients, e.g., nitrogen, phosphate, or oxygen), several microorganisms redirect the acetyl-CoA flux from biomass formation towards accumulation of PHA [140,141].

At present, these biopolymers are synthesized in pure cultures, using synthetic pure substrates (as monosaccharides and organic acids), in macronutrient limiting conditions (N, O, P). Because of substrate weights on whole process cost at 40%, it becomes necessary to search alternative cheaply available source materials for PHA production, as food by-products (e.g. cheese whey).

In the last two decades, a broad number of studies were related to the production of PHA from milk whey permeate using pure cultures of wild type microorganisms or recombinant ones. In several of these works, whey lactose was hydrolyzed by lactase: poly-(3HB-3HV) was produced with Ralstonia eutropha DSM545 on whey permeate and inverted sugars [142], poly-3-(hydroxybutyrate-co-hydroxyvalerate) was produced by P. hydrogenovora with hydrolyzed whey permeate and sodium valerate [143].

In [141] Hydrogenophaga pseudoflava DSM1034 was reported as unique example of wild type microorganism able to synthesize PHA directly from lactose, three possible routes from whey lactose to PHA have been suggested: direct conversion of lactose to PHA, hydrolysis of lactose (chemically or enzimatically) and conversion of glucose and galactose to PHA, lactose fermentation to lactic acid and then conversion of lactic acid to PHA. In reference [144] a recombinant E. coli strain GCSC 6576 and whey powder was used in a pH-stat fed-batch fermentation. After 47 h of fermentation, it was obtained a dry cell weight and P(3HB) concentration of 109 g L−1 and 50 g L−1, respectively. Further, with the same organism and using pH-stat fed-batch culture and concentrated whey solution containing 210 g L−1 lactose as nutrient feed, a dry cell weight of 87 g L−1 and P(3HB) concentration of 69 g L−1 containing 87% dcw P(3HB) was achieved. Using the recombinant E. coli strain (K24K), the authors successfully produced 70.1 g L−1 biomass containing 51.1 g L−1 P(3HB) in a pH controlled fed-batch fermentation at pH 7.20 with whey and corn steep liquor as carbon and nitrogen sources, respectively[145]. Previously, PHB production by Methylobacterium sp. ZP24 on lactose and whey with similar values of biomass polymer content was described [146].

Poly-β-hydroxybutyrate production was also described in lactic acid bacteria belonging to Lactococcus, Lactobacillus, Pediococcus and Streptococcus genera [147].

Lactic acid producing bacteria such as Lactobacillus lactis [148], Propionibacterium [149], L. delbrueckii [150-152] and C. necator have been also used in a co-culture fermentation system: LAB converted sugars into lactic acid which was later taken up by C. necator to produce PHAs. Generally, it has been demonstrated that co-culture fermentations resulted in increased yield, improved control of product qualities. A further advantage in the application of co-cultures is the possibility of utilizing secondary products (e.g. whey) cheaper than glucose as substrates for production of PHAs [153]. Bacteria that have a "generally recognised as safe" (GRAS) status for PHA-production such as lactic acid bacteria and bacilli belonging to probiotic species [147,154] might constitute an added value to these biotechnological processes [155].

3.2 Lactic acid

Acetic, propionic, lactic, lactobionic, citric, gluconic and itaconic acids can be obtained from lactose/whey fermentation. Among organic acids that find applications in specialty chemicals, lactic acid is the most important from an economical point of view [156].

Lactic acid is used in food and chemical industries (pharmaceutical products, textiles, leather), primarily as a preservative and as acidulant [157-159]. Also it has applications as a biodegradable plastic component (polylactide, polymers, polyhydroxybutyrate) [159]. Notably, the industrial demand for lactic acid (LA) has been increasing considerably in recent years, owing to the promising applications of its polymer, the polylactic acid (PLA), as an environment-friendly alternative to plastics derived from petrochemicals [160].

Polylactic acid (PLA) is a biodegradable polyester made by condensation of lactic acid (LA) monomers [161]. It can be worked with conventional facilities and techniques to produce implant devices and internal sutures [162]. Due to its low toxicity, it is classified as GRAS and can be also used in food packaging [161,163], which is one of the widest fields of plastics market. PLA is completely biodegradable under compostable conditions: however, if disposed in landfills, it will last in the environment for years, likewise oil-based plastics [164].

Cheese whey effluents have been used in fermentation processes to produce lactic acid [165-168,158,169,159]. Microorganisms used in lactic acid production are Lactobacillus casei [166,157,158,169], Lactobacillus helveticus [170-172,168,156,159]; Lactobacillus acidophilus [167]; Lactobacillus delbrueckii [165,167,168]; Lactobacillus salivarius [169]; Lactococcus lactis [158]. Streptococcus thermophilus [167]; K. marxianus [168]; Leuconostoc and Pediococcus [4].

Among different lactobacilli species employed in lactic acid production, L. helveticus is the generally preferred organism, it is a homolactic LAB that produces LA in racemic mixture (DL) [173]. L. helveticus showed enhanced lactose utilisation and lactic acid production at 42°C and pH 5.8 [159].

Currently, a high fraction of generated CW is managed by membrane processes, mainly, ultrafiltration. The obtained permeate has a low protein content and an elevated lactose and mineral salts concentration, both aspects are advantageous in lactic acid fermentation. As a consequence, several works have been carried out aimed at obtaining lactic acid after ultrafiltration of CW [171,157,172,169]. Highest lactic acid production rate was obtained with L. helveticus cultivated in whey permeate, with corn steep liquor (CSL) as the nitrogen source [174]. Lactic acid productivity of 9.7 g/L/h using L. helveticus strain milano has been obtained in continuous fermentation of whey-yeast extract permeate medium [173,175]. In a work with L. salivarum YE supplementation was replaced by in situ treatment of fermentation medium with proteolytic microorganisms [169].

However, lactic acid production from cheese whey or its permeate obtained without nutrients supplementation [165,168,159,169,] is of limited application to industrial scale because of the low productivity, nutrients supplementation is a key factor limiting the process efficiency.

Some studies report the use of mixed cultures in lactic acid production with synergistic effects [168,158]. Other research groups tried to improve LA production using rE.coli harbouring an inducible expression plasmid containing D-lactate dehydrogenase encoding gene of Lactobacillus plantarum [176] or using metabolic engineering of LAB, fungal or yeast systems [177,178,179], but all of these strategies, if compared with the use of mixed cultures, involve higher costs due to genetic engineering studies and the need of sterilization.

LAB have been immobilised by several methods on different supports (calcium alginate, k-carragenane, agar and polyacrylamide gels) [4] and the immobilised systems have been investigated for lactic acid production from whey. A two-stage process was used for continuous fermentation of whey permeate medium with L. helveticus immobilised in k-carrageenan/ locust bean gum, which resulted in high lactic acid productivity (19–22 g L−1 h−1) [172].

In [4] L. casei was immobilized in Ca pectate gel. A high lactose conversion (94.37%) to lactic acid (32.95 g L−1) was achieved and the cell system was found highly stable, no decrease in lactose conversion to lactic acid was observed up to 16 batches.

3.3 Exopolysaccharides

Exopolysaccharides (EPS) are long-chain polysaccharides (4*104-6*106 Da). They are produced by bacteria and microalghe and can be divided in homopolysaccharides (Homo-EPS) and heteropolysaccharides (Hetero-EPS) [180]. Homo-EPS consist of either D-glucose (glucans) or D-fructose (fructans) residues, with different types of linkage and branching degree. Hetero-EPS are constructed from multiple copies of an oligosaccharide and show little structural similarity to one another: glucose, galactose, xylose, mannose, arabinose and rhamnose are the most represented sugars but also amino-sugars and polyols can be occasionally present as well as glucuronic acid. They are often highly branched with different binding types [181].

The physiological role of EPS is related to microbial cell protection from toxic compounds, phagocytosis, antibiotics effect and osmotic stress [182]. In the last years, their prebiotic role was identified and described [183,184]. In addition to role as food additives and prebiotics, LAB synthesized EPS have been implicated as anti-tumor agent, immuno-stimulator and blood cholesterol lowering agent [185,184].

On the basis of their rheological properties, EPS are applied as stabilizers and emulsifiers in food industry, particularly for yogurt, fermented milk and mozzarella production [186,187]. EPS used in food industry must be considered additives and, consequently, must ensure safety qualification. EPS synthesize by LAB, generally used in food production, possess this safety characteristics [184]. The GRAS and probiotic status of some lactobacilli give to them more preference for consumable EPS production. Main drawbacks limiting their industrial expansion are their low yields of production [188]. Approximately 30 species of lactobacilli are described as EPS producers, among them, the best known are L. casei, L. acidophilus, L. brevis, L. curvatus, L. delbrueckii bulgaricus, L. helveticus, L. rhamnosus, L. plantarum. They are principally cultivated between 30 and 37 °C on rich media as Man Rogosa Sharp (MRS), milk or milk derivatives [189,188]. But Lactobacillus sp. are not the best polysaccharide producers compared to some soil bacteria, as Xanthomonas campestris [190,188]. It is known that different strains of X. campestris can produce xanthan gums of different composition, viscosity and yield. In reference [191], cheese whey was used as substrate in the production of xanthan gum with optimised high gum production in a bioreactor and at a wide range of viscosity values. With the strain X. campestris C7L, [192] a production level of 14.7 g kg−1 was obtained, in a 40 g L−1 whole milk whey-based medium (powdered cheese whey) with the addition of 0.1 g L−1 magnesium sulphate and 5 g L−1 potassium phosphate. In [193], the feasibility of using cheese whey as carbon source for xanthan gum production was investigated using two different strains: Xanthomonas campestris pv mangiferaeindicae 1230 and X. campestris pv manihotis 1182. At 72 h of fermentation, using cheese whey as sole carbon source, in presence of 0.1% (w/v) MgSO4*7H2O and 2.0% (w/v) of K2HPO4, maximum xanthan gum productions were observed. Although the xanthan gum concentration was similar for the two strains, approximately 25 g L-1, chemical composition and ionic strength presented several differences.

A modelling approach was used to describe the influence of temperature and pH on the kinetics of both growth and EPS production of Streptococcus thermophilus ST 111 in milk-based medium; addition of whey protein hydrolysate to milk medium resulted in an increased growth and EPS production [194].

Growth and EPS production during free and immobilized cell chemostat culture of Lactobacillus rhamnosus RW-9595M D was described in [195]. Whey permeate powder was used, both very high EPS production (1800 mg L-1) and volumetric productivity (542 mg L-1 h-1) were obtained during chemostat culture with free cells for a D of 0.3 h-1.

3.4 Bacteriocins

Bacteriocins are antimicrobial peptides synthesized, at ribosomal level, by various Gram positive and and Gram negative bacteria, acting towards strictly correlated bacteria (growth inhibition). Generally, bacteriocins are produced at the end of the exponential growth-phase and their spectrum of action can vary, depending on the producing specie [196].

Positive effect of bacteriocins was spotlights in different types of food: dairy and bakery products, meat and fishing products, fruit and vegetables, beverage

[197]. Moreover, further application are known, in medical, pharmaceutical and veterinary fields [198].

LAB are particularly prolific in bacteriocins production and can biosynthesize different types of antagonistic molecules. Many LAB are able to synthesize different classes of bacteriocins, currently, nysin is the only bacteriocin industrially produced and which use in food is allowed and authorized [199,200]. Over the last two decades, there has been an explosion of basic and applied research on lactic acid bacteria (LAB) bacteriocins, primarily due to their potential application as biopreservatives in food and food products to inhibit the growth of food-borne bacterial pathogens [196]. Different experimental models demonstrated efficiency on pathogens and spoilage microrganisms growth control, both adding bacteriocins in food [201,202] and using bacteriocins in situ synthesized [203].

Although bacteriocins can be produced in the food matrix during food fermentation (in situ), bacteriocins by LAB can be produced in much higher amounts during fermentations under optimal physical and chemical conditions [204-207].

A recent review [208] summarized information on nisin production by L. lactis in batch cultures utilizing skimmed milk or whey as inexpensive medium. Nisin biosynthesis occurs during the exponential growth phase, several cultural factors influence nisin production: producer strain, media composition, pH and temperature values, agitation and aeration. Nisin production with Lactococcus lactis ATCC 11454 was carried out in two different media with pasteurized milk whey, filtered or not. In filtered milk whey nisin titer was 11120.13 mgL-1 up to 1628-fold higher than the filtered milk whey. The higher nisin concentration was probably related to insoluble proteins released into the media [204]. In [209] the utilization of milk whey was studied in batch cultures, a higher nisin production was observed in diluted whey (mixed with wash waters), 22.9 BU mL-1 (BU – bacteriocin units) in relation to concentrated whey (liquid remaining after the first cheese pressing), 8.3 BU mL-1.

In reference [205], the production of nisin by L. lactis UQ2 in a bioreactor using supplemented sweet whey (SW) was optimized by a statistical design of experiments and response surface methodology (RSM). A 2nd-order model built from a CCD experiment predicted a maximum nisin production of 178 IU/mL at 6 h of incubation in the bioreactor, leading to a productivity of 0.74 mg nisin/([L][h]), which increased 1.62 times when using controlled pH (6.5) fermentation.

Continuous production of nisin in whey permeate, supplemented with casein hydrolysate, was investigated using a packed-bed bioreactor. Lactococcus lactis subsp. lactis ATCC 11454 was immobilized by natural attachment to fiber surfaces and entrapment in the void volume within spiral wound fibrous matrix. Optimal conditions for continuous nisin production were pH5.5, 31°C, 10–20 g/l casein hydrolysate, and D 0.2 h-1. A maximum nisin titer of 5.1 × 104 AU/ml was observed. The bioreactor was operated continuously for 6 months without encountering any clogging, degeneration, or contamination problems [206].

Previously, nisin-Z production was studied during repeated-cycle pH-controlled batch (RCB) cultures using Lactococcus lactis subsp. Lactis biovar. diacetylactis UL719 immobilized in k-carrageenan/locust bean gum gel beads in supplemented whey permeate [210,211].

Bacillus bacteriocins are increasingly becoming more important due to their sometimes broader spectra of inhibition (as compared with most LAB bacteriocins), which may include Gram-negative bacteria, yeasts or fungi, in addition to Gram-positive species, some of which are known to be pathogenic to humans and/ or animals. Bacteriocins from Bacillus species offer a much broader spectrum of potential applications compared with LAB bacteriocins. The use of Bacillus bacteriocins in food preservation is just starting to be investigated [212]. A BLIS with a broad spectrum of activity against pathogenic and spoilage bacteria (L. monocytogenes, B. cereus and clinical isolates of Streptococcus spp.) was produced by B. licheniformis P40 when cheese whey concentration in the growth medium was about 70 g L-1 [213].

Pediocin production on whey by Pediococcus acidilactici NRRL B-5627 was described in [214], pediocin was obtained both in batch and re-alkalized fed-batch fermentations on diluted whey supplemented with 2% (w/v) yeast extract.

3.5 Single cell protein and single cell oil

The term SCP refers to dried cells of microorganisms such as algae, actinomycetes, bacteria, yeast, molds, and higher fungi grown in large-scale culture systems for use as protein source in human food or animal feed. Among suitable microorganisms to be grown on whey lactose as a substrate for SCP production, more research work has been carried out on yeasts, in particular K. marxianus [215-217]. The use of lactose or whey as a carbon source for the production of yeast biomass is a simple treatment process for increasing the value of food industry co-products [14,218].

Biomass produced from both batch and continuous processes [219-221] is mostly used as animal feed supplement [215] but also in production of baker's yeast [217].

In reference [221] deproteinized sweet and sour cheese whey concentrates were investigated as substrates for the production of SCP with Kluyveromyces marxianus CBS6556 up to a 100-l scale. Biomass concentrations up to 50 g l-1 (Yx/s 0.52) for sweet whey and 65 g l-1 (Yx/s 0.48) for sour whey concentrates were obtained.

Use of whey or buttermilk supplemented with YE for the growth of thermophilic LAB was reported in [222]; cell yields and kinetic parameters obtained were comparable or better than those obtained from control media, which appear to be too expensive for growing LAB on an industrial scale.

High added value probiotic biomass from deproteinized and non-supplemented milk whey was reported in [223]. Growth kinetics of Lactobacillus casei in deproteinized goat milk whey was analyzed in batch, continuous and fed-batch conditions.

In [224] a technology for kefir SCP production using whey was developed, a three-step process to scale-up kefir biomass production at a semi industrial scale pilot plant (100- and 3,000- L bioreactors) has been described.

The biotechnological production of SCO has been focused on the ability of various oleaginous microorganisms to convert agro-industrial wastes or raw materials into specialty lipids [225], or equivalents of plant oils that contain poly-unsaturated fatty acids (PUFAs) [226]. Three Zygomycetes, Mortierella isabellina, Thamnidium elegans and Mucor sp., were tested for their ability of producing biomass and lipid-containing g-linolenic acid (GLA) during their cultivation on cheese whey. All the tested microorganisms presented appreciable microbial growth; GLA concentration presented differences related with the strains and the fermentation time. M. isabellina produced noticeable quantities of SCO resulted in a maximum GLA production of 301 mg/L [227].

3.6 Enzymes

β-D-Galactosidase most commonly known as lactase, is one of the most important enzymes used in food processing, which catalyses the hydrolysis of lactose to glucose and galactose. The enzyme has been isolated and purified from a wide range of microorganisms but most commonly used β-D-galactosidases are derived from yeasts and fungal sources [228,229].

The ultrafiltration technique generates a whey permeate abundant in lactose, including over 80% of the original lactose content present in milk. Advanced methodologies, such as nanofiltration and reverse osmosis, have been established for lactose concentration, pertinent to the confectionery industry and pharmaceutical fermentation processes [7]. Whey permeate, in conjunction with lactose, contains extra nutrients vital for microbial growth; therefore, its potential as a fermentation medium for producing high-value products is a significant opportunity that deserves consideration. Furthermore, whey permeate constitutes a promising source of oligosaccharides for prospective use in human nutrition.

REFERENCES

1. Baldasso C., Barros T.C., Tessaro I.C. Concentration and purification of whey proteins by ultrafiltration. Desalination 2011;278 381–386.
2. Spalatelu C. Biotechnological valorization of whey. Innovative Romanian Food Biotechnology 2012;10 1-8.

3. Koller M., Bona R., Braunegg G., Hermann C., Horvat P., Kroutil M., Martinz J., Neto J., Pereira L., Varila P. Production of Polyhydroxyalkanoates from Agricultural Waste and Surplus Materials. Biomacromolecules 2005;6 561-565.
4. Panesar, P.S., Kennedy J.F., Gandhi D.N., Bunko K. Bioutilisation of whey for lactic acid production. Food Chemistry 2007;105 1–14.
5. Jelen, P. Whey processing. In: H. Roginski, J. W. Fuquay, P. F. Fox (ed.) Encylcopedia of dairy sciences. London: Academic Press. 2003 p2739–2751.
6. Prazeres A.R., Carvalho F., Rivas J. Cheese whey management: A review. Journal of Environmental Management 2012;110 48-68.
7. Atra R., Vatai G., Bekassy-Molnar E., Balint A. Investigation of ultra- and nanofiltration for utilization of whey protein and lactose. Journal of Food Engineering 2005;67 325–332
8. Barile D., Tao N., Lebrilla C.B., Coisson J-D., Arlorio M., German J.B. Permeate from cheese whey ultrafiltration is a source of milk oligosaccharides. International Dairy Journal 2009;19 524–530.
9. Marshall K. Therapeutic Applications of Whey Protein. Altern Med Rev 2004; 9(2) 136-156.
10. Huffman, L.M. Processing whey protein for use as a food ingredient. Food Technology 1996; 5049–52.
11. Mann E.J. Whey products and their uses. Dairy Industries International 2000; 13-14.
12. Pouliot Y. Membrane processes in dairy technology-from a simple idea to worldwide panacea. International Dairy Journal 2008;18(7) 735- 740.
13. deWit J.N. Nutritional and functional characteristics of whey proteins in food products, J. Dairy Science 1998;81 597–608.
14. Siso, M.I.G. The biotechnological utilization of cheese whey: a review. Bioresource Technology 1996;57(1) 1-11
15. Gunasekaran S., Ko S., Xiao L. Use of whey proteins for encapsulation and controlled delivery applications. Journal of Food Engineering 2007;83 31–40.
16. López-Rubio A., Lagaron J.M. Whey protein capsules obtained through electrospraying for the encapsulation of bioactives. Innovative Food Science and Emerging Technologies 2012;13 200–206.
17. Jovanovic S., Barac M., Macej O., Vucic T., Lacnjevac C. SDS–PAGE analysis of soluble proteins in reconstituted milk exposed to different heat treatments. Sensors 2007;7 371–383.

18. Santos M.J., Teixeira J.A., Rodrigues L.R. Fractionation of the major whey proteins and isolation of beta-Lactoglobulin variants by anion exchange chromatography. Separation and Purification Technology 2012;90 133–139.
19. Evans E.W. Uses of milk proteins in formulated foods. In B.J.F. Hudson (Ed.), Developments in Food Proteins. London: Applied Science; 1982.
20. Etzel M.R. Manufacture and use of dairy protein fractions. The Journal of Nutrition 2004;134 Suppl. 996S-1002S.
21. Madureira A.R., Pereira C.I., Gomes A.M.P., Pintado M.E., Malcata F.X. Bovine whey proteins – Overview on the main biological properties. Food Research International 2007;40(10) 1197-1211.
22. Huffman L.M., Harper W.J. Maximizing the value of milk through separation technologies. J. Dairy Sci. 1999;82 2238–2244.
23. Zemel M.B. Role of calcium and dairy products in energy partitioning and weight management. American Journal of Clinical Nutrition 2004;79 907S–912S.
24. Smilowitz J.T., Dillard C.J., German J.B. Milk beyond essential nutrients: The metabolic food. Australian Journal of Dairy Technology 2005;60 77–83.
25. Krissansen G.W. Emerging health properties of whey proteins and their clinical implications. Journal of the American College of Nutrition 2007;26(6) 713S–723S.
26. McIntosh G.H., Royle P.J., Le Leu R.K., Regester G.O., Johnson M.A., Grinsted R.L., Kenward R.S., Smithers G.W. Whey Proteins as Functional Food Ingredients? Int. Dairy Journal 1998;425-434.
27. Shahidi F. Nutraceuticals and functional foods: whole versus processed foods. Trends in Food Science and Technology 2009;20 376-387.
28. Kaur S., Das M. Functional foods: an overview. Food Sci. Biotechnol. 2011;20(4) 861-875.
29. Hardy G. Nutraceuticals and Functional Foods: Introduction and Meaning. Nutrition 2000;16 688-698.
30. Bonnaillie, L.M., Tomasula, P.M., Kinetics, aggregation behavior and optimization of the fractionation of whey protein isolate with hydrochloric acid. Food Bioprod Process 2012;doi:10.1016/j.fbp.2012.01.002
31. Lozano J.M., Giraldo G.I., Romero C.M. An improved method for isolation of β-lactoglobulin. International Dairy Journal 2008;18 55-63.
32. Wang Q., Tolkach A., Kulozik U. Quantitative assessment of thermal denaturation of bovine alfa-lactalbumin via low-intensity ultrasound, HPLC, and DSC. J. Agric. Food Chem. 2006;54 6501–6506.

33. Fernandez A., Menendez V., Riera F.A., Alvarez R. Caseinomacropeptide behaviour in a whey protein fractionation process based on alpha-lactalbumin precipitation. J. Dairy Res. 2011;78 196–202.

34. Zydney, A.L. Protein separation using membrane filtration: new opportunities for whey fractionation. Int. Dairy J. 1998;8 243–250.

35. Smithers G.W., Ballard F.J., Copeland A.D., De Silva K.J., Dionysius D.A., Francis G.L., et al. Symposium: advances in dairy foods processing and engineering. J. Dairy Sci. 1996;79 1454–1459.

36. Almecija M.C., Ibanez R., Antonio Guadix A., Guadix E.M. Effect of pH on the fractionation of whey proteins with a ceramic ultrafiltration membrane. Journal of Membrane Science 2007;288 28–35.

37. Gurgel P.V., Carbonell R.G., Swaisgood H.E. Fractionation of whey proteins with a hexapeptide ligand affinity resin. Bioseparation 2000;9 385.

38. Noppe W., Haezebrouck P., Hanssens I., De Cuyper M. Asimplified purification procedure of alpha-lactalbumin from milk using Ca2+-dependent adsorption in hydrophobic expanded bed chromatography. Bioseparation 1999;8 153.

39. Wang Q., Swaisgood H.E. Characteristics of beta-lactoglobulin binding to the all- trans-retinal moiety covalently immobilized on Celite™. J. Dairy Sci. 1993;76 1895– 1901.

40. Schlatterer B., Baeker R., Schlatterer K. Improved purification of β-lactoglobulin from acid whey by means of ceramic hydroxyapatite chromatography with sodium fluoride as a displacer. J. of Chromatography B 2004;807 223-228.

41. Doultani S., Turhan K.N., Etzel M.R. Fractionation of proteins from whey using cation exchange chromatography. Process Biochemistry 2004;37 1737-1743.

42. Lightfoot E.N., Cockrem M.C.M. What are dilute-solutions. Sep. Sci. Technol. 1987;22 165–189.

43. Heeboll-Nielsen A., Justesen S.F.L., Thomas O.R.T. Fractionation of whey proteins with high-capacity superparamagnetic ion-exchangers, J. Biotechnol. 2004;113 247.

44. Sawyer L. β-Lactoglobulin. Pages 319–386 in Advanced Dairy Chemistry I, 3rd ed. P.F. Fox and P. McSweeney, eds. Kluwer, Amsterdam, The Netherlands; 2003.

45. Kontopidis G., Holt C., Sawyer L. Invited Review: β-Lactoglobulin: Binding Properties, Structure, and Function. Journal of Dairy Science 2004;87(4) 785-796.

46. Ragona L., Catalano M., Zetta L., Longhi R., Fogolari F., Molinari H. Peptide models of folding initiation sites of bovine b-lactoglobulin: Identification of

native like hydrophobic interactions involving G and H strands. Biochemistry 2002;41 2786–2796.

47. Apenten R.K.O., Khokhar S., Galani D. Stability parameters for b-lactoglobulin. Thermal dissociation and unfolding in phosphate buffer at pH 7. Food Hydrocolloids 2002;16 95–103.

48. Perez M.D., Calvo M. Interaction of beta-lactoglobulin with retinol and fatty acids and its role as a possible biological function for this protein: a review. J. Dairy Sci. 1995;78 978–988.

49. Yoshida T., Sasahara Y., Miyakawa S., Hattori M. Reduced T cell response to beta- lactoglobulin by conjugation with acidic oligosaccharides. J. Agric. Food Chem. 2005;53 6851–6857.

50. Adel-Patient K., Ah-Leung S., Creminon C., Nouaille S., Chatel J.M., Langella P., Wal J.M. Oral administration of recombinant Lactococcus lactis expressing bovine beta- lactoglobulin partially prevents mice from sensitization. Clin. Exp. Allergy 2005;35 539–546.

51. Cortes-Perez N.G., Ah-Leung S., Bermudez-Humaran L.G., Corthier G., Wal J.M., Langella P., Adel-Patient K. Intranasal coadministration of live lactococci producing interleukin-12 and a major cow's milk allergen inhibits allergic reaction in mice. Clin. Vaccine Immunol. 2007;14 226–233.

52. Prioult G., Pecquet S., Fliss I. Stimulation of interleukin-10 production by acidic beta- lactoglobulin-derived peptides hydrolyzed with Lactobacillus paracasei NCC2461 peptidases. Clin. Diagn. Lab. Immunol. 2004;11 266–271.

53. Kushibiki S., Hodate K., Kurisaki J., Shingu H., Ueda Y., Watanabe A., Shinoda M. Effect of beta-lactoglobulin on plasma retinol and triglyceride concentrations, and fatty acid composition in calves. J. Dairy Res. 2001;68 579–586.

54. Meisel H. Biochemical properties of peptides encrypted in bovine milk proteins. Curr. Med. Chem. 2005;12 1905–1919.

55. Sipola M., Finckenberg P., Vapaatalo H., Pihlanto-Leppala A., Korhonen H., Korpela R., Nurminen M.L. Alpha-lactorphin and beta-lactorphin improve arterial function in spontaneously hypertensive rats. Life Sci. 2002;71 1245–1253.

56. Yamauchi R., Wada E., Yamada D., Yoshikawa M., Wada K.: Effect of beta-lactotensin on acute stress and fear memory. Peptides 2006;27 3176–3182.

57. Ohinata K., Sonoda S., Inoue N., Yamauchi R., Wada K., Yoshikawa M. beta-Lactotensin, a neurotensin agonist peptide derived from bovine beta-lactoglobulin, enhances memory consolidation in mice. Peptides 2007;28 1470–1474.

58. Yamauchi R., Ohinata K., Yoshikawa M. Beta-lactotensin and neurotensin rapidly reduce serum cholesterol via NT2 receptor. Peptides 2003;24 1955–1961.

59. Barbana C., Pérez M.D. Interaction of a-lactalbumin with lipids and possible implications for its emulsifying properties: A review. International Dairy Journal 2011;21 727-741.

60. Grobler J.A., Wang M., Pike A.C., Brew K. Study by mutagenesis of the roles of two aromatic clusters of a-lactalbumin in aspects of its action in the lactose synthase system. Journal of Biological Chemistry 1994;269 5106-5114.

61. Brew K. alfa-Lactalbumin. In P. F. Fox, & P. L. H. McSweeney (Eds.), Proteins (3rd ed.). Advanced dairy chemistry, Vol. I (pp. 387e421) New York, NY, USA: Kluwer Academic/Plenum Press;2003.

62. Lien E.L. Infant formulas with increased concentrations of a-lactalbumin. American Journal of Clinical Nutrition 2003;77 1555S-1558S.

63. Wijesinha-Bettoni R., Dobson C.M., Redfield C. Comparison of the structural and dynamical properties of holo and apo bovine a-lactalbumin by NMR spectroscopy. Journal of Molecular Biology 2001;307 885-898.

64. Pettersson-Kastberg J., Mossberg A.K., Trulsson M., Yong Y.J., Min S., Lim Y. alfa- Lactalbumin, engineered to be nonnative and inactive, kills tumor cells when in complex with oleic acid: a new biological function resulting from partial unfolding. Journal of Molecular Biology 2009;394 994-1010.

65. Noyelle K., Van Dael H. Kinetics of conformational changes induced by the binding of various metal ions to bovine a-lactalbumin. Journal of Inorganic Biochemistry 2002;88 69-76.

66. Owusu R.K. The effect of calcium on bovine a-lactalbumin conformational transitions by ultraviolet difference and fluorescence spectrophotometry. Food Chemistry 1992a; 43 41-45.

67. Owusu R.K. Thermodynamic analysis of the effect of calcium on bovine a-lactalbumin conformational stability. Food Chemistry 1992b;44 189-194.

68. Stanciuc N., Rapeanu G. An overwiev of bovine α-lactalbumin structure and functionality. Food Technology 2010;34(2) 82-93.

69. Otani H., Mizumoto G.. Suppressive effect of lysozymes and α-lactalbumins on mitogen-induced proliverative responses of mouse lymphocytes. Animal Science Technology 1998;11 1029-1039.

70. Matsumoto H., Shimokawa Y., Ushida Y., Toida T., Hayasawa H. New biological function of bovine α-lactalbumin: Protective effect against ethanol- and stress-induced gastric mucosal injury in rats. Bioscience, Biotechnology and Biochemistry 2001;65 1104-1111.

71. Heine W.E., Klein P.D., Reeds P.J. The importance of α-lactalbumin in infant nutrition- Critical Review. The Journal of Nutrition 1991;121 277-283.

72. Maubois J.L., Ollivier G. Extraction of milk proteins. In S. Damodaran, A. Paraf (Eds.), Food proteins and their applications (pp. 225–256). New York, NY, USA: Marcel Dekker;1997. p225-256.

73. Gustafsson L., Hallgren O., Mossberg A. K., Pettersson J., Fischer W., Aronsson A. HAMLET kills tumour cells by apoptosis: Structure, cellular mechanisms and therapy. Journal of Nutrition 2005;135 1299-1303.

74. Pellegrini A., Thomas U., Bramaz N., Hunziker P., von Fellenberg R., Isolation and identification of three bactericidal domains in the bovine α-lactalbumin molecule. Biochimica et Biophysica Acta- General Subjects 1999;1426(3) 439-448.

75. Teschemacher H., Koch G., Brantl V. Milk protein-derived opioid receptor ligands. Peptide Science 1998;43(2) 99-117.

76. Yamaguchi M. and Uchida M. α-Lactalbumin suppresses interleukin-6 release after intestinal ischemia/reperfusion via nitric oxide in rats. Inflammopharmacology 2007;15 43-47.

77. Gauthier S.F., Pouliot Y., Saint-Sauveur D. Immunomodulatory peptides obtained by the enzymatic olysis of whey proteins-Review. International Dairy Journal 2006;16 1315-1323.

78. de Wit J.N. Functional properties of whey proteins. In Developments of dairy chemistry. In P. F. Fox (Ed.). Functional milk proteins (Vol. 4). London: Applied Science; 1989.

79. Korhonen H., Marnila P., Gill H. Milk immunoglobulins and complement factors: A review. British Journal of Nutrition 2000;84 S1-S7.

80. Mulvihill D.M., Donovan M. Whey proteins and their thermal denaturation – a review. Irish Journal of Food Science and Technology 1987;11 43–75.

81. Sarker S.A., Casswall T.H., Mahalanabis D., Alam N.H., Albert M.J., Brussow H., et al. Successful treatment of rotavirus diarrhoea in children with immunoglobulin from immunised bovine colostra. Peadiatric Infectious Diseases Journal 1998;17 1149–1154.

82. Zenk J.L., Helmer T.R., Kuskowski M.A. The effects of milk protein concentrate on the symptoms of osteoarthritis in adults: An exploratory, randomized, double-blind, placebo-controlled trial. Curr. Ther. Res. 2002;63 430–442.

83. Oona M., Rago T., Maaroos H., Mikelsaar M., Loivukene K., Salminen S., et al. Helicobacter pylori in children with abdominal complaints: Has immune bovine colostrum some influence on gastritis? Alpe Adria Microbiology Journal 1997;6 49–57.

84. Loimaranta V., Laine M., Soderling E., Vasara E., Rokka S., Marnila P., et al. Effects of bovine immune- and non-immune whey preparations on the composition and pH response of human dental plaque. European Journal of Oral Science 1999;107 244–250.

85. Ormrod D.J., Miller T.E. The anti-inflammatory activity of a low molecular weight component derived from the milk of hyperimmunized cows. Agents and Actions 1991;32 160–166.

86. Korhonen H., Syvaoja E., Ahola-Luttila H., Sivela S., Kopola S., Husu J., et al. Bactericidal effect of bovine normal and immune serum, colostrum and milk against Helicobacter pylori. Journal of Applied Bacteriology 1995;78 655–662.

87. Sharpe S.J., Gamble G.D., Sharpe D.N. Cholesterol lowering and blood pressure effects of immune milk. American Journal of Clinical Nutrition 1994;59 929–934.

88. Huang B.X., Kim H., Dass C. Probing three-dimensional structure of bovine serum albumin by chemical cross-linking and mass spectrometry. J. Am. Soc. Mass. Spectrom. 2004;15 1237-1247.

89. Fox P.F. The milk protein system. In Developments in Dairy Chemistry. In P. F. Fox (Ed.). Functional Milk Proteins (Vol. 4). London: Applied Science;1989.

90. Kinsella J.E., Whitehead D.M., Brady J., Bringe N.A. Milk proteins, possible relationships of structure and function. In Developments in Dairy Chemistry. In P. F. Fox (Ed.). Functional Milk Proteins (Vol. 4). London: Applied Science;1989.

91. Laursen I., Briand P., Lykkesfeldt A.E. Serum albumin as a modulator of the human breast cancer cell line MCF-7. Anticancer Research 1990;10 343–352.

92. Lu B., Xiong S., Yang H., Yin X., Zhao R. Mitoxantrone-loaded BSA nanospheres and chitosan nanospheres for local injection against breast cancer and its lymph node metastases II: Tissue distribution and pharmacodynamics. International Journal of Pharmaceutics 2006;307 175–181.

93. Choi J.K., Ho J., Curry S., Qin D., Bittman R., Hamilton J.A. Interactions of very long- chain saturated fatty acids with serum albumin. Journal of Lipid Research 2002;43 1000–1010.

94. Tong L.M., Sasaki S., McClements D.J., Decker E.A. Mechanisms of the antioxidant activity of a high molecular weight fraction of whey. Journal of Agricultural and Food Chemistry 2000;48 1473–1478.

95. Adlerova L., Bartoskova A., Faldyna M. Lactoferrin: a review. Veterinarni Medicina 2008;53 9 457-468.

96. Sharma S., Jasti J., Kumar J., Mohanty A.K., Singh T.P. Crystal structure of a proteolytically generated functional monoferric C-lobe of bovine lactoferrin at 1.9 A resolution. J. Mol. Biol. 2003;331 485-496.

97. Brock J.H. The physiology of lactoferrin. Biochemistry and Cell Biology 2002;80 1–6.

98. Van Snick J.L., Masson P.L., Heremans J.F. The involvement of lactoferrin in the hyposideremia of acute inflammation. The Journal of Experimental Medicine 1974;140 1068–1084.

99. Van Vugt H., Van Gool J., Ladiges N.C., Boers W. Lactoferrin in rabbit bile: its relation to iron metabolism. Quarterly Journal of Experimental Physiology and Cognate Medical Sciences 1975;60 79–88.

100. Suzuki Y.A., Lopez V., Lonnerdal B. Mammalian lactoferrin receptors: structure and function. Cellular and Molecular Life Sciences 2005;62 2560–2575.

101. Legrand D., Elass E., Carpentier M., Mazurier J. Lactoferrin: a modulator of immune and inflammatory responses. Cellular and Molecular Life Sciences 2005;62 2549– 2559.

102. Brock J.H. Lactoferrin in human milk: its role in iron absorption and protection against enteric infection in the newborn infant. Archives of Disease in Childhood 1980;55 417–421.

103. Sherman M.P., Bennett S.H., Hwang F.F., Yu C. Neonatal small bowel epithelia: enhancing anti-bacterial defense with lactoferrin and Lactobacillus GG. Biometals 2004;17 285–289.

104. Rossi P., Giansanti F., Boffi A., Ajello M., Valenti P., Chiancone E., Antonini G. Ca2+ binding to bovine lactoferrin enhances protein stability and influences the release of bacterial lipopolysaccharide. Biochemistry and Cell Biology 2002;80 41–48.

105. Valenti P., Antonini G. Lactoferrin: an important host defense against microbial and viral attack. Cellular and Molecular Life Sciences 2005;62 2576–2587.

106. Ward P.P., Paz E., Conneely O.M. Multifunctional roles of lactoferrin: a critical overview. Cellular and Molecular Life Sciences 2005;62 2540–2548.

107. Yi M., Kaneko S., Yu D.Y., Murakami S. Hepatitis C virus envelope proteins bind lactoferrin. Journal of Virology 1997;71 5997–6002.

108. Damiens E., El Yazidi I., Mazurier J., Duthille I., Spik G., Boilly-Marer Y. Lactoferrrin inhibits G1 cyclin-dependent kinases during growth arrest of human breast carcinoma cells. Journal of Cellular Biochemistry 1999;74 486–498.

109. Damiens E., Mazurier J., El Yazidi I., Masson M., Duthille I., Spik G., Boilly-Marer Y. Effects of human lactoferrin on NK cell cytotoxicity against haematopoetic and epithelial tumour cells. Biochimica et Biophysica Acta1998;1402 277–287.

110. Fujita K., Matsuda E., Sekine K., Iigo M., Tsuda H. Lactoferrin enhances Fas expression and apoptosis in the colon mucosa of azoxymethane-treated rats. Carcinogenesis 2004;25 1961–1966.

111. Yanaihara A., Toma Y., Saito H., Yanaihara T. Cell proliferation effect of lactoferrin in human endometrial stroma cells. Molecular Human Reproduction 2000;6 469–473.

112. He J., Furmanski P. Sequence specificity and transcriptional activation in the binding of lactoferrin to DNA. Nature 1995;373 721–724.

113. Cornish J., Callon K.E., Naot D., Palmano K.P., Banovic T., Bava U., Watson M., Lin J.M., etc. Lactoferrin is a potent regulator of bone cell activity and increases bone formation in vivo. Endocrinology 2004;145 4366–4374.

114. Reiter B., Perraudin J.P. Lactoperoxidase, biological functions. In J. Everse, K. E. Everse, & M. B. Grisham (Eds.). Peroxidases in Chemistry and Biology. Boca Raton, FL: CRC Press;1991. Vol.1 p144–180.

115. Watanabe S., Murata S., Kumura H., Shingo Nakamura S., Bollen A., Moguilevsky N., Shimazaki K. Bovine Lactoperoxidase and Its Recombinant: Comparison of Structure and Some Biochemical Properties. Biochemical and Biophysical Research Communications 2000;274 756–761.

116. Reiter B., Pickering A., Oram J. D., Pope G.S. Peroxidase thiocyanate inhibition of streptococci in raw milk. Journal of General Microbiology 1963;33 12-15.

117. Tenovuo J. Clinical applications of antimicrobial host proteins lactoperoxidase, lysozyme and lactoferrin in xerostomia: efficacy and safety. Oral Diseases 2002;8 23-29.

118. Touch V., Hayakawa S., Yamada S., Kaneko S. Effects of a lactoperoxidase–thiocyanate–hydrogen peroxide system on Salmonella enteritidis in animal or vegetable foods. International Journal of Food Microbiology 2004;93 175– 183.

119. Wolfson L.M. and Sumner S.S. Antibacterial activity of the lactoperoxidase system. A review. J. Food Prot. 1993;56 887-892.

120. de Wit J.N. and van Hooydonk A.C.M. Structure, functions and application of lactoperoxidase in natural antimicrobial system. Netherland Milk Dairy J. 1996;50 227-244.

121. Seifu E., Buys E.M. and Donkinc E.F. Significance of the lactoperoxidase system in the dairy industry and its potential applications: a review. Trends in Food Science & Technology 2005;16 137–154.

122. Reiter B.V., Marshall M.E., Bjorck L., Rosen C.G. Non specific bactericidal activity of the lactoperoxidase – thyocianate – hydrogen peroxide system on milk against Escherichia coli and some Gram negative pathogens. Infect. Immun. 1976;13 800–807.

123. Carlsson J., Iwami Y., Yamada, T. Hydrogen peroxide excretion by oral streptococci and effect of lactoperoxidase- thiocyanate-hydrogen peroxide. Infection and Immunity 1983;40 70–80.

124. Tenovuo J., Makinen K.K., Sievers G. Antibacterial effect of lactoperoxidase and myeloperoxidase against Bacillus cereus. Antimicrobial Agents and Chemotherapy 1985;27 96–101.

125. Pruitt K.M., Kamau D.N. The lactoperoxidase system of bovine and human milk. In D. S. Robinson, & N. A. M. Eskin (Eds.), Oxidative enzymes in foods. London: Elsevier Applied Science;1991. p133–174.

126. Garcia-Graells C., van Opstal I., Vanmuysen S.C. M., Michiels C.W. The lactoperoxidase system increases efficacy of high-pressure inactivation of foodborne bacteria. International Journal of Food Microbiology 2002;81 211–221.

127. Seifu E., Buys E.M., Donkin E.F., Petzer, I.M. Antibacterial activity of the lactoperoxidase system against foodborne pathogens in Saanen and South African Indigenous goat milk. Food Control 2004b;15 447–452.

128. Jacob B.M., Antony E.K., Sreekumar, B., Haridas M. Thiocyanate mediated antifungal and antibacterial property of goat milk lactoperoxidase. Life Sciences 2000;66 2433– 2439.

129. Belding M. E., Klebanoff S. J., Ray G.C. Peroxidase mediated virucidal systems. Science 1970;167 195–196.

130. Yamaguchi Y., Semmel M., Stanislawski L., Strosberg A.D., Stanislawski M. Virucidal effects of glucose oxidase and peroxidase or their protein conjugates on human immunodeficiency virus type 1. Antimicrobial Agents and Chemotherapy 1993,37 26– 31.

131. Pourtois M., Binet C., Van Tieghem N., Courtois P., Vandenabbeele A., Thiry L. Inhibition of HIV infectivity by lactoperoxidase-produced hypothiocyanite. Journal de Biologie Buccale 1990;18 251–253.

132. Kosikowsky F.V. Whey utilization and whey products. J. Dairy Science 1979;62(7) 1149-1160.

133. Zafar S., Owais M. Ethanol production from crude whey by Kluyveromyces marxianus. Biochemical Engineering J. 2006;27 295-298.

134. Sansonetti S., Curcio S., Calabrò V., Iorio G. Bio-ethanol production by fermentation of ricotta cheese whey as an effective alternative non-vegetable source. Biomass Bioenergy 2009;33(12) 1687-1692.

135. Guimarães P.M.R., François J., Parrou J.L., Teixeira J.A., Domingues L. Adaptive evolution of a lactose-consuming Saccharomyces cerevisiae recombinant. Applied Environmental Microbiology 2008;74(6) 1748-1756.

136. Guimarães P.M.R., Teixeira J.A., Domingues L. Research review paper: fermentation of lactose to bio-ethanol by yeasts as part of integrated solutions for the valorisation of cheese whey. Biotechnology Advances 2010;28(3) 375-384.

137. Anderson A.J., Dawes E.A. Occurrence, Metabolism, Metabolic Role, and Industrial Uses of Bacterial Polyhydroxyalkanoates. Microbiological Reviews 1990;54(4) 450-472.

138. Khanna S., Srivastava A.K. Recent advances in microbial polyhydroxyalkanoates. Process Biochemistry 2005;40 607-619.

139. Reddy C.S., Ghai R., Rashmi, Kalia V.C. Polyhydroxylakanoates: an overview. Bioresource Technology 2003;87 137-146.

140. Sudesh K., Abe H., Doi Y. Synthesis, structure and properties of polyhydroxyalkanoates: biological polyesters. Progress Polymer Science 2000;25 1503-1555.

141. Koller M., Hesse P., Bona R., Kutschera C., Atlic A., Braunegg G. Potential of various archae- and eubacterial strains as industrial polyhydroxyalkanoate producers from whey. Macromol. Biosci. 2007;7 218-226.

142. Marangoni C., Furigo A. Jr., de Aragão G.M.F. Production of poly(3-hydroxybutyrate-co-3-hydroxyvalerate) by Ralstonia eutropha in whey and inverted sugar with propionic acid feeding. Process Biochemistry 2002;38 137-141.

143. Koller M., Bona R., Chiellini E., Grillo Fernandes E., Horvat P., Kutscher C., Hesse P., Braunegg G. Polyhydroxyalkanoate production from whey by Pseudomonas hydrogenovora. Bioresource Technology 2008;99 4854-4863.

144. Wong H.H., Lee S.Y. Poly(3-hydroxybutyrate) production from whey by high density cultivation of recombinant Escherichia coli. Applied Microbiology Biotechnology 1998;50 30–33.

145. Nikel P.I., Almeida A.D., Melillo E.C., Galvagno M.A., Pettinari M.J. New recombinant Escherichia coli strain tailored for the production of poly(3-hydroxybutyrate) from agro-industrial by-products. Applied Environmental Microbiology 2006;72 3949–3954.

146. Yellore V., Desai A. Production of poly-3-hydroxybutyrate from lactose and whey by Methylobacterium sp. ZP24. Letters in Applied Microbiology 1998;26 391-394.

147. Aslım B., Çalışkan F., Beyatli Y., Gündüz U. Poly-β-hydroxybutyrate production by lactic acid bacteria. FEMS Microbiology Letters 1998;159 293–297.

148. Tanaka K., Katamune K., Ishizaki A. Fermentative production of poly-(β-hydroxybutyric acid) from xylose via L-lactate by a two stage culture method employing Lactococcus lactis IO-1 and Alcaligenes eutrophus. Canadian J. Microbioliology 1995;41 257– 261.

149. Tohyama M., Patarinska T., Qiang Z., Shimizu K. Modeling of the mixed culture and periodic control for PHB production. J. Biochemical Engineering 2002;10 157–173.

150. Tohyama M., Shimizu K. Control of a mixed culture of Lactobacillus delbrueckii and Ralstonia eutropha for the production of PHB from glucose via lactate. J Biochemical Engineering 1999;4 45–53.

151. Tohyama M., Takagi S., Shimizu K. Effect of controlling lactate concentration and periodic change in DO concentration on fermentation characteristics of a mixed culture of Lactobacillus delbrueckii and Ralstonia eutropha for PHB production. J. Bioscience Bioengineering 2000;89 323–328.

152. Patnaik P.R. Perspectives in the modelling and optimization of PHB production by pure and mixed cultures. Critical Reviews in Biotechnology 2005;25 153–171.

153. Bader J., Mast-Gerlach E., Popovic M.K., Bajpai R., Stahl U. Relevance of microbial co-culture fermentations in biotechnology. Journal of Applied Microbiology 2010;109 371–387.

154. Yilmaz M., Soran H., Beyatli Y. Determination of poly-b-hydroxybutyrate (PHB) production by some Bacillus spp. World J. Microbiology Biotechnology 2005; 21 565-566.

155. Defoirdt T., Boon N., Sorgeloos P., Verstraete W., Bossier P. Short-chain fatty acids and poly-β-hydroxyalkanoates: (New) Biocontrol agents for a sustainable animal production. Biotechnology Advances 2009;27 680–685.

156. Yang N., Silva E.M., Novel products and new technologies for use of a familiar carbohydrate, milk lactose. J. Dairy Science 1995;78 2541–2562.

157. Pauli T., Fitzpatrick, J. J. Malt combing nuts as a nutrient supplement to whey permeate for producing lactic by fermentation with Lactobacillus casei. Process Biochemistry 2002;38 1-6.

158. Roukas T., Kotzekidou P. Lactic acid production from deproteinized whey by mixed cultures of free and coimmobilized Lactobacillus casei and Lactococcus lactis cells using fedbatch culture. Enzyme and Microbial Technology 1998;22(3) 199-204.

159. Tango M.S.A., Ghali A.E. Effect of temperature on lactic acid production from cheese whey using Lactobacillus helveticus under batch conditions. Biomass and Bioenergy 1999; 16(1) 61-78.

160. John P.R., Nampoothiri K.M., Pandey A. Fermentative production of lactic acid from biomass: an overview on process developments and future perspectives. Applied Microbiology Biotechnology 2007;74 524-534.

161. Dana R. Biodegradable packaging materials for food products. Scientific and Technical Bulletin of University "A.Vlaicu" Arad 2005;10 88-93.

162. Lim L.T., Auras R., Rubino M. Processing technologies for poly(lactic acid). Progress in Polymer Science 2008;33: 820-852.

163. Weber C.J., Haugaard V., Festersen R., Bertelsen G. Production and applications of biobased packaging materials for the food industry. Food Additive Contaminations 2002;19 172-177.

164. Comstock K., Farrell D., Godwin C., Xi Y. From hydrocarbons to carbohydrates: food packaging of the future. 2004; Website: http://depts.washington.edu/poeweb/gradprograms/envmgt/mposium/GreenPackagingReport.pdf (Accessed 3/28/06).

165. Arasaratnam V., Senthuran A., Balasubramaniam K. Supplementation of whey with glucose and different nitrogen sources for lactic acid production by Lactobacillus delbrueckii. Enzyme and Microbial Technology 1996;19 482-486.

166. Mostafa N.A. Production of lactic acid from whey with agar immobilized cells in a continuous packed tubular reactor. Energy and conversion management 1996;37(3) 253-260.

167. Pescuma M., Hebert E.M., Mozzi F., de Valdez G.F. Whey fermentation by thermophilic lactic acid bacteria: Evolution of carbohydrates and protein content. Food Microbiology, 2008;25 442-451.

168. Plessas S., Fischer A., Koureta K., Psarianos C., Nigam P., Koutinas A.A. Application of Kluyveromyces marxianus, Lactobacillus delbrueckii ssp. bulgaricus and L. helveticus for sourdough bread making. Food Chemistry 2008;106(1) 985-990.

169. Vasala A., Panula J., Neubauer P. Efficient lactic acid production from high salt containing dairy by-products by Lactobacillus salivarius ssp. salicinius with pre-treatment by proteolytic microorganisms. Journal of Biotechnology 2005;117 421–31.

170. Gonzalez M.I., lvarez S. A., Riera F., Alvarez R. Economic evaluation of an integrated process for lactic acid production from ultrafiltered whey. Journal of Food Engineering 2007;80 553-561.

171. Kulozik U., Wilde J. Rapid lactic acid production at high cell concentration in whey ultrafiltrate by Lactobacillus helveticus. Enzyme and Microbial Technology 1999;24(5-6) 297-302.

172. Schepers A.W., Thibault J., Lacroix C. Lactobacillus helveticus growth and lactic acid production during pH-controlled batch cultures in whey permeate/ yeast extract medium. Part I: multiple factor kinetic analysis. Enzyme and Microbial Technology 2002;30 176-186.

173. Roy D., Goulet J., LeDuy A. Batch fermentation of whey ultrafiltrate by Lactobacillus helveticus for lactic acid production. Applied Microbiol. Biotechnol. 1986; 24 206-13.

174. Amrane A., Prigent Y. Influence of yeast extract concentration on batch cultures of Lactobacillus helveticus: growth and production coupling. World J. of Microb. Biot. 1993;14 529-34.

175. Roy D., Goulet J., LeDuy A. Continuous production of lactic acid from whey permeate by free and calcium alginate entrapped Lactobacillus bulgaricus. J. Dairy Science 1987;70 506-13.

176. Taguchi H., Ohta T. Cloning, sequencing and expression in Escherichia coli of the D- lactate dehydrogenase gene of Lactobacillus plantarum: D-lactate dehydrogenase is a member of the D-isomer-specific 2-hydroxyacid-dehydrogenase family. Journal of Biological Chemistry 1991;266(12) 588-594.

177. Adachi E., Mikiko T., Sugiyama M., Nikawa J.I., Schimuzu K. Modification of metabolic pathways of Saccharomyces cerevisiae by the expression of lactate dehydrogenase and deletion of pyruvate decarboxylase genes for the lactic acid fermentation at low pH value. Journal of Fermentation and Bioengineering. 1998;86 284-289.

178. Porro D., Bianchi M.M., Brambilla L., Meneghini R., Bolzani D., Carrera V. Replacement of a metabolic pathway for large-scale production of lactic acid from engineered yeast. Applied and Environmental Microbiology 1999;65 4211-4215.

179. Bai D.M., Zhao X.M., Li X.G., Xub S.M. Strain improvement of Rhyzopus oryzae for over-production of L-(+)-lactic acid and metabolic flux analysis of mutans. Biochemical Engineering Journal 2004;18 41-48.

180. De Vuyst L., Degeest B. Heteropolysaccharides from lactic acid bacteria. FEMS Microbiology Review 1999;23 153-177.

181. van Hijum S.A.F.T., Kralj S., Ozimek L. K., Dijkhuizen L., van Geel-Schutten I.G. Structure-function relationships of glucansucrase and fructansucrase enzymes from lactic acid bacteria. Microbiol. Mol. Biol. Rev. 2006;70 157–176.

182. De Vuyst L., De Vin F., Vaningelgem F., Degeest B. Recent developments in the biosynthesis and applications of heteropolysaccharides from lactic acid bacteria. International Dairy J. 2001;11 687–707.

183. Krinos C. M., Coyne M. J., Weinacht K.G., Tzianabos A.O., Kasper D.L., Comstock L.E. Extensive surface diversity of a commensal microorganism by multiple DNA in versions. Nature 2001;414 555–558.

184. Patel S., Majumder A., Goyal A.. Potentials of Exopolysaccharides from Lactic Acid Bacteria. Indian J. Microbiology 2012;52(1) 3–12

185. Makino S., Ikegami S., Kano H., Sashihara T., Sugano H., Horiuchi H., Saito T., Oda M. Immunomodulatory effects of polysaccharides produced by Lactobacillus delbrueckii spp. bulgaricus OLL1073R-1. J. Dairy Science 2006;89 2873-2881.

186. Broadbent J.R., McMahon D.J., Welker D.L., Oberg C.J., Moineau S. Biochemistry, Genetics, and Applications of Exopolysaccharide. Production in Streptococcus thermophilus: A Review. J. Dairy Science 2003;86 407–423.

187. Laws A.P., Marshall V.M. The relevance of exopolysaccharides to the rheological properties in milk fermented with ropy strains of lactic acid bacteria. International Dairy J. 2001;11 709–721.

188. Badel S., Bernardi T., Michaud P. New perspectives for Lactobacilli exopolysaccharides. Biotechnology Advances 2011;29 54–66.

189. Looijesteijn P.J., Hugenholtz J. Uncoupling of growth and exopolysaccharide production by Lactobacillus lactis subsp. cremoris NIZO B40 and optimization of its synthesis. Journal of Bioscience and Bioengineering 1999;88 178–182.

190. Sutherland I.W. Polysaccharides for microbial exopolysaccharides. Carbohydrate Polymer 1999;38 319–328.

191. Mesomo M., Silva M.F., Boni G., Padilha F.F., Mazutti M., Mossi A., de Oliveira D., Cansian R.L., Di Luccio M., Treichela H. Xanthan gum produced by Xanthomonas campestris from cheese whey: production optimisation and rheological characterization. J. Science Food Agric. 2009;89 2440–2445.

192. Nitschke M., Rodrigues V., Schinatto L.F. Formulacao de meios de cultivo a base de soro de leite para a producao de goma xantana por X. campestris C7L. Cienc. Tecnol. Alim. 2001;21 82–85.

193. Silva M.F., Fornari R.C.G., Mazutti M.A., Oliveira D., Padilha F.F., Cichoski A.J. Production and characterization of xantham gum by Xanthomonas campestris using cheese whey as sole carbon source. J. Food Eng. 2009;90 119–123.

194. Vaningelgem F., Zamfir M., Adriany T., De Vuyst L. Fermentation conditions affecting the bacterial growth and exopolysaccharide production by Streptococcus thermophilus ST 111 in milk-based medium. Journal of Applied Microbiology 2004;97 1257– 1273.

195. Bergmaier D., Champagne C.P., Lacroix C. Growth and exopolysaccharide production during free and immobilized cell chemostat culture of Lactobacillus rhamnosus RW-9595M D. Journal of Applied Microbiology 2005;98 272–284.

196. Beshkova D., Frengova G. Bacteriocins from lactic acid bacteria: Microorganisms of potential biotechnological importance for the dairy Industry. Eng. Life Sci. 2012;12(4) 1–14.

197. Luchansky J. B. Overview on applications for bacteriocin-producing lactic acid bacteria and their bacteriocines. Antonie van Leeuwenhoek 1999;76 335.

198. Ross R.P., Galvin M., McAuliff O., Morgan S.M., Ryan M.P., Twomey D.P., Meaney W.J., Hill C. Developing applications for lactococcal bacteriocins. Antonie van Leeuwenhoek 1999;76 337–346.

199. Cheigh C.-I., Pyun Y.-R., Nisin biosynthesis and its properties. Biotechnology Letters. 2005;27 1641–1648.

200. Cleveland J., Montville T. J., Nes I. F., Chikindas M. L. Bacteriocins: safe, natural antimicrobials for food preservation. International J. Food Microbiology 2001;71 1–20.

201. Messens W., De Vuyst L. Inhibitory substances produced by lactobacilli isolated from sourdoughs - a review. International J. Food Microbiology 2002;72 31–43.

202. Neysen P., De Vuyst L. Kinetic and modeling of sourdough lactic bacteria. Trends Food Science Technology 2005;16 95–103.

203. Delgado A., Brito D., Peres C., Arroyo-Lopez F.N., Garrido-Fernandez A. Bacteriocin production by Lactobacillus pentosus B96 can be expressed as a function of temperature and NaCl concentration. Food Microbiology 2005;22 521–528.

204. Juncioni de Arauz L., Jozala A. F., Pinheiro G.S., Gava Mazzola P., Pessoa A. Junior, Vessoni Penna T.C. Nisin expression production from Lactococcus lactis in milk whey medium. Chemical Technology Biotechnology 2008;83 325–328.

205. Gonzalez-Toledo S.Y., Domınguez-Domınguez J., Garcıa-Almendarez B.E., Prado- Barragan L.A., Regalado-Gonzalez C. Optimization of Nisin Production by Lactococcus lactis UQ2 Using Supplemented Whey as Alternative Culture Medium. Journal of Food Science 2010;75(6) M347-M353.

206. Liu X., Chung Y.K., Yang S.T., Yousef A.E. Continuous nisin production in laboratory media and whey permeate by immobilized Lactococcus lactis. Process Biochemistry. 2005;40 13–24.

207. Juncioni de Arauz L., Jozala A.F., Baruque-Ramos J., Gava Mazzola P., Pessoa A. Junior, Vessoni Penna T.C. Culture medium of diluted skimmed milk for the production of nisin in batch cultivations. Ann. Microbiol. 2012;62 419–426.

208. Juncioni de Arauz L., Jozala A. F., Gava Mazzola P., Vessoni Penna T.C. Nisin biotechnological production and application: a review. Trends in Food Science & Technology 2009;20 146-154.

209. Guerra N.P., Rua M.L., Pastrana L. Nutritional factors affecting the production of two bacteriocins from lactic acid bacteria on whey. International J. Food Microbiology 2001;70 267–281.

210. Bertrand N., Fliss I., Lacroix C. High nisin-Z production during repeated-cycle batch cultures in supplemented whey permeate using immobilized Lactococcus lactis UL719. International Dairy Journal. 2001;11 953–960.

211. Desjardins P., Meghrous J., Lacroix C. Effect of aeration and dilution rate on nisin Z production during continuous fermentation with free and immobilized Lactococcus lactis UL719 in supplemented whey permeate. International Dairy Journal 2001;11 943–951.

212. Abriouel H., Franz C.M.A.P., Omar N.B., Galvez A. Diversity and applications of Bacillus bacteriocins FEMS Microbiol. Review. 2011;35 201–232

213. Cladera-Olivera F., Caron G.R., Brandelli A. Bacteriocin production by Bacillus licheniformis strain P40 in cheese whey using response surface methodology. Biochemical Engineering J. 2004;21 153–58.

214. Guerra N.P., Fajardo Bernardez P., Pastrana Castro L. Fed-batch pediocin production on whey using different feeding media. Enzyme and Microbial Technology 2007;41 397–406

215. Ghaly A.E., Kamal M.A. Submerged yeast fermentation of acid cheese whey for protein production and pollution potential reduction. Water Research 2004;38 631–644.

216. Ghaly A.E., Kamal M., Correia L.R. Kinetic modelling of continuous submerged fermentation of cheese whey for single cell protein production. Bioresource Technology 2005;96 1143–1152.

217. de Palma Revillion J.P., Brandelli A., Záchia Ayub M.A. Production of yeast extract from whey using K. marxianus. Braz. Arch. Biol. Technol. 2003;46(1) 121-127.

218. Gardner D. New technologies in the conversion of whey to high protein products. Mod. Dairy 1989; 68 15–17.

219. Champagne C.P., Gioulet J., Lachance R.A. Production of baker's yeast in cheese whey ultrafiltrate. Applied Environmental Microbiology 1990;56 425–430.

220. Sandhu D.K., Waraich M.K. Conversion of cheese whey to single-cell protein. Biotechnology Bioengineering. 1983;25 797–803.

221. Schultz N., Chang L., Hauck A., Reuss M., Syldatk C. Microbial production of single- cell protein from deproteinized whey concentrates. Applied Microbiology Biotechnology 2006;69 515–520.

222. Vigliengo E., Reinheimer J. Use of whey and buttermilk based media to obtain biomass of thermophilic LAB. International J. of Dairy Technology 2009;62(3) 431-437.

223. Aguirre-Ezkauriatza E.J., Aguilar-Yáñez J.M., Ramírez-Medrano A., Alvarez M.M. Production of probiotic biomass (Lactobacillus casei) in goat milk whey: Comparison of batch, continuous and fed-batch cultures. Bioresource Technology 2010;101 2837– 2844.

224. Koutinas A.A., Athanasiadis I., Bekatorou M., Iconomopoulou G. Blekas Kefir Yeast Technology: Scale-Up in SCP Production Using Milk Whey. Biotechnology and Bioengineering 2005;89(7) 788-796.

225. Ykema A., Kater M. M., Smit H. Lipid production in whey permeate by an unsaturated fatty acid mutant of the oleaginous yeast Apiotrichum curvatum. Biotechnology Letters 1989;11 477–482.

226. Fakas S., Papanikolaou S., Batsos A., Galiotou-Panayotou M., Mallouchos A., Aggelis G. Evaluating renewable carbon sources as substrates for single cell oil production by Cunninghamella echinulata and Mortierella isabellina. Biomass & Bioenergy 2009; 33(4) 573-580.

227. Vamvakaki A.N., Kandarakis I., Kaminarides S., Komaitis M., Papanikolaou S. Cheese whey as a renewable substrate for microbial lipid and biomass production by Zygomycetes. Engeneering Life Science 2010;10(4) 348–360.

228. Bansal S., Oberoi H.S., Dhillon G.S., Patil R.T. Production of β- galactosidase by Kluyveromyces marxianus MTCC 1388 using whey and effect of four different methods of enzyme extraction on β- galactosidase activity. Indian J. Microbiology 2008;48 337– 341.

229. Panesar P.S., Panesar R., Singh R.S., Kennedy J.F., Kumar H. Microbial production, immobilization and applications of β-D-galactosidase. J. Chemical Technology Biotechnology 2006;81 530–543.

230. Ramirez-Matheus A.O., Rivas N. Production and partial characterization of β-galactosidase from Kluyveromyces lactis grown in deproteinized whey. Arch. Latinoam. Nutr. 2003;53194–201.

231. Ramana-Rao M.V., Dutta S.M. Production of betagalactosidase from Streptococcus thermophiles grown in whey. Applied Environmental Microbiology 1977;34 185–188.

232. Vasiljevic T., Jelen P. Production of β-galactosidase for lactose hydrolysis in milk and dairy products using thermophilic lactic acid bacteria. Innov. Food Sci. Emerg. Technol. 2001;2 75–85

233. Rech R., Ayub M.A.Z. Simplified feeding strategies for fed-batch cultivation of Kluyveromyces marxianus in cheese whey. Process Biochemistry 2007;42 873–877.

234. De Leon-Rodrıguez A., Rivera-Pastrana D., Medina-Rivero E., Flores-Flores J.L., Estrada-Baltazar A., Ordonez-Acevedo L.G., Barba de la Rosa A.P. Production of penicillin acylase by a recombinant Escherichia coli using cheese whey as substrate and inducer. Biomolecular Engineering 2006;23 299–305.

235. Romero F.J., Garcıa L.A., Salas J.A., Dıaz M., Quiros L.M. Production, purification and partial characterization of two extracellular proteases from Serratia marcescens grown in whey. Process Biochemistry 2001;36 507–515

236. Ustariz F., Laca A., Garcıa L.A., Dıaz M. Mixed cultures of Serratia marcescens and Kluyveromyces fragilis for simultaneous protease production and COD removal of whey. J. of Applied Microbiology 2007;103 864–870.

25 Chapter

THE ANTIOXIDANT AND EMULSIFYING PROPERTIES OF MODIFIED SUNFLOWER LECITHIN AS DETERMINED BY FRACTIONATION WITH ETHANOL-WATER MIXTURES

1. INTRODUCTION: AN OVERVIEW

Lecithins comprise a blend of acetone-insoluble phospholipids, predominantly including phosphatidylcholine (PC), phosphatidylethanolamine (PE), phosphatidylinositol (PI), phosphatidic acid (PA), along with minor constituents such as carbohydrates and triglycerides [1-3]. The production of sunflower oil in Argentina is critically significant from an economic perspective. In this country, sunflower lecithin may serve as an alternative to soybean lecithin due to its classification as a non-GMO product, aligning with the preferences of certain consumers [5].

Lecithins, whether in their original or modified forms, are utilized across various industrial sectors, including dietary, pharmaceutical, food, and cosmetics. This oil industry by-product serves as a multifunctional component in the production of chocolate, baked goods, instant food, margarine, and mayonnaise, owing to the properties of its phospholipids [6-8].

Modifications in the relative concentration of the original phospholipid composition of lecithin can provide enriched fractions of specific phospholipids with varying physicochemical and functional qualities for various industrial applications [9-11].

The fractionation procedure utilizing ethanol or ethanol:water mixtures exploits the varying solubility of phospholipids in this solvent. PC is very soluble in ethanol, but PI and PA are nearly insoluble. Phosphatidylethanolamine is present in both fractions. This process can be carried out alone or in combination with other techniques such as chromatography as a further purification step, especially for pharmaceutical, cosmetic and dietetic industry [12-13].

The main application of lecithin at the food industry is associated with its rol as emulsifying agent for dispersions or emulsions [14]. Emulsions are thermodynamically unstable systems from a physicochemical point of view. In virtue of that, it is important to characterize their behaviour against different destabilization processes (flocculation, coalescence, creaming, etc.) [15]. PC enriched fraction, due to its high PC/PE ratio and the lamellar phase structure of the PC at the interface between oil in water, is recognized to be a good oil-in-water (O/W) emulsifier [12,16,17].

Conversely, PLs can enhance the oxidative stability of fats and oils. Diverse antioxidative pathways have been suggested for the actions of phospholipids. The amino functionalities of PC, PS, or PE, as well as the sugar moiety of PI, have demonstrated metal-chelating activities, while PC and PE exhibited a synergistic impact with phenolic antioxidants such as tocopherols and flavonoids [18, 19].

This study aimed to assess the antioxidant and emulsifying capabilities of modified sunflower lecithins obtained through fractionation using ethanol-water combinations. This study aims to provide the food industry with valuable insights into the development of customized surface-active emulsifiers.

2. MATERIALS AND METHODS

2.1 Materials

Native sunflower lecithin was provided by a local oil industry (Vicentin S.A.I.C.).. This lecithin present a phospholipid composition of 43.1% (PC 16.2%, PI 16.5%, PE 5.3%, minor PLs 5.1%), 23.5% other compounds (glycolipids, complex carbohydrates), 33.4% oil.

Sunflower lecithin was deoiled with acetone, according to AOCS Official Method Ja 4-46, procedures 1–5 [20], obtaining the deoiled sunflower lecithin (DSL) (Figure 1). Then, DSL was stored at 0 ºC. This modified lecithin was used as control sample. Deoiling procedure was performed in duplicate.

2.2. Sunflower lecithin fractionation

Fractionation process was performed to 30 g of native sunflower lecithin with the addition of extraction solvents with different ethanol/water ratio (96:4, 100:0) using an ethanol/lecithin ratio of 3:1. These samples were incubated in a water bath at 65 °C during 60 min with moderate agitation (60 rpm), and then centrifuged at 1880 g, 10 ºC during 10 min. Afterwards, the corresponding ethanolic extracts were obtained and ethanol was eliminated by evaporation under vacuum [17].

Ethanol soluble phases were further deoiled with acetone, according to AOCS Official Method Ja 4-46, procedures 1–5, obtaining the different PC enriched fractions (PCF 96, PCF 100) (Figure 1). Then, both fractions were stored at 0 ºC. Fractionation and deoiling procedures were performed in duplicate.

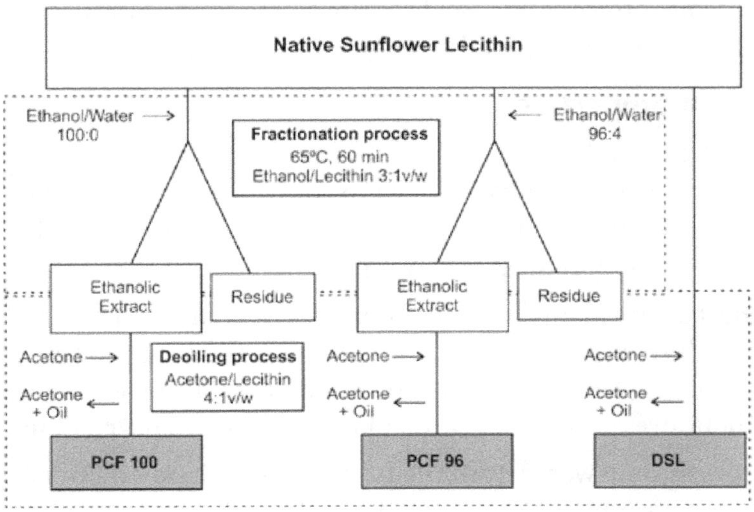

Figure 1. Flow diagram for deoiling and ethanol fractionation of native sunflower lecithin

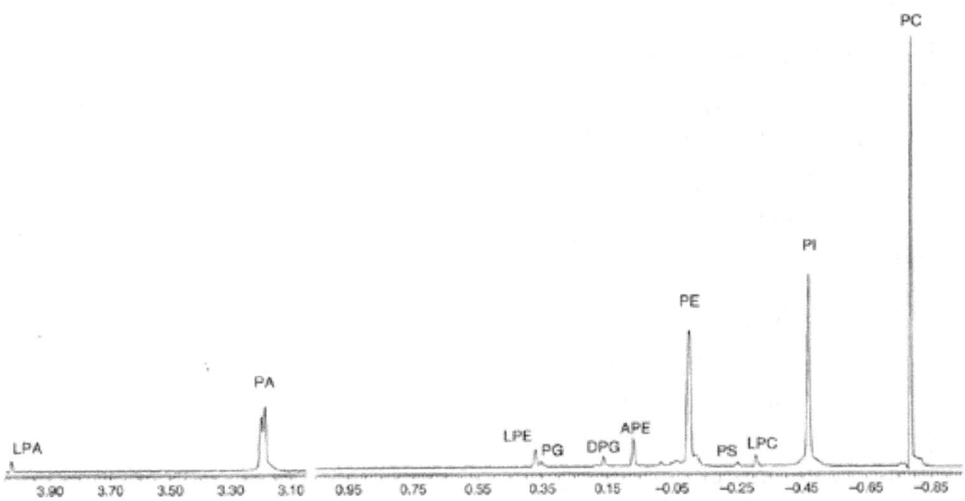

Figure 2. Spectrum of a phospholipidic sample

2.3 Phospholipid composition

Phospholipid composition of samples obtained after different modification processes was determined by 31P NMR analysis in a Bruker Avance 600 MHz automatic spectrometer, using triphenyl phosphate as internal standard (Spectral

Service GmbH, Köln, Germany) (Figure 2) [21-23]. For this purpose, 100 mg of each modified lecithin were diluted in 1 mL of deuterated chloroform, 1 mL of methanol and 1 mL of 0.2 M Cs-EDTA (pH 8.0). The organic layer was separated after 15 min shaking, and analyzed by the described spectroscopic technique. Phospholipid composition of the different modified sunflower lecithins (MSLs) was expressed in terms of molar concentration (mol / 100 mol lecithin) [17].

2.4 Antioxidant properties

The antioxidant properties of the MSLs were evaluated by the Rancimat (Mod 679, Metrohm) method. 5 g of sunflower oil were added with different concentration of the analyzed samples (500-2000 ppm), heated at 98 °C, air flow 20 L/h. Stability was expressed as the induction time, according to Gutiérrez [24].

Also, the highest antioxidant concentration was selected for each modified lecithin. The same procedure was carried out with previous thermal treatments at 120 and 160 °C for 1 h, adding the modified lecithins before heating. Refined sunflower oil with and without previous heat treatments were used as control samples.

Oil tocopherol content was determined by normal phase HPLC using a Hewlett Packard chromatography system (HPLC Hewlett Packard 1050 Series, Waldbronn, Germany) equipped with a fluorescence detector Agilent 1100 Series (Agilent Technology, Palo Alto, CA, USA) following the procedures described in IUPAC 2.432 [25] and AOCS Ce8-89 [20].

2.5 Oil-in-Water (O/W) emulsions preparation

Refined sunflower oil was used to prepare oil-in-water (O/W) emulsions with a formulation of 30:70 wt/wt. Emulsions were prepared at room temperature in an Ultra-Turrax T25 homogenizer using S 25 N–10 G dispersing tool (7.5 mm rotor diameter) at 10,000 rpm for 1 min, according to Cabezas et al. [26] with the addition of MSLs in a range of 0.1-2.0% (wt/wt).

2.6 Optical characterization of emulsions

The backscattering of light was measured using a QuickScan Vertical Scan Analyzer (Coulter Corp., Miami, FL). The backscattering of monochromatic light (λ = 850 nm) from the emulsions was determined as a function of the height of the sample tube (ca. 65 mm) in order to quantify the rate of different destabilization processes during 60 min. This methodology allowed to discriminate between particle migration (sedimentation, creaming) and particle size variation (flocculation, coalescence) processes [27]. The basis of the multiple light scattering theory has been exhaustively studied by Mengual et al. [28].

2.7 Statistical analysis

Data were evaluated by analysis of variance (ANOVA) using the software Systat® 12.0 [29]. For this purpose, differences were considered significant at $p < 0.05$.

3. RESULTS

3.1 Phospholipid composition

The quantitative analysis of phospholipids was performed by 31PNMR which represents a modern and the most sophisticated methodology for evaluating the composition of lecithins, since it is possible to obtain a separate signal for each phospholipid class [9, 21]. The results presented in Table 1, evidenced the high solubility of the phosphatidylcholine in the different ethanolic solvents assayed. 31PNMR determinations of the different enriched PC fractions (PCF 96 and PCF 100) exhibited an important concentration of PC (>71.0 %) as well as low values of PI (<8%) in comparison with the native and deoiled sunflower lecithin assayed.

3.2 Rancimat analysis

The oxidative stability of the refined sunflower oil (control) and the activity of different modified lecithins were evaluated determining the induction time (ti) by a Rancimat analysis. This methodology can be used to evaluate the efficiency of various synthetic or natural antioxidants to stabilize fats and oils against an accelerated oxidation test [30]. Figure 3 shows ti for the modified sunflower lecithins used at different concentrations.

Table 1. Percentage phospholipid composition (mol PL / mol total PL) of deoiled sunflower lecithin and PC-enriched fractions analyzed by 31PNMR

PL	DSL a	PCF 96 a	PCF 100 a
PC	36.7	71.2	76.3
1-LPC	< 0.1	< 0.1	< 0.1
2-LPC	1.5	3.3	2.8
PI	35.3	7.4	3.6
LPI	< 0.1	< 0.1	< 0.1
PE	15.1	12.2	10.6
LPE	< 0.1	0.8	0.4
PA	5.0	1.3	1.4
LPA	< 0.1	< 0.1	< 0.1
Others PL	6.4	3.8	4.9

a Values represent means (n = 2). The coefficient of variation was lower than 6%

The antioxidant addition increased the ti of the control as a function of increasing concentration. MSLs did not show a marked difference of the respective ti values at concentration of 500 ppm. However, at high concentrations (1000 - 2000 ppm), both PC fractions had a high significant effect ($p < 0.01$) on the oxidative stability of the control system, in relation to the values recorded by DSL addition. In this sense, 2000 ppm of DSL increased the ti value of the control oil of 49.1%, while similar concentrations of PCF 96 and PCF 100 enhanced this value 167.4% and 108.3%, respectively (Table 2).

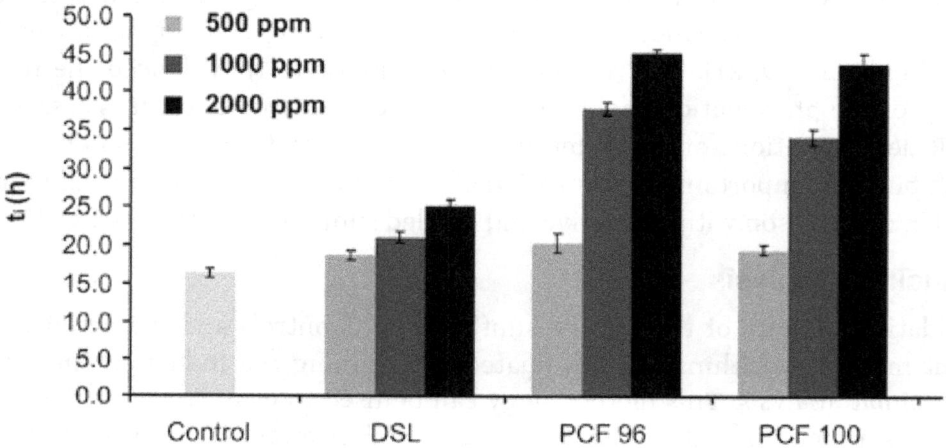

Figure 3. Induction times (ti) of refined sunflower oil (control) added with different concentrations of modified sunflower lecithins (Rancimat Mod 679, Metrohm). Mean values (n = 3) ± sd

Table 2. Percentage induction times increase of refined sunflower oil added with different modified sunflower lecithins (MSLs) with or without thermal pre-treatment

	Δ ti WT / ti control WT (%)	Δ ti T120 / ti control T120 (%)	Δ ti T160 / ti control T160 (%)	Δ ti T120 / ti control WT (%)	Δ ti T160 / ti control WT (%)
Control	0.0	0.0	0.0	-26.6	-25.8
DSL	49.1	56.9	19.0	15.1	-11.7
PCF 96	167.4	211.9	113.9	128.8	58.6
PCF 100	157.7	177.5	95.4	103.6	44.9

control, refined sunflower oil; WT, samples without thermal pre-treatment; T120, samples with thermal pre-treatment 120°C, 1h; T160, samples with thermal pre-treatment 160°C, 1h

Induction times of oils added with the different MSLs without thermal pre-treatments, showed higher values than those obtained by Pan et al. [31], who reported under similar test conditions that 2000 ppm of native sunflower lecithin produced a ti increase of 12%. The introduction of changes in the original phospholipids concentration of this native lecithin by different modification processes (deoiling, ethanol fractionation) allow to obtain modified lecithins with better physicochemical and functional properties with respect to the starting material.

A concentration of 2000 ppm of the different modified lecithins showed the highest antioxidant activity in this assay. Therefore, refined sunflower oil added with this concentration of MSLs was previously subjected to different heat treatments at 120 and 160 ° C for 1 h. The ti values of the oils added with PCF samples had a high significant difference ($p < 0.01$) in relation to those added with DSL (Table 2). PC enriched fractions reduced the negative effect of the heat treatments over the oxidative stability of refined sunflower oil. These pre-treated samples showed higher ti values respect to the control system, even without thermal pre-treatment. In this sense, oils added with PCF 96 and PCF 100, subjected to a heat treatment at 160°C presented induction times of 58.6 and 44.9% higher than the untreated oil, respectively.

It is interesting to note that PCF 96 recorded the best antioxidant capacity producing an increase of 211.9% (120°C, 1h) and 113.9% (160°C, 1h) in relation to those recorded for the respective control oils. This fact indicates that the fractionation process carried out with ethanol 96 ° not only allows to obtain fractions enriched in PC with antioxidant activity, but also this activity is greater than those exhibited using DSL and the PCF obtained through fractionation with absolute ethanol (PCF 100).

The ability of phospholipids to inhibit lipid oxidation in bulk oils has been known for several decades, but the mechanism of stabilization still remains controversial [32]. However, many research works have proposed different antioxidant mechanisms for these compounds. Particularly, PC and PE have been shown to have metal-chelating and scavenging properties. Also, this type of phospholipids present a synergistic effect with the different tocopherols (α-tocopherol 512.84 μg/g; β-tocopherol 4.55 μg/g) regenerating the oxidized tocopherol molecule by donation of a hydrogen atom of their amino function [18, 33]. This fact and the high PC and PE concentrations are in correlation with the better antioxidants characteristics observed for both PCF assayed.

3.3 Optical characterization of O/W emulsions

Stability of the different O/W emulsions (30:70 wt/wt) was studied recording the backscattering (BS) profiles as a function of the cell length and time, by a vertical scan analyzer (QuickScan). These profiles were analysed in different zones of the emulsion: Zone I (10-20 mm) to visualize the destabilization process by migration of the oil droplets from the bottom towards the top of the tube (creaming), Zone II (40-45 mm) characterized by the accumulation of oil droplets after the creaming process (cream phase) and Zone III (50-60 mm), to analyze the destabilization of the cream phase [34]. For instance, Figure 4 shows a typical profile obtained for an O/W emulsion with the addition of 0.5% of DSL.

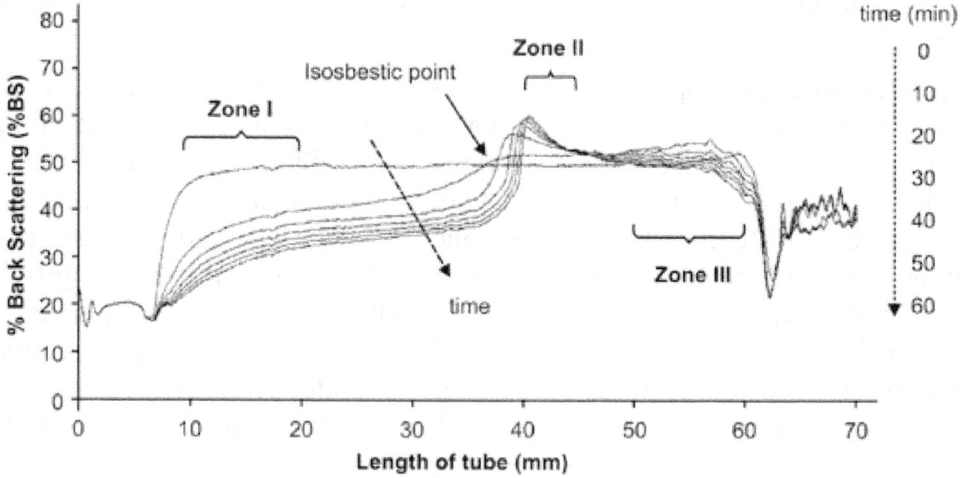

Figure 4. Backscattering (%BS) profile of a O/W emulsion (30:70 wt/wt) with the addition of 0.5% of DSL

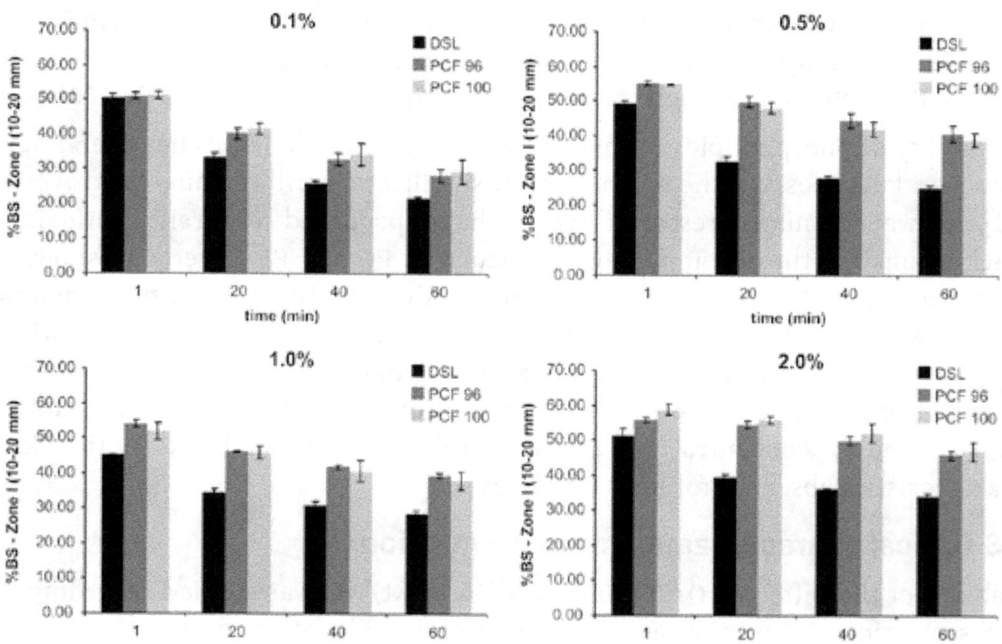

Figure 5. Backscattering (%BS) values of O/W emulsions (30:70 wt/wt) with the addition of different modified sunflower lecithins, Zone I (10-20 mm). Mean values (n = 3) ± sd

The creaming destabilization process (i.e. migration of oil particles to the upper portion of the tube) is evidenced by a decrease of %BS values at the bottom of the tube and the appearance of an isosbestic point [34]. This point separates the zones with lower (left) and higher (right) values than the initial %BS (Figure 3). The QuickScan profiles corresponding to the zone I (10-20 mm) showed an increase of the emulsion stability against the creaming process, as a function of increasing concentration of different modified lecithins (Figure 5). In particular, the PC enriched fractions (PCF 96 and PCF 100) produced a high stability in O/W emulsions than DSL, over the range of concentration studied. In this sense, it should be noted that QuickScan profiles of the different PCF did not show significant variations of %BS values for 2.0% during 40 minutes. These results are in concordance with those previously reported by Wu and Wang [12], related to the emulsifying activity of PC enriched fractions from soy lecithin.

Moreover, O/W emulsions with 0.1-0.5% of DSL showed a sharp decrease of %BS in the Zone I. This behaviour indicates a rapid destabilization of these emulsions by creaming.

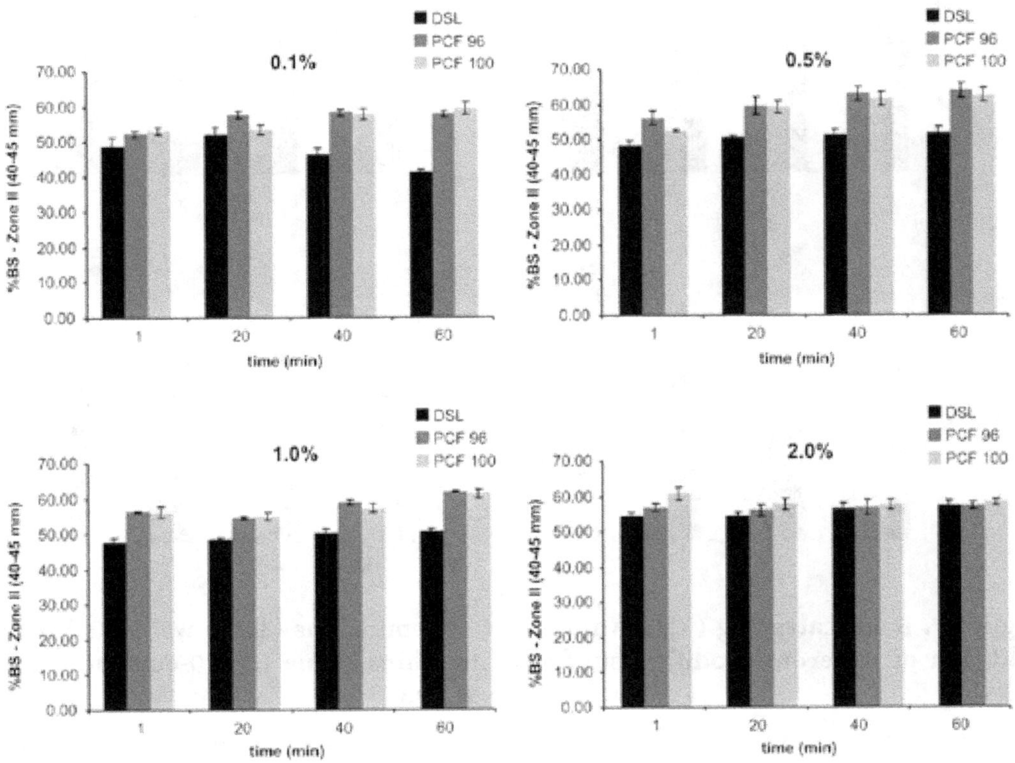

Figure 6. Backscattering (%BS) values of O/W emulsions (30:70 wt/wt) with the addition of different modified sunflower lecithins, Zone II (40-45 mm). Mean values (n = 3) ± sd

The tube zone between 40-45 mm (Zone II) is characterized by the accumulation of oil droplets after the creaming process (cream phase). Figure 6 shows the %BS values vs. time in Zone II. Emulsions formulated with PC enriched fractions presented higher %BS values than those obtained using DSL, for all concentrations studied. The higher levels of %BS and the greater stability of these emulsions formulated with a high concentration of phosphatidylcholine would be associated with the formation of dense cream phases with a lower proportion of continuous phase inside [26].

The stability of the different cream phases were analyzed by the evolution of the % BS values vs. time in Zone III (50-60 mm) (Figure 7). %BS values remained constant at concentrations of emulsifier above 0.1% of the different modified lecithins. This behaviour confirms the stability provided by these MSLs against the coalescence process, in these conditions.

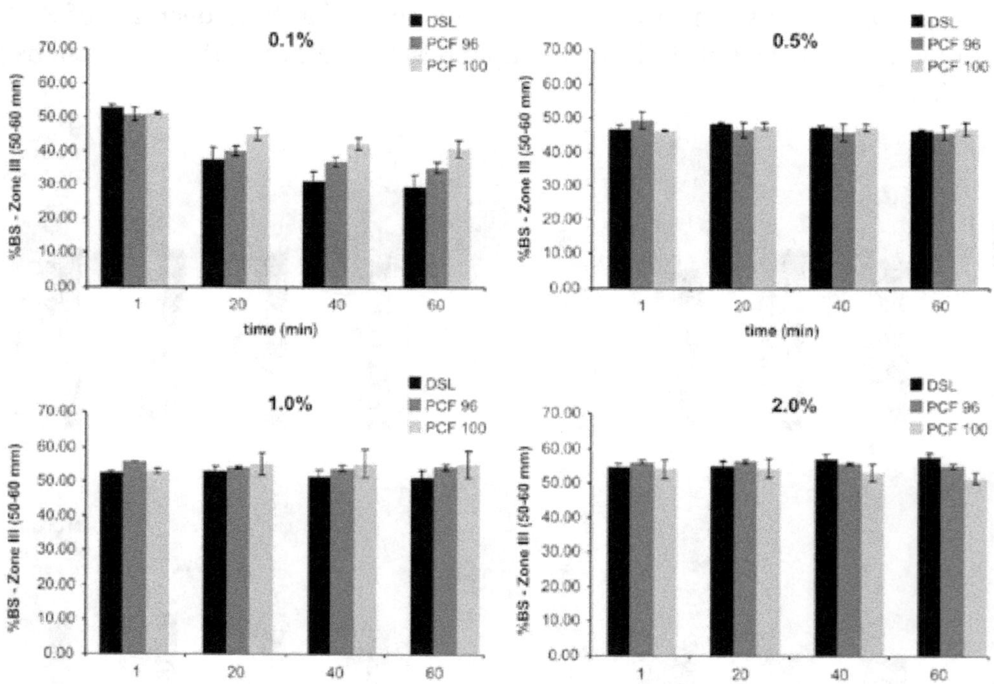

Figure 7. Backscattering (%BS) values of O/W emulsions (30:70 wt/wt) with the addition of different modified sunflower lecithins, Zone III (50-60 mm). Mean values (n = 3) ± sd

However, emulsion with 0.1 of DSL did not allow the formation of the cream phase. These results are related to the rapid decrease of %BS, the formation of an oil layer in the upper part of the tube and the absence of an isosbestic point (data not shown) ; suggesting the occurrence of a cream phase destabilization by coalescence [27].

The hydrophilic-lipophilic balance value (HLB) is often used in order to explain the performance of emulsifiers [15]. The high concentration of phosphatidylcholine (hydrophilic phospholipid) increases this empirical value in the PC enriched fractions. In this sense, the best properties of these modified lecithins as emulsifying agents in O/W emulsions was according with Carlsson [35], who has determined that lecithin with high HLB values present best O/W emulsifying properties.

Also, the phase structure at the interface of the different phospholipids influences the emulsion formation and stability [16]. PC forms a lamellar phase at the interface between oil and water with well ordered mono- and bi- layers. This structure has a great importance for the stabilisation of O/W emulsions.

4. CONCISE SUMMARY

The analysis of induction times revealed a statistically significant difference ($p < 0.01$) in the antioxidant activity displayed by the various PCF concerning the incorporation of DSL. Furthermore, these fractions facilitated the formation of more stable O/W emulsions (30:70 wt/wt) compared to those supplemented with DSL at varying concentrations (0.1-2.0%), regarding kinetic degradation as indicated by alterations in backscattering values over time. The results indicated that PC-enriched fractions (PCF 96 and PCF 100) provide a viable alternative as emulsifying agents for the food sector.

REFERENCES

1. Krawczyk T. Lecithin: consider the possibilities. International News on Fats, Oils and Related Materials 1996; 7, 1158-1167.
2. Wendel A. Lecithin: the first 150 years Part I: From discovery to early commercialization. International News on Fats, Oils and Related Materials 2000; 11, 885 – 892.
3. Wendel A. Lecithin: the first 150 years Part II: Evolution to a global pharmaceutical industry. International News on Fats, Oils and Related Materials 2000; 11, 992 – 997.
4. Cabezas DM, Diehl B, Tomás MC. Sunflower Lecithin: Application of a Fractionation Process with Absolute Ethanol. Journal of the American Oil Chemists' Society 2009; 86, 189 – 196.
5. Cabezas DM, Madoery R, Diehl B, Tomás MC. Emulsifying properties of hydrolized sunflower lecithins by phospholipases A2 of different sources. In: Valdez B. (ed.) Food industrial processes - methods and equipment. Rijeka: Intech; 2012. p39-50. Available from http://www.intechopen.com/books/food-industrial-processes-methods-and-equipment/emulsifying-properties-of-hydrolized-sunflower-lecithins-by-phospholipases-a2-of-different-sources (accessed 7 August 2012).

6. Prosise WE. Commercial Lecithin Products: Food Use of Soybean Lecithin. In: Szuhaj BF, List G (eds.) Lecithins. Champaign: AOCS Press; 1985. p163-181.
7. Schmidt JC, Orthoefer FT. Nonfood Uses of Lecithin. In: Szuhaj BF, List G (eds.) Lecithins. Champaign: AOCS Press; 1985. p183-200.
8. van Nieuwenhuyzen W. The Industrial uses of special lecithins. Journal of the American Oil Chemists' Society 1981; 58, 886–888.
9. van Nieuwenhuyzen W, Tomás MC. Update on vegetable lecithin and phospholipid technologies. European Journal of Lipid Science and Technology 2008; 110, 472-486.
10. van Nieuwenhuyzen W. Fractionation of Lecithins. The European Food and Drink Review Process Technology 1999; 27 – 32.
11. Kullenberg FW. Lecithin in animal health and nutrition. In: Szuhaj BF (ed.) Lecithins: Sources, Manufacture & Uses. Champaign: AOCS Press; 1989. p237–252.
12. Wu Y, Wang T. Soybean lecithin fractionation and functionality. Journal of the American Oil Chemists' Society 2003; 80, 319–326.
13. Sosada M. Optimal Conditions for Fractionation of Rapeseed Lecithin with Alcohols. Journal of the American Oil Chemists' Society 1993; 70, 405-410.
14. Hernández E, Quezada N. Uses of phospholipids as functional ingredients. In: Gunstone FD (ed.) Phospholipid Technology and Applications. Bridgwater: The Oily Press; 2008, p83-94.
15. McClements DJ. Food emulsions: principles, practice and techniques. Florida: CRC Press; 1999.
16. van Nieuwenhuyzen W, Tomás MC. Update on vegetable lecithin and phospholipid technologies. European Journal of Lipid Science and Technology 2008; 110, 472-486.
17. Cabezas DM, Diehl BWK, Tomás MC. Effect of processing parameters on sunflower PC enriched fractions extracted with aqueous-ethanol. European Journal of Lipid Science and Technology 2009; 111, 993-1002.
18. Saito H, Ishihara K. Antioxidant activity and active sites of phospholipids as antioxidants. Journal of the American Oil Chemists' Society 1997; 74, 1531-1536.
19. Shukla VKS, Wanasundara PKJPD, Shahidi F. Natural antioxidants from oilseeds. In: Shahidi F (ed.) Natural Antioxidants, Chemistry, Health Effects, and Applications. Illinois: AOCS Press; 1997, p97-132.

20. AOCS. Official methods and recommended practices of the American Oil Chemists' Society, 5th ed. Champaign: AOCS Press; 1998.

21. Diehl BWK, Ockels W. Quantitative Analysis of Phospholipids. In: Cevc G, Paltauf F (eds.) Characterization, Metabolism and Novel Biological Applications. Proceeding of the 6th International Colloquium Phospholipids. Champaign: AOCS Press; 1995, p29-32.

22. Diehl BWK. Multinuclear high-resolution magnetic resonance spectroscopy. In: Hamilton RJ (ed.) Lipid Analysis in Oils and Fats. London: Blackie Academic & Professional; 1997, p87-135.

23. Diehl BWK. High resolution NMR spectroscopy. European Journal of Lipid Science and Technology 2001; 103, 830-834.

24. Gutiérrez F. Determinación de la estabilidad oxidativa de aceites de oliva vírgenes: comparación entre el método A.O.M. y el método Rancimat. Grasas y Aceites 1989; 40, 1–5

25. International Union of Pure and Applied Chemistry. Standard Methods for the Analysis of oils, Fats and Derivates, 7th edn. Oxford: Blackwell Scientific; 1992.

26. Cabezas DM, Madoery R, Diehl BWK, Tomás MC. Emulsifying properties of different modified sunflower lecithins. Journal of the American Oil Chemists' Society 2012; 89, 355-361.

27. Pan LG, Tomás MC, Añón MC. Effect of sunflower lecithins on the stability of water in Oil (W/O) and Oil in Water (O/W) emulsions. Journal of Surfactants and Detergents 2002; 5, 135 – 143

28. Mengual O, Meunier G, Cayré I, Puech K, Snabre P. Turbiscan MA 2000: multiple light scattering measurement for concentrated emulsion and suspension instability analysis. Talanta 1999; 50, 445-456.

29. Systat version 12. Systat Software Inc. California; 2007

30. Verleyen T, Van Dick S, Adams CA. Accelerated stability test. In: Kamal-Eldin A, Pokorný J (eds.) Analysis of lipid oxidation. Champaign: AOCS Press; 2005, p21-228

31. Pan LG, Tironi VA, Tomás MC, Añón MC. Actividad Antioxidante del Extracto de Romero (Rosmarinus officinalis L.) y Lecitinas de Girasol en Aceite de Girasol. Proceedings of XIII Seminario Latinoamericano y del Caribe de Ciencia & Tecnología de los Alimentos, 12-16 Oct 2004, Montevideo, Uruguay; 2004

32. Cardenia V, Waraho T, Rodriguez-Estrada MT, Mc Clements DJ, Decker EA. Antioxidant and prooxidant activity behavior of phospholipids in stripped soybean Oil-inWater emulsions. Journal of the American Oil Chemists' Society 2011; 88, 1409–1416

33. Judde A, Villeneuve P, Rossignol-Castera A, Le Guillouc A. Antioxidant effect of soy lecithins on vegetable oil stability and their synergism with tocopherols. Journal of the American Oil Chemists' Society 2003; 80, 1209-1215

34. Cabezas DM. Modificaciones tecnológicas de lecitinas de girasol. Aplicaciones en el área alimentaria. PhD Thesis. Centro de Investigación y Desarrollo en Criotecnología de Alimentos (CIDCA). Facultad de Ciencias Exactas, Universidad Nacional de La Plata (FCE-UNLP); 2010

35. Carlsson A. Physical properties of phospholipids. In: Gunstone FD (ed.) Phospholipid technology and applications, Chapter 5. Bridgwater: The Oily Press; 2008, p104

SECTION 5
ECONOMIC AND SOCIAL ASPECTS

26 Chapter

AN AGRO-ECOLOGICAL CHALLENGE FOR SUSTAINABLE RURAL DEVELOPMENT IN NORTHERN ITALY: CABANNINA CATTLE BREEDING

1. INTRODUCTION: AN OVERVIEW

Intensive farming is an agricultural production system marked by the extensive use of external inputs, including capital, machinery, infrastructure, pesticides, and chemical fertilizers, which impacts the natural environment and rural communities. This agricultural strategy has been the primary answer to population growth, as it enables increased food production on a specific land area. Intensive animal farming systems, while allowing for the rearing of numerous animals in confined spaces, necessitate substantial quantities of food, water, medical interventions, capital-intensive machinery, energy, and fuel. The objective of every industrial farm is to choose animals that efficiently convert feed into milk and meat; nevertheless, this often results in a decline in reproductive performance and product quality. Consequently, contemporary issues in the dairy cow sector are readily apparent. The Friesian breed has experienced a decline in reproductive performance globally, adversely affecting cow robustness and longevity due to heightened stress, udder health issues, and locomotor abnormalities, resulting in detriments to the physiological parameters characteristic of healthy cows.

Notwithstanding the aforementioned issues related to conventional agriculture, numerous positive advancements are emerging. Numerous alternative projects are already thriving throughout the Italian peninsula to advance ecological agriculture, safeguard small farmers' livelihoods, produce healthy, safe, and tradition-based foods, and localize distribution, trade, and marketing. These typologies of traditional agriculture offer promising models for marginal areas as they promote biodiversity, thrive without agrochemicals, and sustain year-round yields. Technological methods are encouraged if they enhance yield gains grounded in agro-ecological principles, highlighting diversity, synergy, recycling, and the integration of animals, soil, and water; additionally, social processes that involve and empower communities are also welcomed. In the north-western Italian region of Liguria, Petramartina farm exemplifies a pioneering endeavor by adhering to agro-ecological principles to establish a traditional and sustainable dairy operation centered on indigenous Cabannina cattle.

2. CONVENTIONAL DAIRY FARMING IN THE "RIVER PO PLAIN" AND ITS CRITICAL POINTS

The River Po Plain is characterized by an industrial breeding system whose bovine herds at present are fewer, larger, more specialized and more capital intensive than they were 60 years ago.

Revealing data are briefly shown in the following table (Table 1).

Table 1. Italian National Data about number of herds, total cows, and average number of cows per herd.

NATIONAL DATA			
YEAR	CONSISTENCY		COWS/HERD
	HERDS	COWS	
2002	15.106	1.088.178	72
2003	14.984	1.107.701	74
2004	14.823	1.100.543	74
2005	14.317	1.101.657	77
2006	14.069	1.102.655	78
2007	13.818	1.100.401	80
2008	13.510	1.101.868	82
2009	13.327	1.103.453	83
2010	13.164	1.113.859	85
2011	12.922	128.626	87

As clearly shown, since the last 10 years the Italian dairy situation has followed the international trend for which farmers, have to increase the number of animals in their herds to sustain their businesses.

Thus, in recent years the dairy system has been mainly concerned with both improving animal performances and with increasing the number of heads per farm in order to reduce fixed costs that represent, together with raw materials necessary to compose animal feed formulations, the most important passive voice of a farmer's budget.

The following table (Table 2), gives an overview of Italian Friesian cow milk production and their reproductive situation in the last years.

Table 2. An overview of Italian Friesian cow official production by AIA (Italian Breeders Association)

Year	Milk Production (q/cow/year)	First Calving/ Total Cows (%)	Days open Mean	Number of Lactation/Cow Mean
2004	85,9	34,8	147	2,48
2005	89,1	35,6	136	2,47
2006	89,4	36,1	141	2,45
2007	90,3	35,5	143	2,46
2008	89,8	36,2	143	2,46
2009	88,7	35,6	141	2,46
2010	91,3	36,1	141	2,44
2011	91,9	36,4	144	2,39

As shown, the increased number of heads and of milk production amount were not followed by any improvement in reproductive efficiency. In fact, Table 2 clearly demonstrates how the reproductive physiology of the animals was damaged. And mainly in Friesian breeds, cows' robustness and longevity negatively affected by increasing stress, udder health disturbances and locomotion disorders [1], are an issue of major concern to farmers. From a physiological point of view, Holstein cows reached a very critical situation having missed their good reproductive efficiency characters (e.g. calving interval and conception rate [2], excellent longevity in farm (only 2,39 lactations in their life), resistance to stress and diseases (metabolic syndromes, ketosis, mastitis and foot diseases) [1,3,4], whereas they dramatically increased "energy and financial voracity" (diet based on starch and protein meals, great health and structural investments due to several highly-recurring diseases [4-6]. A recent study conducted on an Italian dairy farm, located in the River Po Plain, highlighted the extreme financial voracity affecting a conventional farm of good quality in order to run it and make it survive [7]. The Authors showed that even an excellent business consisting of 208 productive cows, 70% of which are healthy animals, can lose more than 36000 €/year on including both direct and indirect costs due to animal diseases such as veterinary therapies, decrease in productive performances, milk removed, price-cutting for high milk quality, etc. These data are worrying in the consideration that in Italy milk production represents one of the most important sectors of its agro-industry,. In fact, in Europe Italy is the most important cheese producer and exporter, with its about 460,000 tons of products and almost 3 billion Euros (data from www.clal.it, www.ismea.it) derived from PDO (Protected Designation Origin) and PGI (Protected Geographical Indication) production. Its most representative cheese is certainly Parmigiano Reggiano and Grana Padano that recently increased by 9.8% its export trend to Germany, the United States, France, Switzerland and the United Kingdom.

Italian breeders have to care about all the important milk quality parameters involved in cheese making processes such as milk fat, proteins and somatic cell count.Recently, Italian researchers demonstrated that milk whose somatic cell content is greater than 400,000 cells/ml evidences a scarce aptitude to rennet coagulation and cannot be generally considered suitable for cheese production, with particular reference to Grana cheese production [8].

Having in mind the difficulties characterizing the present background and in the aim to confirm Italian national data, some analytical studies were performed in five different Friesian herds bred in the River Po Plain.

The following parameters were investigated: milk production (herds A and B), reproduction and sanitary situation (herds C, D, E).

In herds A and B from an area south of Milan (Lodi) fat and protein in the milk yield as well as total cheese yield were monitored. Tables 3 and 4 show key results from this investigation.

Table 3. Milk Production Data in Herd A

YEAR	LACTATIONS CLOSED	MILK YIELD Kg	FAT		PROTEIN		CHEESE YIELD Kg
			%	Kg	%	Kg	
2007	239	10.608	3,74	397	3,57	379	825
2008	243	10.639	3,68	392	3,48	370	811
2009	254	10.184	3,78	385	3,56	363	793
2010	232	9.991	3,87	387	3,56	356	785
2011	228	10.193	3,52	359	3,48	348	765

Table 4. Milk Production Data in Herd B

YEAR	LACTATIONS CLOSED	MILK YIELD Kg	FAT		PROTEIN		CHEESE YIELD Kg
			%	Kg	%	Kg	
2007	54	10.691	3,37	360	3,3	353	768
2008	55	9.385	3,53	331	3,39	318	695
2009	48	9.344	3,68	344	3,29	307	690
2010	52	9.380	3,67	344	3,27	307	690
2011	55	8.536	3,45	294	3,28	280	616

Clearly, milk production and milk aptitude to produce cheese underwent a dramatic decrease causing economic losses to breeders.

In herds C, D, and E from an area west of Milan (Abbiategrasso), days open (DO), number of insemination for conception (AI), culling cows (CC) together with sanitary problems and treatments were monitored and key results from this investigation are represented in table 5.

Table 5. Average data in Herds C, D, and E.

HERD	COWS	DO	AI	CC
C	688	154	2,53	30%
D	438	147,5	2,09	36,50%
E	345	130,7	3,18	29,03%
TOTAL	1471			
AVERAGE		144,1	2,53	32%

C,D,E = herds in Abbiategrasso (MI); DO = days open; AI =number of artificial insemination per conception; CC= number of culling cows.

Such data broadly confirm official data related to Italian national situation. Among critical points to be underlined both nationally and internationally, we should consider that not only animal intensive breeding dramatically increase worldwide, but vegetal monoculture as well. Thus, while intensive farming and agriculture can be considered extremely competitive and the only way to feed a fast-growing human population, they also cause different economic, environmental and social problems [9]. In fact, from an ecological point of view, consequences of industrialization are manifold [10] as detailed below:

a. Cycles of nutrients, energy, water and wastes become more open rather than closed as occurring in a natural ecosystem;
b. The natural habitat of most wild creatures can be limited or destroyed by industrial agro-farm, and most soil can suffer erosion;
c. A large amount of energy input is needed to produce, transport, and apply chemical fertilizers;
d. Dwindling and ever more expensive fossil fuels are used to derive agrochemicals, and to work fuel-based mechanization and irrigation operations, the core of industrialized agriculture;
e. Air pollution can be prompted by chemical fertilizers that were recently implicated in the destruction of the ozone layer and in global warming;
f. Animal and vegetal biodiversity are severely affected as species more susceptible to diseases are selected.

Moreover, an excessive use of fertilizers is linked to acidification/salinization of soils and to a higher incidence of insect pests and diseases through mediation of negative nutritional changes in crop plants [11]. It also implies several harmful effects on the health of workers spreading them, of people living nearby or downstream/downwind an area of application, and of consumers swallowing leftover pesticides from food.

Last, but not least, a genetic selection is required in both vegetal and animal conventional production to maximize production, which implies biodiversity decrease. Thus, the amount of crop diversity per unit of arable land decreased and croplands tended towards concentration. In fact, in order to serve as many markets as possible some political and economic forces influence a general trend to devote larger and larger areas to monoculture [10].

3. AGRO-ECOLOGICAL PRINCIPLES AS A POTENTIAL SOLUTION TO EXTREMELY INTENSIVE BREEDING

Nonetheless, excessive intensive breeding can be successfully contrasted and new ecological concepts and principles can be applied to design, development and management of sustainable agricultural systems. In fact, agro-ecology can be a sensible solution.

Agro-ecology is the science that studies the relationship between agricultural production, land and traditions regarding a given territory [11]. It is a theoretical and practical discipline derives from different experiences related to organic agricultural production, characterizing the last and present centuries.

It investigates the elements of an agricultural ecosystem and their interactions carefully and provides principles and methods of work. Principles and methods that do not merely consider production but also the ecological, technical, socio-economic and cultural spheres of the agro-ecosystem.

Major aims of this science are:

1. to increase both functionality and productivity of a business and an ecosystem;
2. to deal with biodiversity conservation and nutrient recycling;
3. to optimize local resources usage and economic viability of a farm.

Following these objectives, some practices - such as crop rotation and use of local varieties - essential to the microbiological and mineral balance of the soil can be recovered together with techniques well-known since past times but never applied on a large scale, e.g. the on- farm production of natural preparations and fertilizers [11]. Even planting local varieties of fruit trees and hedges, a good practice swept away by the advent of industrial agriculture, can help to increase biodiversity within a farm and promote a more balanced agro-ecosystem. A subsoil rich in organic and mineral elements and a greatly diversified topsoil make vegetation more durable and more resistant against diseases and insects.

Present industrialized agriculture, as Altieri underlines, is no longer sustainable. In fact, conceiving a constant supply of water, some cheap power and climate exploitation with no abrupt changes is absolutely utopian nowadays.

For instance, the old system based on mechanization, e.g. forage drying, cannot work any longer because of the continuous rise in price of fossil fuels. Not to say about the global use of herbicides that counts about 2.6 million tons per year and costs 25 billion euros. Being excessive and uncontrolled, they negatively affects wildlife and pollinators, the quality of water and fishing activity, and pose a serious risk of poisoning humans and animals. In addition, arthropods and weeds show high adaptation to these substances, while abuse of monocultures indirectly select pests and makes them more resistant [10]. Table 6 illustrates the fundamentals of agro-ecology.

Table 6. Principles of Agro-ecology by Altieri (2012) [11].

Enhancing biomass recycling, with a view to optimizing organic matter decomposition and nutrient cycling over time.
Strengthening the " immune system" of agricultural systems by enhancing functional biodiversity, natural enemies, antagonists, etc.
Providing the most favorable soil conditions for plant growth, particularly by managing organic matter and by improving soil biological activity.
Minimizing losses of energy, water, nutrients and genetic resources by enhancing conservation and regeneration of soil and water resources and agro-biodiversity.
Diversifying species and genetic resources.
Improving beneficial biological interaction and synergies among the components of agro-biodiversity, thereby promoting key ecological processes and services.

As seen in Table 7, in a recent study Koohafkan et al. [12] formulated some questions indicating what can be defined as an agro-ecological farm.

Table 7. Questions to define an agro-ecological farm. (Koohafkan, 2011) [12]

Do they reduce poverty?
Are they based on rights and social equity?
Do they reduce social exclusion, particularly for women, minorities and indigenous people?
Do they protect access and rights to land, water and other natural resources?
Do they favor redistribution of productive resources?
Do they substantially increase food production and contribute to household food security and improved nutrition?
Do they enhance families water access and availability?
Do they regenerate and conserve soil and increase soil fertility?
Do they reduce soil loss / degradation and enhance soil regeneration and conservation?

Do practices maintain or enhance organic matter and the biological life and biodiversity of the soil?
Do they prevent pest and disease outbreaks?
Do they conserve and encourage agro-biodiversity?
Do they reduce greenhouse gas emissions?
Do they increase income opportunities and employment?
Do they reduce variation in agricultural production under climatic stress conditions?
Do they enhance farm diversification and resilience?
Do they reduce investment costs and farmer dependence on external inputs?
Do they increase the degree and effectiveness of farmer organizations?
Do they increase human capital formation?
Do they contribute to local / regional food sovereignty?

4. CABANNINA BREEDING IN AVETO VALLEY

4.1 Territory

Aveto Valley represents the natural settlement area of Cabannina cows. When a war correspondent, Ernest Hemingway wrote in his diary in 1945: "I have just passed across the most beautiful valley in the world". In fact, it is one among the narrowest valleys of the northern side of the eastern Ligurian Apennine, where both the climate and the lands have unique characteristics thanks to the favorable altitude, which ranges from about 350 to 1800 meters, and the proximity to the sea. Rain is much more abundant than in the rest of the region, with about 2500 mm per year as peak value, especially in autumn and spring. In winter, snowfalls are stable, usually abundant, and temperatures can drop below 10 °C. In summer, the valley climate is humid and can become cooler at high altitude.

Figure 1. Aveto valley panorama, from Scabbiamara area.

The river Aveto flows through the valley floor; it springs in Prato Lungo, an area just below the village of Acquapendente near Mount Caucaso. Slowly winding, it creates different landscapes where pastures alternate to woods, meadows and rocky canyons. Then it passes through the province of Genoa, to reach the province of Piacenza. In Confiente place (a term that means "confluence") it gets into the river Trebbia, a tributary of the Po river.

In this context, the Cabannina, a great grazer even in extreme conditions, perfectly adapted to the territory and evolved. In fact, it can climb to the ridges and resist cold temperatures as well as snow without any problem like other northern breeds, such as the Highlands of Scotland. Thanks to these features this breed could spread to other areas of Liguria where the animals can graze free on green pastures surrounded by woods all summer long and where many businesses breed only one race.

4.2 A short historical account

Modern cattle breeds derive from two domestication events of the ancestral "Bos Primigenius" occurred in South West Asia and in Asia, which gave rise to the "Bos Taurus" and "Bos Indicus" respectively [13-15]. Hence, it is believed that these animals may result from a particular population of Uri that lived in the Southern Alps area [16]. Their dairy characteristics and adaptation to the territory still show their ties with the Bos Primigenius.

In the past the breeder families of Aveto Valley gathered in consortia; alternately, every producer received milk by the other members and made cheese. Every family was given a wooden stick where some notches were carved to mean milk given to community, a form of barter locally named 'cangiu', i.e. 'exchange'. The animals were let loose in the pastures; for most of the year they were fed only with fresh forage and homemade feed supplements such as boiled potato skins, cooking water and bran [17]. When necessary, transhumance to Tigullio coast occurred. At the beginning of the twentieth century there were about 40,000 specimens that in the postwar period underwent a dramatic reduction in their number as countryside was abandoned, people migrated towards cities and intensive farms arose.

In February 1963, Cabannina cows risked their extinction on account of Law No. 126, "Regulation of Bovine Reproduction", - now repealed- requiring their forced slaughter and replacement with apparently more productive breeds. When this law became effective the Aveto breeders, strongly determined to preserve Cabannina, started their resistance against the Provincial Inspectorate of Agriculture. Attempts were made to combine different races, such as Bruno-Alpine, to the native heads, but with little results. In this regard, Maimone studied these two races years later in 1981 [18]. Combining the data collected by Orefice in 1978 [19], it was certified that

Cabannina reproductive efficiency was better than Brown one and its economic life was greater thanks to the early age of its first birth and the reduced inter-calving period (84 days versus 140). Brown also appeared to have an inferior capacity to pasture, especially in villages where Cabannina was farmed, and could not stand the scarce food resources in the area showing phenomena of piroplasmosis which Cabannina breed could better fight against. In fact, originally Aveto Valley was a large swamp and over time native breed may have developed a natural resistance to the protozoa responsible for the disease.

Since 1974, the studies by Usai (1974), Orefice (1978) and Maimone (1981) [18-20] contributed to the promotion of Cabannina breed. In 1978 Orefice proved that in the town of Rezzoaglio there were about 500-550 heads of this typical breed. In 1981 when a census was made, out of 759 heads left only 158 were phenotipically Cabannina. Thus, an exemption to law No. 126 was decided in order to use Cabannina reproductors, and in 1985 the "Anagraphic Register of indigenous cattle populations and ethnic groups of limited diffusion " was established.

Despite various interventions Cabannina heads have steadily declined. From a survey conducted in 2010 by the Provincial Association of Breeders, only 254 heads are left on the Ligurian territory, mostly concentrated in Aveto valley. In the same year Cabannina breed has got one of the 193 Slow Food Presidia on the Italian territory. The network of Slow Food Presidia was established in 1998; it protects and preserves traditional products made according to old techniques, often present in remote areas of our country, and makes them distinguishable by a logo.

Figure 2. U'Cabanin trademark of Cabannina raw milk production.

In our case it is not the product to be preserved, but the breed as a part of it, similarly to other endangered species such as the Black Pig Nebrodi of Sicily and the Black Cock of Val di Vara in Liguria (www.slowfood.it). In 2007 a cheese made only from Cabannina raw milk began to be produced under the trademark "U Cabanin". Compared to the 1981 census the number of Cabannina heads has slightly increased, but the situation remains critical, being this breed still at high risk of extinction.

4.3 Cabannina cow

Breed standards required to register animals were defined in a specific Registry for Breeds under Limited Distribution. A recent dissertation from the Faculty of Veterinary Medicine in Parma [21], collected biometric data on Cabannina cows; such data prove how their average measurements have changed according to observations performed from mid'70s. In fact, compared against race specifications, height at withers, set from 1.18 to 1.20 m for females, and from 1.25 to 1.30 m for males, appears to be about 10 cm higher, as well as other parameters related to both sexes. Circumference of front shank, as well as head and rump bis-articular width remained unchanged.

Figure 3. Pasturing Cabannina cows in Scabbiamara area.

Natural breed evolution and feed were considered to be factors influencing the named changes, but the investigation even considered the influence from crossings with Brown Alpine cows after Law 126 came into force when they began to replace the crossings. Moreover, while in the past the cows were fed exclusively on forage and leftover food, such as potato peels, the recent introduction of feed concentrates, even not properly balanced ones, may have boosted the growth of Cabannina breed. Actually, its small size should be considered as a positive feature to keep because it makes these cows excellent grazers even under extreme conditions, and low feed consumers, which means a substantial economic advantage for breeders [21].

4.3.1 Rusticity

Cabannina breed shows a very low incidence of diseases, virtually no hoof problems, or mastitis. No dystocical parturitions and on average calving-conception occurs after 80 days [21].

4.3.2 Physiological reproductive features

Under normal conditions and in the absence of any drug treatment, the involution of the uterus, is completed within 4-6 weeks post-partum. A recent work about uterine involution [22], shows that the reproductive physiology of Cabannina is characterized by a rapid recovery of ovarian activity. In fact, the onset of first estrus can be observed 20 days after birth and the fertilizing opportunity occurs in the following cycle, at about day 40. Features that allow farmers to achieve the advantage of a calf per year, i.e. maximum productivity, maximum reproductive efficiency and excellent mammary functionality.

4.3.3 Physiological productive features

Milk from Cabannina cows was carefully examined at the laboratory of Veterinary Physiology - University of Milan – [22-23] as to the dimensional characteristics of its fat globules, which make it a product of great value. In fact, it was proven the presence of small fat globules resulting in a wide specific surface area (SSA); thus, intestinal lipases can easily attack them and milk results to be highly digestible [24]. Moreover, as fat globules interact with casein curd, their size also influence the processes of lipolysis, bacterial colonization and cheese ripening [25]. High levels of unsaturated fatty acids and desaturase indices were highlighted, with important antimicrobial and biological effects and influences on human health as described in recent literature [26-30].

During cheese making processes, titrable acidity is one among basic parameters to control as it indicates good quality and good preservation of the product. Its normal values range between 3.20 and 3.80 °SH/50 ml. in fresh milk [21]. In 1987, Zanetti et al. [31]found that a low titrable acidity represents a condition technologically limiting optimal cheese making. In fact, hypoacidic milk, takes longer to turn into cheese when rennet coagulation is used, thus affecting dairy processing negatively. Finally, even though both remained within normal range values, milk from Cabannina showed average values of SH ° definitely higher than milk from Friesian breed, which can positively influence yield in cheese making processes [32].

5. PETRAMARTINA FARM

5.1 Farm structure and management

Petramartina farm is located at Scabbiamara in Aveto Valley at an altitude of 1000 m above sea level. (GPS coordinates: 44 ° 31’60 “N - 09 ° 22’60” E). It was established in 2009; since then, the owners and two workers have looked after the animals and manufactured dairy products. The farm represents a modern rural reality carefully merged with traditions. In fact, its products are sold directly to consumers or to small local retailers (EEC Reg. 852/04) according to the principles of short food chain. The business is articulated into several buildings: an animal shed, a service area, a dairy product manufacturing room, a seasoning room. An area devoted to direct sale and a classroom to provide educational services for students and adult parties have already been designed and will be realized to complete the farm.

5.2 Cattleshed, animals and their reproduction

The old cattleshed is under the house of the herdsman/milker as it used to be in the past when they derived heat from nearby animals. It is characterized by stalls for the cows, used only to milk them and shelter them in winter, when they rest on chipboard.

As shown in Figure 5, the cattleshed is built according to an old concept: with no separating bars or "educators", as they are not essential where human-animal relation is still strong. Manure removal is not mechanized; the pit where it is kept to be spread over crops the farm

can use is just close to the stable. In spring, the animals can graze freely. They are usually let loose day and night and lead to the cattleshed only for milking. During the night cows in lactation are kept in the pastures closer to the shed, whereas animals in dry and heifers can graze in the fields farther away.

Figure 4. Petramartina Farm Logo.

Figure 5. The cattleshed.

Milking is performed at 6.00 am and 6.00 pm by a cart machine; it starts with tail binding and then pre-dipping follows, i.e. a neutral liquid detergent (foam on) characterized by a sanitizing and softening action is applied. Such a procedure makes it easier to milk cows. To finish, post-dipping is carried out using a highly viscose solution of benzyl alcohol to protect nipples; in fact, it forms a barrier against bacteria, thus reducing mastitis and related problems for Cabannina cows at lowest level possible.

Table 8. Animals in the stable in August 2011

NUMBER	COW NAME	AURICULAR BRAND	STATUS
1	BIONDA	IT010990001714	Production
2	MARTINA	IT010990009965	Production
3	NORA	IT010990009966	Production
4	NEBBIA	IT018990022884	Production
5	BELLA	IT010990013162	Production
6	PETRA	IT010990013163	Production
7	ESMERALDA	IT010990012274	Production
8	PICOLA	IT018990042730	Production
9	LILLY	IT018990032152	Production
1	RAYA	IT010990013806	NORA
2	LENA	IT010990013807	NEBBIA
3	SUSY	IT010990015551	ESMERALDA
4	MIRA	IT010990013805	PICOLA

As shown in Table 8, productive animals are currently limited to 9 but their number is constantly increasing when compared with previous years. Figure 6 proves that trend related to number of Cabannine cows assigned to milk production is becoming larger and is equal to 2.5 subjects/year, which can prompt towards implementing their relative abundance in the area, sustaining their reintegration within typical Ligurian breeding and, last but not least, ensuring biodiversity in cattle. Year 2011 was a greatly positive as 4 female cows directed to production were born.

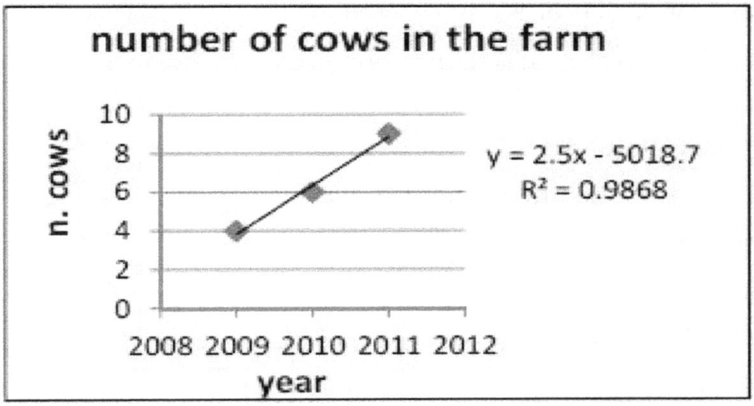

Figure 6. Cabannina cows increase from 2009 to 2011. Trend for forthcoming years.

Management of reproduction represents one among breeders' tasks: to increase genetic variability artificial insemination is performed using Cabannino bull paillettes preserved by APA (Association of Provincial Breeders) since 20 years ago, which is essential to recover a cattle breed. Bulls are chosen carefully to keep blood lines as much as possible divided. This is important primarily because deleterious inbreeding must be avoided and, secondly, because genetic and phenotypic data about bull semen have not been collected so far.

Table 9. Insemination data and birth calendar.

NAME	INSEMINATION DATA	BULL	DELIVERY DEADLINE	DELIVERY
PICOLA	20-05-2010	PIPPO CAB	26-02-2011	06-03-2011
NORA	19-07-2010	PIP	19-04-2011	06-05-2011
NEBBIA	01-09-2010	BIMBOTTO	01-06-2011	08-06-2011
ESMERALDA	26-09-2010	LEO	26-06-2011	06-07-2011
BIONDA	05-11-2010	MANDRIN 35992	05-08-2011	23-08-2011
MARTINA	29-12-2010	FARFALLO	29-09-2011	-
LILLY	07-05-2011	LEO	-	-

Table 9 details dates about each insemination performed and deliveries occurred. As to inseminations, only one is necessary per animal, which can decrease reproduction management costs. As to calving-conception interval, data from the farm under investigation showed compatible with the breed trend, reaching about 80 days, against current 145 days related to Italian Friesian cows [21].

5.3 Dairy, seasoning room and dairy products

At Petramartina farm, dairy consists of two distinct working units: one devoted to cheese cooling where there is a milk cooling tank, the other for its processing with a boiler, a table for curd processing and a press.

Nowadays the farm, one among Slow Food Presidia, manufactures several dairy products:

- U Cabanin: matured cheese (60 days at least) from raw milk
- Nina:, a fresh 'caciotta' from raw milk
- Velleja: a fresh ricotta (sort of cottage cheese)
- Yocab: whole white yoghurt or with honey from the farm apiary
- Mamma d'oro: a creamy cheese to spread, particularly suitable for children
- Primo sale of Cabannina a refreshing light cheese
- Squacquero of Cabannina a creamy cheese

Figure 7. Petramartina dairy.

Since when named among Slow Food Presidia, Petramartina farm has been able to produce many different dairy products from just one ingredient, milk, which meets with the recent consumers' demand for products with a characteristic taste on which they can indulge. Physiologically, Cabannina breed cannot compete with milk production from Frisian. Being its quantity of milk per lactation equal to 26-30 kg, Cabannina raw milk sale would not be economically relevant. Nonetheless, the percentages of fat (about 3.7%), proteins (about 3.3%) and lactose (5.3%) make it particularly suitable for producing very good cheese, as recently confirmed by a recent study [11]. Briefly, Cabannina breeding is now possible thanks to its highly differentiated and tasty dairy products.

5.4 Pasture management and diet

The Ligurian hinterland is characterized by small sloping plains among its woods. Cabannina cows have very strong and resistant hoofs and a low live weight (about

4.5 quintals), and can move on this territory without great difficulties. For their correct management, a breeder should divide pasture into different areas where the animals can alternatively rotate to be provided with a constant supply of organic material, to avoid excessive exploitation of vegetation and to ensure a good cleaning of the whole undergrowth. Being highly adaptive, animals of this breed graze looser most part of the year, but from February to March and from October to November, depending on the weather.

The typical diet of Cabannina cows consists of pasture forage. During winter, when kept inside in stalls, the farmer must provide their whole daily ration, based only on hay originated from areas in Aveto Valley. Lactating cows are given an integration all year long at milking time, i.e. a feed supplement specifically made for this breed derived from raw materials typical of pastures in these areas (table 10).

Such a feed has an average protein content; it is to be administered daily with fodder as much as 2% of each animal body (30% maximum according to U Cabanin production protocols).

So far no studies on Aveto Valley forage and related nutritional value have been conducted, which makes it impossible to consider how the Cabannina diet is balanced. Nonetheless, such forage is pivotal to the exploitation of this breed whose production results to be constant even though it only feed on fodder from the pasture they graze.

Table 10. Raw materials from which Cabannina special feed is derived.

Raw materials	Analytical components T.Q.:
wheat meal	protein 15,5%
(small/young) broad beans	Oils and fat 3%
wheat bran	crude cellulose 7%
maize	ash 7%
barley	
sugar cane treacle	
calcium carbonate	
Bicalcium phosphate	
Magnesium carbonate	
Sodium bicarbonate	
farina glutinata di mais	
vitamine e provitamine (vit.A, vit.D3, vit.E, vit.B1)	
oligoelementi (Mn,Zn,Fe,Cu,I,Cb,Se)	

6. CONCISE SUMMARY

Petramartina farm exemplifies a contemporary agro-ecological model, as evidenced by research that directly compare it to the principles of agro-ecology established by Altieri. (2012) [11] (table 11):

Table 11. Principles of Agro-Ecology by Altieri (2012) [11]

Pertramartina Farm	Yes	No
Enhancing biomass recycling, with a view to optimizing organic matter decomposition and nutrient cycling over time.	X	
Strengthening the " immune system" of agricultural systems by enhancing functional biodiversity, natural enemies, antagonists, etc.	X	
Providing the most favorable soil conditions for plant growth, particularly by managing organic matter and by improving soil biological activity.	X	
Minimizing losses of energy, water, nutrients and genetic resources by enhancing conservation and regeneration of soil and water resources and agro-biodiversity.	X	
Diversifying species and genetic resources.	X	
Improving beneficial biological interaction and synergies among the components of agro-biodiversity, thereby promoting key ecological processes and services.	X	

The farm totally follows the above mentioned agro-ecological principles.

Similarly, we can compare it with the questions proposed by Koohafkan (2011) [12] (table 12):

Table 12. Questions to define an agro-ecological farm (Koohafkan, 2011) [12]

Petramartina Farm	Yes	No
Do they reduce poverty?	X	
Are they based on rights and social equity?	X	
Do they reduce social exclusion, particularly for women, minorities and indigenous people?	X	
Do they protect access and right to land, water and other natural resources?	X	
Do they favor redistribution of productive resources?	X	
Do they substantially increase food production and contribute to household food security and improved nutrition?	X	
Do they enhance families water access and availability?		X
Do they regenerate and conserve soil and increase soil fertility?	X	
Do they reduce soil loss/ degradation and enhance soil regeneration and conservation?	X	
Do practices maintain or enhance organic matter and the biological life and biodiversity of the soil?	X	

Do they prevent pest and disease outbreaks?	X	
Do they conserve and encourage agro-biodiversity?	X	
Do they reduce greenhouse gas emissions?	X	
Do they increase income opportunities and employment?	X	
Do they reduce variation in agricultural production under climatic stress conditions?		
Do they enhance farm diversification and resilience?	X	
Do they reduce investment costs and farmer dependence on external inputs?	X	
Do they increase the degree and effectiveness of farmer organizations?	X	
Do they increase human capital formation?	X	
Do they contribute to local/ regional food sovereignty?	X	

The merits of Petramartina farm are obvious, with particular reference to its strong relation with territory, animals and humans. The main economic returns of this business undoubtedly are due to:

1. a low use of water for field irrigation and for animal management in the stables;
2. a reduced consumption of fossil fuel during different stages of feed preparation and preservation.

Some critical points can be identified:

1. in the current lack of direct production of forage; in fact, for the winter period it has to be purchased from other local businesses;
2. hay integration through the use of a feed supplement does not fully meet the agro-ecological principles, but the issue is absolutely irrelevant when compared to its irrational use in industrial breeding.

Noticeably, Petramartina farm is soon going to implement its properties by acquiring new pastures in the aim of improving the territorial hydrogeological structure. Moreover, farming is also designed to become an essential element in promoting environmental education at any school level for the east Ligurian territory through adequate didactic structures. A further step towards preservation and spreading of Cabannina breed and its derived products, as well as the creation of new jobs in the farming and marketing areas.

REFERENCES

1. Roxström A, Strandberg E, Berglund B, Emanuelson U, Philipsson J. Genetic and environmental correlations among female fertility traits and milk production in different parities of Swedish Red and White dairy cattle. Acta Agriculturae Scandinavica, Section A - Animal Science 2010a;51(1) 7-14. http://www.

tandfonline.com/doi/abs/ 10.1080/090647001300004745 (accessed August 19, 2012)

2. Sørensen AC, Lawlor T & Ruiz, F. A survey on fertility in the Holstein populations of the world. In: Proceedings of the Int. Conf. on Fertility in dairy cows "EAAP Satellite Meeting", pp. 1:17, 30-31 August 2007 Liverpool Hope University, UK; 2007.

3. Roxström A, Strandberg E, Berglund B, Emanuelson U, Philipsson J. Genetic and environmental correlations among female fertility traits, and the ability to show oestrus, and milk production in dairy cattle. Acta Agriculturae Scandinavica, Section A - Animal Science 2001b; 51(3) 192-199. http://www.tandfonline.com/doi/abs/10.1080/09064700118617 (accessed August 19, 2012)

4. Carlén E, Strandberg E & Roth A. Genetic Parameters for Clinical Mastitis, Somatic Cell Score, and Production in the First Three Lactations of Swedish Holstein Cows. Journal of Dairy Science 2004;87(9) 3062-3070. http://download.journals.elsevierhealth.com/pdfs/journals/0022-0302/PIIS0022030204734396.pdf (accessed August 19, 2012)

5. Ingvartsen KL, Dewhurst RJ & Friggens NC. On the relationship between lactational performance and health: is it yield or metabolic imbalance that causes diseases in dairy cattle? A position paper. Livestock Production Science 2003;83 277-308. http:// (accessed August 19, 2012)

6. Collard BL, Boettcher PJ, Dekkers JCM, Petitclerc D & Schaeffer LR. Relationships between energy balance and health traits of dairy cattle in early lactation. Journal of Dairy Science 2000;83 2683- 2690. http://download.journals.elsevierhealth.com/pdfs/ journals/0022-0302/PIIS0022030200751629.pdf (accessed August 19, 2012)

7. Sgoifo, C.A.; Franciosi, M.; Compiani, R.; Caprarotta, L.; Vandoni, S.(2010). Malattie delle lattifere, ma quanto ci costano? L'Allevatore magazine. Anno LXVI, 2:53-61.

8. Sandri S, Summer A, Tosi F, Mariani MS, Pecorari M, Franceschi P, Formaggioni P, Pederzani D & Malacarne M. Influence of somatic cell content on dairy aptitude of milk. Scienza e Tecnica Lattiero-Casearia 2010;61(1): 5-18.

9. Conway GR, Pretty JN. Unwelcome harvest: agriculture and pollution. Earthscan publisher, London; 1991.

10. Altieri MA. Agroecology: the science of sustainable agriculture. Westview press, boulder, CO; 1995

11. Altieri MA. The scaling up of agroecology: spreading the hope for food sovereignty and resiliency. 2012. ftp://ftp.fao.org/ag/agp/ca/CA_CoP_June12/Rio20-Altieri.pdf (accessed August 19, 2012)

12. Koohafkan P, Altieri MA, Gimenez EH. Green Agriculture: Foundations for Biodiverse, Resilient and Productive Agricultural Systems. International Journal of Agricultural Sustainability 2011;10(1) 61-75. http://www.tandfonline.com/doi/pdf/10.1080/14735903.2011.610206 (accessed August 19, 2012)

13. Achilli A, Bonfiglio S, Olivieri A, Malusa A, Pala M, Kashani BH, Perego UA, Ajmone-Marsan P, Liotta L, Semino O, Bandelt H, Ferretti L, Torroni A. The Multifaceted Origin of Taurine Cattle Reflected by the Mitochondrial Genome, Plos One 2011;4(6) 5753.http://www.plosone.org/article/info%3Adoi%2F10.1371%2Fjournal.pone.0005753 (accessed August 19, 2012)

14. Loftus RT, MacHugh DE, Bradley DG, Sharp PM, Cunningham P. Evidence for two independent domestication of cattle, Proceedings of the National Academy of Sciences USA;1994. 91 757-2761. 1994. http://www.pnas.org/content/91/7/2757.full.pdf +html (accessed August 19, 2012)

15. Troy CS, MacHugh DE, Bailey JF, Magee DA, Loftus RT, Cunningham P, Chamberlain AT, Sykes BC, Bradley DG. Genetic evidence for Near-Eastern origins of European cattle, Nature 2001; 410 1088-1099. http://www.nature.com/nature/journal/v410 n6832/abs/4101088a0.html (accessed August 19, 2012)

16. Achilli A, Olivieri A, Pellecchia M, Uboldi C, Colli L, Al-Zahery N, Accetturo M, Pala M, Hooshiar Kashani B, Perego UA, Battaglia V, Fornarino S, Kalamati J, Houshmand M, Negrini R, Semino O, Richards M, Macaulay V, Ferretti L, Bandelt H, Ajmone-Marsan P, Torroni A. "Mithocondrial genomes of extinct aurochs survive in domestic cattle", Current Biology 2008; 18(4) R157-R158. http://www.sciencedirect.com/science/article/pii/S0960982208000651 (accessed August 19, 2012)

17. Picone L. Dove pascola un erede del Bos Primigenius, La Casana, 2010;3 34-37. http://books.google.it/books?hl=it&lr=&id=XpJfGzhbdLwC&oi=fnd&pg=PA34&dq=picone+2010+Dove+pascola+un+erede+del+Bos+Primigenius&ots=KC14XfE9O1&sig=RPq-tB1al9bWmNIX-MKtbNSNqEQ#v=onepage&q&f=false (accessed August 19, 2012)

18. Maimone M. Studio sulla valorizzazione di due razze bovine autoctone dell'Appennino settentrionale: Cabannina e Ottonese-Varzese. Agriculture degree thesis, University of Milan; 1981.

19. Orefice G. Studio dei caratteri morfologici e produttivi di una razza bovina autoctona della Valle D'Aveto. Veterinary Medicine degree thesis, University of Pisa, 1978.

20. Usai G. Ricerche biometriche sulla razza bovina Cabannina in provincia di Genova, Specialization thesis, University of Milan, 1974.

21. Guida S.. Caratterizzazione morfologica e studio della curva di lattazione della razza bovina cabannina. Veterinary Medicine degree thesis. University of Parma, 2011.

22. Communod R, Faustini M, Chiesa LM, Torre ML, Lazzati M, Vigo D. Milk biodiversity: future perspectives of milk and dairy products from autochthonous dairy cows reared in northern Italy. In Food Production - Approaches, Challenges and Tasks. Rijeka: Intech; 2012. P169-184. Avaible from http://cdn.intechweb.org/pdfs/26522.pdf (accessed August 19, 2012)

23. Communod R. Bovine Lactation and Reproduction Physiology: A Survey on Some Fundamental Features In Autochthonous Dairy Breeds Reared in Northern Italy, PhD thesis. University of Milan; 2012.

24. Berton A, Rouvellac S, Robert B, Rousseau F, Lopez C, Crenon I. Effect of the size and interface composition of milk fat globules on their in vitro digestion by the human pancreatic lipase: Native versus homogenized milk fat globules, Food Hydrocolloids 2012, 29: 123-134.

25. http://www.sciencedirect.com/science/article/pii/S0268005X12000471 (accessed August 19, 2012)

26. Lopez C. Focus on the supramolecular structure of milk fat in dairy products, ReproductionNutritionDevelopment2005,45:497-511.http://rnd.edpsciences.org/index.php?option=com_article&access=standard&Itemid=129&url=/articles/rnd/pdf/ 2005/04/r5404.pdf (accessed August 19, 2012)

27. Isaacs C.E. The antimicrobial function of milk lipids, Advances in Food and Nutrition Research 2001, 10: 271–285. http://ukpmc.ac.uk/abstract/MED/11795045/ reload=0;jsessionid=3oVWMaB4iKWbHa7RWIkr.0 (accessed August 19, 2012)

28. Elwood PC, Pickering JE, Givens DI, Gallacher JE. The consumption of milk and dairy foods and the incidence of vascular disease and diabetes: an overview of the evidence.Lipids2010,45(10):925-939.http://www.springerlink.com/content/ k2475344067821um/ (accessed August 19, 2012)

29. Clément M, Tremblay J, Lange M, Thibodeau J, Belhumeur P. Whey-derived free fatty acids suppress the germination of Candida albicans in vitro, FEMS Yeast Research 2007, 7: 276–285. http://onlinelibrary.wiley.com/doi/10.1111/j.1567-1364.2006.00166.x/ full (accessed August 19, 2012)

30. Desbois PA, Smith JV. Antibacterial free fatty acids: activities, mechanisms of action and biotechnological potential, Applied Microbiology and Biotechnology 2010, 85: 1629-1642. http://www.springerlink.com/content/2324w3453025n730/ (accessed August 19, 2012)

31. Iguchi K, Okumura N, Usui S, Sajiki H., Hirota K, Hirano K. Myristoleic Acid, a Cytotoxic Component in the extract from Serenoa repens, induces apoptosis and necrosis in Human Prostatic LNCaP Cells, The Prostat Journal 2001, 47: 59-65. http:// onlinelibrary.wiley.com/doi/10.1002/pros.1047/abstract (accessed August 19, 2012)

32. Zanetti G, Martelli P, Gorreri M, Pecorari M, Mariani P. Contributo alla conoscenza della patogenesi della sindrome "Ipoacidità del latte": rilievi emato e galattochimici, Atti Società Italiana di Buiatria 1987, 19: 195-202.

33. Michalsky M, Gassi JY, Famelart MH, Leconte N, Camier B, Michel F, Briard V. The size of native milk fat globules affects physico-chemical and sensory properties of Camembert cheese, Le Lait 2003, 83(2): 131-143. http://lait.dairy-journal.org/index.php?option=com_article&access=standard&Itemid=129&url=/articles/lait/abs/ 2003/02/L3203/L3203.html (accessed August 19, 2012)

27 Chapter
THE SFIN INNOVATION SYSTEM - INSIGHTS ON FOOD CLUSTER ADMINISTRATION

1. INTRODUCTION: AN OVERVIEW

The Skane Food Innovation Network aims to facilitate innovation within the food industry, fostering improved food quality, enhanced competitiveness for regional food enterprises, and a more robust culinary identity for the area. This chapter seeks to elucidate the management of a Triple Helix cluster project in relation to innovation. Skane Food Innovation Network (SFIN) is a triple helix cluster initiative situated in southern Sweden. SFIN signifies a novel form of innovation system. Government assistance to individual firms is prohibited by EU rules, necessitating a transformation in the structure and administration of Swedish public innovation support. Currently, such money is allocated to regional industry clusters rather than to individual enterprises. This has stimulated the advancement of the Swedish innovation system. SFIN is regarded as a leader in this advancement due to its sophisticated and coherent triple helix model of innovation (OECD, 2012).

The chapter examines the functionality of the innovation system, both in reality and in theory. It relies on extensive institutional and industry legitimacy, twelve distinct networks centered around various nodes in the value system, the interaction between systemic innovation meetings and the development of pilot projects, and importantly, the active participation of entrepreneurs and customers, along with the shared value generated through their collaborative efforts. SFIN has adopted a systematic approach to invention—detect, develop, and diffuse—advancing towards a new innovation system.

The chapter expands on the concept of "bridging regimes" (Jönsson et al, 2011) and posits that innovation may be viewed as a collaborative cognitive activity. This scenario illustrates the significance of recruiting intermediaries capable of bridging the disparate regimes inherent in the actors, thereby engaging them in the development of creative solutions. Innovation inside an existing organization is particularly described and analyzed (Govindarajan and Trimble, 2010; Kotter, 1985). By innovation itself is meant any process or product that sells and creates new value in a market (Edquist, 2002). In theory, the innovation system may be described as a collective cognitive process. The chapter connects theoretical underpinnings from strategy theory and market theory with models of change, institutionalization and culture. Strategy is discussed as a consistent and long-term set of activities (Porter, 1996) and a clear and understandable modus operandi (Miles and Snow, 1984; Drucker, 1994). The system creates a "liquid environment" and increases the "adjacent possible" (Johnson, 2010), where ideas are detected and further developed in a market, defined as any self-referencing group with certain needs in common (Moore, 1991). Such a market view is consistent with the idea that any innovation established through replication (Christensen et al., 2006) in order to survive. Once the change is made (Kotter, 1996)– the developed new value creation is accepted, widespread and taken for granted, the new solution is institutionalized (Veblen, 1904; Hodgson, 1988; Scott, 1995) and part of a specific culture on one or more levels in society (Schein, 2010). The innovation process involves connecting regimes, fundamentally controlling stakeholder perceptions through the three phases of detection, development, and diffusion across the three tiers of individual, organization, and society.

1.1 Shaping the shaping of the future food industry

How do you shape the future? As a matter of fact, you don't have to do anything. It will shape itself around you, with or without your involvement or approval. But if you have an idea of what you want to see in the future, there is a challenge ahead. The Skane Food Innovation Network works to help the food industry innovate, to make the future happen in a way that includes better food, increased competitiveness for the food companies within the region and a stronger culinary profile of the region. This chapter aims at presenting and discussing how a Triple Helix cluster initiative can be managed towards innovation.

This chapter is about regional economic development and innovation is a key concept. It is used in the meaning of any process or product that sells and creates new value in a market (Edquist, 2002). The text revolves around the development of Skane Food Innovation Network (SFIN). In 2003 - winning the competition Vinnväxt - SFIN was boosted by national funding the coming 10 years. The objective was to systematically facilitate innovation in the since long established food cluster in the south of Sweden, also known as Skane. Reading the official program description today, almost ten years later, it is strikingly void of normative guidelines of how to go about creating the innovation system. The purpose of the program, called Regional growth through dynamic innovation systems, is

"to promote sustainable growth in the regions based on international competitive ability, by successively developing or further developing the functioning, dynamics and effectiveness of innovation systems in functional regions at an international level. A prerequisite for the programme is the active participation of players from the business community, research organisations, politics and public administration."

The concept of Triple Helix is mentioned in the original directions. The collaboration between business firms, universities and public sector organisation, innovation will result.

This view of the innovation process could be interpreted in two opposite ways. On one hand, you could expect innovation as a result of the "adjacent possible" in a "liquid environment" (Johnson, 2010), where the mere coexistence of different individuals gives the future new combinations of ideas, DNA, artefacts or other things. This is what Edquist and Hommen (1999) call the "systems oriented" view on innovation processes. On the other hand, there is a strong contender in the "linear" perspective on innovation. Innovation is unidirectional process, where research and development efforts is initiated by public authorities or private firms, which is expected to produce new technology, which in turn provides new solutions to market needs.

1.2 Purpose and approach

The purpose of this chapter is to describe and interpret the development and the dynamics of the way SFIN facilitate innovation.

By innovation is meant any process or product that sells and creates new value in a market (Edquist, 2002). In theory, the model may be understood in terms of strategy as a consistent and long-term set of activities (Porter, 1996) and the need for a clear and understandable modus operandi (Miles and Snow, 1984; Drucker, 1994). The system creates a "liquid environment" and increases the "adjacent possible" (Johnson, 2010). A market is any self-referencing group with certain needs in common (Moore, 1991), in which any innovation has to be established through replication (Christensen, 2006) in order to survive.

A major impediment for innovation tends to be the idea that things should be done the way they always have been done. Innovations therefore often occur in settings where actors with different backgrounds join up with a mutual interest to solve the same problem. However, things can also go terribly wrong if the different backgrounds clash instead of fertilizing each other. Sidney Winter and Richard Nelson (1982) have discussed the importance of studying regimes in order to understand why or why not innovations tend to happen. Winter defines regimes in a sector as a specific set of not only regulative institutions and norms but regimes also regulate codified formal as well as tacit informal habits and routines related to common collective and individual practices and beliefs. These practices and beliefs shape and coordinate actions between various groups, individuals, and organizations in the sector. An important role for organisations working with facilitating innovation may be to work for "bridging regimes", as argued in Jönsson et al, (2011).

We are both active in social sciences, business administration and ethnology. We use our set of models of concepts as "temporary walking sticks [to] aid sense making as we go along" (Ghoshal, 2005). Ghoshal questions the pretense of knowledge, resulting from the idea that social sciences should be formed, tested and verified in the same way as natural science and causal theories are. Thus, our eclectic approach is a start trying to understand and verbalise the shaping of the shaping of the future. The approach is to use an eclectic frame of reference, where different models and concepts in the field of business administration are put together in order to show how concepts of strategy, markets, institutions and culture have human understanding and acceptance in common.

Our effort could be labelled action research (Lewin, 1946, Rapaport, 1970, Susman and Evered, 1978). Action research is defined by Rapaport (ibid.:499): *"Action research aims to contribute both to the practical concerns of people in an immediate problematic situation and to the goals of social science by joint collaboration within a mutually acceptable ethical framework."*

The empirical account of the chapter is based on our direct experiences of working inside SFIN since 2004. We both started off leading different projects for a few years. Since 2008 we have been part of the management team of SFIN, each heading a separate large operation. In 2010, the Skåne Regional Council, one of the major stakeholders of SFIN, decided to allocate the initiative Taste of Skane inside SFIN. The assignment is to develop Skane in terms of food experiences - tourism, restaurants, food education, and much more. In 2011, the same organisation gave SFIN the assignment of innovating the hospital meals, starting with the small regional hospital in Trelleborg where a role model is to be presented in early 2014.

So, from an eclectic frame of reference and first hand empirical experiences, this story will serve two general purposes - making sense of cluster management in SFIN and advocating the use of individual and common understanding of the subject of change. Individual come collective understanding, as we aim to show, is at the core of innovation.

2. SKANE FOOD INNOVATION NETWORK (SFIN)

2.1 The development phases of SFIN

During most part of the 20th century, the Swedish food market has been protected from international competition. The idea of national self-subsistence used to guide Swedish food policy. International competitiveness was not a major issue, since surplus production was limited. During the last decades, though, the Swedish food industry has been rapidly changing from a sheltered national industry into an industry exposed to strong international competition (Lagnevik 2006). In 2003, VINNOVA, the state agency for innovation and systems- oriented research, launched a regionally oriented program for research, technology and deployment/ demonstration (RTD). The aim was to promote upgrading and renewal of local innovation and R&D capabilities and skill building in certain important growth areas with strong regional profiles. One initiative within the Vinnväxt (Winn-Growth) program was granted to the Skane Food Innovation Network, a triple Helix network formed in Skane, Sweden's "bread basket" with approximately half of the nation's food production and R&D.

After winning the Vinnväxt competition in 2003, this first period of the "cluster initiative" was research oriented. This "research phase" saw a large share of the funding funnelled into research and PhD student projects. It was characterised by a linear view of the innovation process. As a triple helix organisation, the reactions from the business part of the stakeholders gradually expressed doubts about how the process should gain their companies. The inherent tension between long-term research and short-term profitability pressures of larger companies needed to be resolved.

The former CEO of SFIN retired in 2006, and a new one was recruited. From being an anonymous "industry body", SFIN now got a person with a strong background working with one of the most successful and innovative Swedish consumer products in a long time. Around 2007, "the entrepreneurial phase" began. Now there was an increased support of entrepreneurs, seen as innovation synapses. It could be regarded a reflex of large companies' urge for here-and-now innovation. SFIN now changed into a rhythm of making more frequent decisions. Each decision, though, concerned smaller amounts of money. In retrospect, the development could be described as the recognition of the systems view on innovation. The typical decision was about a business plan or project plan, and a subsequent need for financial support for implementing it. This kind of decision-making turned out to be very difficult, sometimes almost ad hoc, and the organisation then was struggling for criteria from which to evaluate the proposals. The result, and the problem, was that an inconsistency emerged, no other logic than stimulating individual entrepreneurs in order to create innovation, new firms and more jobs.

The orientations seen in this first half of the Vinnväxt program are direct reflections of what was evaluated and, hence, expected from SFIN by Vinnova. The new CEO, slightly flabbergasted, turned to the board for directions. None was given. Instead, the board and CEO together began articulating a strategic orientation for SFIN. The approach used was recognised from straightforward strategic planning and the traditional management of business. State a vision, formulate a mission and set goals and strategies in order to fulfil the mission. The process was clarifying the ideas and purposes of SFIN. A new organisation structure was created and still holds sway today.

2.2 The shape of the SFIN innovation system

The Skane Food Innovation Network is registered as a non-profit organisation. Its constituents are a number of partners, today about 40, and members, today more than 100. Partners pay an annual fee of about 4 000 euro, whereas members pay 300 euro. Partners qualify for board representation, members do not. The board mirrors the triple helix view - large and small companies, universities, and public authorities are all represented by the eleven board members. The Governor of Skane is chairman of the board.

The board has formulated a so called ”VAMS”, an acronym for Vision, Business Idea (Swedish: Affärsidé), Objectives (Swedish: Mål) and Strategies.

The vision states that SFIN is shaping the future food industry and meal experiences. The business idea is to offer the best network for cooperation of different competences, stage pilot projects and gain credibility from a concrete track record of innovation. From the high credibility, SFIN sustain and develop its network and receives funding from national and international authorities.

The objectives revolve around maintaining the networks in an inclusive manner, and hence increase the attractiveness and deepen and widen the competence base of the industry. Strategies are concerned with deep understanding of future needs and questions in the industry, creating a clear and attractive offer to participate, attracting people with vision and drive, organising meetings between companies, entrepreneurs, universities and industry organisations, focusing commercialisation of innovations, taking the role as the network hub and constantly communicating our work and existence.

The operating core is organised in six different areas, each headed by one person. The areas cover as follows:

- Future Strategy: trend spotting, stakeholder relations, governance issues, finance and funding
- Career: student activities, trainee program, attraction of the food industry
- Entrepreneurship: coaching entrepreneurs in the industry and the adherent support systems
- Public Meals: increasing the status and competence of the public meal sector
- Packaging: develop innovative food packaging, small-scale, local producers etc
- Regional Food and Tourism: develop small-scale food producers, food tourism etc

The CEO and the area managers form a management team, also including the communications manager and one representative each from the two largest public stakeholders and funders, Region Skane and the County Council. The management team meet regularly three hours once a week.

From the different areas, various networks are formed. This model is an offspring from the initial CEO network, created by the new CEO in 2006. These networks are gatherings around specific topics, functions or roles in the system. Currently, 12 different networks meet more or less regularly. Apart from the CEO network, SFIN operates a communications network, a public meal network, a retailers' network, an HR network and seven more networks.

The logic is straightforward. In the networks, different ideas surface. Vital ideas, surviving the initial scrutiny of the network itself, are brought to the management team. The management team discuss the innovation potential of the idea. Today, there is a clear-cut evaluation criteria. The innovation has to be of a systemic character. There has to be a dedicated entrepreneur or coalition ready to test it in a pilot project. Finally, there has to be realistic plans of how the pilot project could reproduce itself, on commercial merits.

In summary, SFIN has over the past three years been compressed into six "business areas", a coherent network structure, a management control model and three distinct activities undertaken in order to create the future food industry and meal experiences: we detect a need or potential for innovation, we develop the solution in pilot projects, and we diffuse them to a wider market. This is the SFIN innovation system. This particular "modus operandi" has produced a wide variety of both short-term projects and long-term signature co-operations. The first one was an industry trainee program in open innovation. Another is the training program for elderly homes, "Meal Pleasure for Elderly". A third example is the retail concept "Locally Produced and Carefully Selected".

On a deeper level, these different examples represent a maturation and consolidation of the early years of wayward - yet legitimate - experimentation. The strategic planning process undertaken in 2008 apparently missed out on the emotional aspect of change by avoiding making a mission statement. The organisation probably was not ready for it, but now the board and management team increasingly speak in terms of social responsibility, sustainability and social capital. The fourth example of this, and a strong confirmation of the credibility that SFIN has earned over time, is the most recent pilot project situated at the Trelleborg hospital. In the next section, the case of Meal Pleasure at the Trelleborg Hospital, will serve as a more detailed look into the dynamics of the SFIN innovation system.

3. THE CASE OF MEAL PLEASURE AT THE TRELLEBORG HOSPITAL

3.1 The background

In 2009, the public foodservice sector was slowly introduced into the SFIN operations. It was obvious that this was a forgotten part of the Swedish food industry. Half of the industry is foodservice, half is retail and consumer-oriented. Half of the foodservice sector is private restaurants and service sector offerings (gas stations, Seven Elevens and the likes), the other half is public sector meals. The major part is served in schools, hospitals and elderly care. Thus, this forgotten part constitutes one quarter of the entire demand in sales value, even more in terms of number of people being served. It is a rather consistent segment from a supplier's point of view, where public procurement law has streamlined the procurement of food and meal solutions. The guests, though, is a completely different thing. The elderly cannot be segmented by age, nor can school kids or patients in hospitals. This is a very demanding part of the food industry with diverse preferences and, on average, weak buying power.

The public foodservice sector is furthermore signified by a low average educational level, a predominantly female labour force, low pay, low rate of training and education on the job and the use of traditional cooking techniques. The most obvious example is the use of high tech ovens, which in fact are small "computerised food factories". Such an oven is expensive. Still it is often used merely as a traditional oven, a few hours a day, five days a week, due to lack of education and incentive.

One particular problem, unique to the public foodservice sector, is the obligation to follow the public procurement law. One important purpose of the law is to optimise the use of tax- payers money and avoid nepotism. This is done by organised procurement using the competitive forces in a marketplace. This works, unless there is a dysfunctional market. In the Swedish food industry, both the retail and the foodservice market are oligopolistic markets. In the foodservice case, there are two large dominating suppliers. These have been continuously intimidating municipalities and counties by frequent and systematic legal actions and court over-rulings of public procurement processes, a tactic aiming at coercing the public organisations to behave in a certain way (likely as to benefit large-scale suppliers).

The educational level of the staff and the status of the foodservice operations are related to the problem of public procurement. Increasing occupational status and educational level could make the public foodservice customers more demanding, in that way increasing the innovation pace in the Swedish foodservice industry. That is the hypothesis that motivated SFIN to start working with the public sector.

3.2 The rationales of the Trelleborg project

Region Skane, the county council, is responsible for healthcare, transportation, cultural and regional development of Skane. It has offered continuous support of

SFIN since the start in 1994. Gradually, the confidence in SFIN has increased. In 2009, with 2012 looming, the need for a new procurement of hospital meals for about two thirds of the healthcare organisation was pressing the politicians. The same year, a hospital food process investigation was initiated. The result was a vision for making Skane a role model in food and meals in Northern Europe by 2025.

It was an ambitious vision and it was necessary to take action. A senior foodservice "celebrity" working for SFIN saw it coming and used her leverage in the political sphere in Region Skane to suggest a pilot project to "walk the talk". Said and done, a pilot project in the small local hospital of Trelleborg; was decided upon. There was an existing kitchen ready to use. "Just dust it off and get it running", we were told.

Apart from starting up a dormant kitchen and start cooking tasty food, the short-term goal was, and still is, to innovate the hospital meal experience and to integrate the meal into the healthcare operation. In other words, we got an opportunity to give the meal a higher status in hospitals. The longer-term goal was set to disseminate the idea, to diffuse the model. This is done in SFIN by way of entrepreneurial incentive. This was decided in the early summer of 2011, just before the holidays. In August that year, the adventure started.

3.3 The SFIN networks paying off

SFIN got the assignment from the board of Region Skane, along with funding for two years. This funding covered SFIN expenses for the project, not for investments or employment of staff or other Region Skane-related costs. The line-up from SFIN was possible to achieve as a direct result of the network organisation. Various SFIN projects in the past had made it possible for a rather unique and unorthodox set of project members to form.

The common denominator of this group is a genuine interest in making a difference. Normally, there is an obvious attraction to public funding as "easy money", but these people all work long-term with SFIN on reasonable market-based terms.

Initially, the SFIN manager Knutsson worked as a project owner in SFIN alongside with two experienced and highly respected persons. Together, these three people formed a project core group and started to write a project plan. The project plan was accepted by the responsible politicians in late August 2011.

The project organisation consisted of a political steering committee, a managing steering committee, a work group and a reference group. SFIN participated in all groups. The rest of the participants varied, but represented either the Region Skane corporate level, the Region Skane business development, the Trelleborg hospital or Regionservice (the facility management part of Region Skane coordinating real estate, food, transportation, laundry, and postal services).

At the time in Region Skane, each hospital manager was responsible for all service operations, including the food process, and the organisation of the healthcare. Regionservice had the role of an in-house service integrator, coordinating the service activities in the entire organization.

There were three distinct stakeholders. The hospital of Trelleborg, Regionservice and SFIN. In this context, SFIN was a fairly unknown organisation of ("self-entitled?", some whispered) experts in the field, who had been given a powerful position in reshaping the hospital foodservice in Trelleborg and, more or less explicitly expected, the rest of the Region Skane hospitals. Adding to that, Regionservice had a rather bad reputation in the hospitals, not meeting expectations. Naturally, Regionservice representatives were suspicious and - perhaps - offended by the political decision to invite an innovation organisation from the outside "that no-one ever has heard of". Moreover, the project was running against the clock, as the contract with the external food supplier was to expire 31 August 2012. Regionservice, the coordinator and recipient of the SFIN innovations, had 15 months to go.

3.4 Colliding regimes – The first year's theme

The first meeting with a group from Regionservice and the Trelleborg hospital was held before the summer, before there was a project plan written. This was the first step in cooperating. The meeting was marked by confusion. There was no previous relations in the group to rely on. Knutsson, chairing the meeting, had nothing but a generally formulated political decision to hold on to. Focus was naturally put on the "dusting off" the hospital kitchen.

SFIN, represented by an experienced, renowned and respected chef, made an initial sketch of the kitchen layout on the basis of available blueprints. The Regionservice real estate division appointed a project leader to organise the starting up the kitchen. This was a young and inexperienced person, as fresh as they come, who got this assignment as her first project. The tight time schedule and the goal to create something entirely new clashed. Soon it was discovered that the previous check of the building, about one year old, had been rough and, as it showed, insufficient. The original estimation was 11 MSEK. It rapidly grew to speculations of 70 MSEK. The needed capacity was 500 meals per day. Any sound investment calculus became impossible. Thereafter, Regionservice got silent and started its own investigation of the Klippan kitchen without the participation of SFIN.

There was obvious tension between Regionservice and SFIN. The director of support and service functions in Region Skane was contacted and introduced to the need to strengthen the Regionservice commitment to the collaborative nature of the project. A collaboration document was written and acknowledged. But the Regionservice investigation was already on its way and the door was not opened to SFIN. At that time, it was also decided by the board of Region Skane that Regionservice should take on the direct responsibility for all support functions. That meant that the hospitals were relieved of the support functions, effective January 1, 2012. This stirred things up in Region Skane and the pilot project did not go unaffected. Regionservice got more powerful and the door to SFIN seemed firmly shut. In late January, the door came ajar, though. Suddenly, the real estate project leader forwarded a number of detailed questions from their kitchen architect and kindly asked that we respond the same day.

In the meantime, two people – one dietician and one gastrome – built a remarkable base of knowledge and learnings about the meals in the Trelleborg hospital. The explicit goal was to observe and analyse the potential and need for change in the meal organisation and the quality of the patient's meal experience. They worked in an exploratory fashion and became more and more independent of the project leader, working increasingly with the kitchen issue and the collaboration between Regionservice and SFIN. The first half of 2012 was a period where the project goals first got under real scrutiny. What does it mean to "innovate the hospital meal experience and to integrate the meal into the healthcare operation"? Frequently the phrase "establish a new norm for hospital food" was used in SFIN. That did not bring light to the more and more blurred question - "What do we have to do?" The pressure increased further from an increasing interest and high expectations of the project. And here we were, fumbling for structure and some guidance. And what about the food, where did that go in this huge project?

A hospital kitchen is an important part of the infrastructure. There is a certain minimum efficient scale, given the choice of building, production methods and technology. In the spring of 2012, there was hardly enough expected volume to guarantee an acceptable cost per meal. The projected daily volumes included 300-350 meals to the hospital and about 150 meals to a new psychiatric unit in Trelleborg (RPC), starting up in 2016. 500 meals per day rendered an unacceptable production cost and the entire credibility of the project was being questioned. In that situation, the municipality of Trelleborg caught the project's interest. In the municipality, there was a commonly known need to invest in new elderly care meal solutions. Contacts were made and a letter of intent was written in the spring. Left with a need to find out ways to share the meal organisation, this was a crucial step that in one blow could double the projected production volume in the kitchen. Also in the late spring of 2012, the political decision was made to invest in total 50,1 MSEK in the Trelleborg kitchen. And the project is delayed by one whole year. With the decision finally at hand, the building process got off to a new start. There was a new building meeting and in that meeting, there was a complete breakdown of communication between SFIN and Regionservice.

The demand and expectations from the hospital, RPC and seven different elderly homes in the town of Trelleborg turned out to be difficult to capture and align. The picture of the total demand was very blurred and the design of the kitchen and logistics turned out to be a bad compromise. A calm and focused foodservice manager in Regionservice now entered the project in order to strengthen the project in the eyes of Regionservice and, in some respect, mediate between SFIN and the real estate project group. In order to get hold of the production model, it became necessary to take a stand on the food issue - do we transport hot or cold food, or both? The decision was made by the experienced chef representing SFIN, in mutual understanding with the foodservice representative from Regionservice, to go with the cold alternative. The decision to work with chilled food was supported by the politicians, and opened up a possibility to solve the problem of supplying food to other hospitals in Skane. If the Trelleborg kitchen was expanded, it could serve so much more than merely the few patients and persons in Trelleborg. This, of course, had ramifications on the building project. But the political negotiations were kept secret. Knutsson was allowed to break the news of going for chilled food in a construction meeting. The impression was that SFIN had surrendered and resorted to the cook/chill method. In the meantime, a plan was outlined for a larger solution, which the politicians initially at first supported. However, one week later there was a 180° turn, and the old plan was only reconfirmed by the politicians. All of this wheeling and dealing went unmentioned to the real estate project group and it was no wonder that SFIN lost credibility in that group. That wasn't enough. The hospital management in Trelleborg had been completely forgotten in this intense and extremely pressured situation and was not happy.

3.5 Current state

As the last steering committee before the summer ended, no decision had been made to reorganise the project. As it happened, the project leader met with the service director/project owner the same morning. Referring the meeting, the project leader got an immediate positive response and the service director offered himself to chair the steering committee and to supervise a revising of the project plan and the project organisation.

So, as the summer is almost gone, the project is now divided into four distinct parts of the project (construction, food, meals, and continuous operations). In this scheme, SFIN has now got an explicit development and innovation responsibility for on one hand the kitchen design and a distinct food strategy, and on the other hand the meal experiences of patients, staff and visitors. The construction part of the project now resides directly under the service director. Articulated goals are now present in each respective part. All in all, order has been reinstated and roles and responsibilities are now much clearer.

In Trelleborg, base groups are formed within each clinic, and the hospital management is engaged in the formulation of the Trelleborg policy of patient meals. In that policy, the integration of hospital meals into the healthcare operation resides. The construction planning is underway, and a new, healthy menu is currently planned.

Of course, it remains to be seen, but by early 2014, SFIN and Regionservice will present a new way of working with food in the hospitals of Region Skane. The meal organisation is revamped, the status of the meals is increasing and the quality of the hospital food is dramatically improved by new production and distribution methods. Moreover, the cost of hospital meals is lower than before and the new kitchen turns the conception of high quality foodservice production upside down. From the rest of the country and from other parts of Europe, people start to benchmark and copy the new norm of hospital food, established by SFIN in tight cooperation with Region Skane. Fingers crossed.

3.6 Case study learnings

This case is a story where we want to show more details of the innovation process. We will dwell somewhat upon the difficulties we have experienced from working side by side, with an established organisation.

The story began with the importance of confronting the basic concept of value creation. Change stemmed from an understanding of (the potential for) value creation, in this case supported by external knowledge and experience, but also a strong political awareness and desire to improve the meals served at the Region Skane hospitals. The staff working in the areas in need of change did not, however, recognise the need to the same extent. This was an important learning in the project. Every organisation is shaped to produce a certain outcome. Ongoing operations create, over time, an inertia and inability to adapt to change. One of the participants in the project, a former foodservice manager, drily summed up the operation in one kitchen by saying "this is like it was 25 years ago when I left it".

This pointed to the need for implementing the wish for change to the staff. The success of an organisation is contextual, every organisation is part of a larger system. The supply chain of the hospital food operation relies on education, procurement, IT cooperation between different occupational groups and so on. To isolate and dissect a single part of a system in order to change only that particular part is doomed - the surrounding system will rein in the deviant.

Innovation is based, the case shows, on every individual's understanding, acceptance and estimation of the value of the "new thing". It was most difficult to convey to large groups of people that we shouldn't talk about just food, we had to take the entire meal into consideration. Hospital staff may have longed for "finally getting better tasting food" but the foodservice organisation had a clear view that "cook/chill food is lousy!". This was the real challenge of innovation, finding the motivation and mutual understanding among the kitchen staff and hospital staff at one hand, and the external innovation facilitating actors at the other. We needed to initiate and encourage change agents in a long-term process of changing attitudes and conceptions of how the meals in hospitals are valued and, hence, organised.

We tried to derive changing attitudes, values and conceptions to individuals influencing individuals. The case reveals the importance of tasting sessions, to organise "base groups" at every single clinic and to align top management support, middle management involvement and floor management dedication.

In essence, this is an argument for treating the innovation process as a cognitive process rather than the physical implementation of new work routines. The innovation in this case is the reconceptualisation of a hospital meal.

4. INNOVATION IN THE LIGHT OF STRATEGY, MARKETS AND INSTITUTIONS

Looking at the Trelleborg case from a strategic point of view may give further insight on how actions can be planned in order to succeed in innovation facilitation. First of all, the way that SFIN runs its course follows the strategic planning paradigm. But, just to conceptualise a management system in terms of a list of words will not suffice. From theory we can deduce a certain frame of reference, guiding both sense-making and action-taking. The argument is the following:

- strategy is about understanding how you create value to a customer
- the flipside of a good strategy is inertia and a compromised ability to adapt to change
- success is contextual, every organisation is part of a larger system
- innovation is based on every individual's understanding, acceptance and estimation of the value of the new thing
- individuals are influenced by other individuals;
- innovation, defined as "any process or product that sells and creates new value in a market" (Edquist, 2002) is built on a collective cognitive process;
- The SFIN innovation system is built around a systems view on innovation and concerns the management of cognitive processes triggered by the meeting between people.

4.1 Innovation and strategy

Strategy has an everyday connotation as a plan. It is implicitly rational, built on analysis and take its form as an ex ante decision to make something special to happen. There is a wide literature on the concept of strategy, but a convergence could be discerned from a number of scholars in the past. In 1975, the Swedish scholar Richard Normann (1975) defined the way a company could dominate competitors, i.e. strategy, as an interplay between the organisation, its products and its markets. Strategy, Mintzberg (1978) has defined as "a pattern in a stream of decisions". A continuation of that definition is recognised in Miles and Snow (1984) and Porter (1996), who define strategy as a matter of activity fit. What counts is what you do, in total.

Miles and Snow (ibid.) present fit in terms of integration of an organisation's strategy, structure and management processes. Misfit breaks down an organisation, whereas tight fit is "the underlying causal dynamic producing sustained, excellent performance and a strong corporate culture" (1984:10). Early fit is the discovery and articulation of new patterns of strategy, structure and process and fragile fit signals vulnerability to change. Tight fit, thus, is the objective for adaptation. Tight fit is built upon

- discovery and establishment of a seemingly easy way to work together towards a chosen goal,
- simplicity and increased understanding by incumbents
- reduced need for elaborate coordinating mechanisms, creating slack resources to be reallocated.
- the causal association between the way to work and achieved performance.

Miles and Snow combine the classic thoughts of Selznick (1957) with similar ideas of Normann (1975, 2001) and Porter (1980, 1985, 1996). The easy way to work, well understood by incumbents, could be regarded as a distinctive competence (Selznick, ibid.) and an interplay between the mental models of people, institutionalised organisational behaviour and the way a company runs its business. Normann (2001) indicates that a "business landscape", i.e. an industry structure and its prevailing business logic, could be a result of the "mental maps" of people. Drucker (1994) calls the same thing "the theory of the business". A later heir of these ideas is Jim Collins (Collins, 2001, Collins and Hansen, 2011), who in a massive empirical material derives an explanation of the success of certain companies. It is from disciplined people, thoughts and actions superior performance stems. In the healthcare organization of Region Skane, this is expressed in distinct occupational roles and a strong hierarchical organization. Even though people have to eat, and our wellbeing depends very much on what we eat, people in the medical profession seldom considers how the patient's eating affects its recovery. The "mental map" of healthcare needs to be redrawn.

4.2 Strategy, inertia and the difficulty of surprise

When, in SFIN, we speak about getting the unexpected to happen, there is one problem with innovation and strategy. Strategy is a matter of pattern, routines and familiarity. March (1991), among many others, is preoccupied with the choice a firm has to make between exploiting "old certainties" and exploring new possibilities: "The essence of exploitation is the refinement and extension of existing competences, technologies, and paradigms. Its returns are positive, proximate, and predictable. The essence of exploration is experimentation with new alternatives. Its returns are uncertain, distant, and often negative. Thus, the distance in time and space between the locus of learning and the locus for the realization of returns is generally greater in the case of exploration than in the case of exploitation, as is the uncertainty." (1991:85)

This frames the challenge of SFIN and other innovation organisations: how do you help an organisation close the gap in time and space, i.e. between innovation experiments and profitable growth of business? The wider the gap, the less interested an organisation is to bet on something other than the proven way. Christensen (1997) distinguish between disruptive and sustaining innovation, where sustaining innovation is an incremental development of the existing products or methods. Disruptive innovation changes the game, for the customer or the organisation or both. Digging further down the theoretical underpinnings of innovation, Argyris and Schon (1978) introduced the idea of single and double loop learning to a wider public. Learning in a single loop is to learn to do better what you already do. Double- loop learning occurs when you question why you are doing what you do in the first place - you test the assumptions, values and policies underlying the particular action.

So, the sustaining innovation is a matter of improvement of the existing solution to a problem. You run your thoughts through a single loop - how can we do this better? The degree of surprise by the unexpected is limited. Disruptive innovation, though, is by definition something other than the existing thing. Then, when you enter a "going concern", the accountant's label on an organisation expected to live long, and you should get the unexpected to happen, you could expect resistance.

In more practical terms, Govindarajan and Trimble (2010) differentiate between the main organisation, what they call the "performance engine", and the innovation organisation, the "dedicated team". In order to succeed with exploration, call it innovation, the dedicated team has to work alongside with the people in the performance engine. A partnership has to be established between the exploring and exploiting parts of the organisation.

The main issue, Govindarajan and Trimble argue, is the work relationships among the participating members and decision-makers. Work relationships are analysed in terms of depth, power balance and operating rhythm. Deep relationships are defined as a continuous and tight relationship where two or more people work together with a specific problem. Power balance is a matter of influence and having the final say. It doesn't rely on individual personal authority, but rather the primacy of certain organisational functions over other. Last, the operating rhythm concerns the pace and intervals in which the cooperating organisations work - if one part work in a monthly budgetary cycle and the other in a three-year development process, the operating rhythm will cause the work relation to wither.

What Govindarajan and Trimble teach us in SFIN is, at the very least, that as soon as we enter an existing organisation with an innovation mission, we must take good care of the work relationships we establish. Infrequent, casual meetings with imbalanced power distribution - one way or the other - and differing time horizons, we are by and large bound to fail to innovate. If we don't get to understand each other, the existing view of reality will prevail. In order to redraw the mental maps of the hospital staff to include the hospital meals as an integrated part of the healthcare, we need to get our "partners" in the "performance engine" Regionservice and the Trelleborg hospital to take the patient meals in a double loop – why do we serve meals in the first place? We have to do this in close work relations, where we need to engage deeply in the realities of kitchen staff and healthcare staff, so that we can better understand how people think about and value the work they do. The change doesn't start from our standpoint, it starts from theirs.

4.3 Innovation, markets and other institutions

The problem of innovating inside an existing organisation could be further discussed in terms of diversity and interdependence in the organisation (Kotter, 1985). Kotter claims that the larger the organisation, the harder it is to run it efficiently. If we look closer to his explanation, again work relations between people is the object of interest. His argument goes that a smaller organisation, with a narrow scope of technology, products and markets, creates less confusion and disputes than in larger organisation. The larger organisation depends on many people working together, so when interests divert among these interdependent individuals, the organisation gets stiff and bureaucratic. Kotter ponders:

"it is not by chance that the economist's traditional model of a firm, where only 'rational' economic decision making occurred, and where power struggles and politics were nonexistent, was a small and technologically simple organization that operated in an environment without large customers, suppliers, unions, or governmental regulators, and that employed a relatively homogeneous labor force in a simple organization structure." (Kotter, 19851). Thus, as soon as the organisation grows, the diversity and interdependence in existing relations have to be managed and led.

Max de Pree in his book "Leadership is an art" (1987) made a classic statement about leading an organisation - "the first responsibility of a leader is to define reality. The last is to say thank you. In between, the leader is a servant." In case of an organisation needing to change, the leader has to spell it out. Kotter (1996) has designed an eight-step model of change, claiming the need to explain, motivate and sustain the change. The now and here must be described, but also the then and there: where are we and where are we going. A small core of people, the "guiding coalition", has the initiative, but in order to succeed, more and more people have to be involved. One important motivator for others to engage is "short-term wins", necessary to underscore the meaning of and reward from participating. Kotter ends his list with the need to persevere and to gradually institutionalise the changes. New behaviour has to be established. If we choose to define "institution" as a pattern of routinized behaviour (Veblen, 1904, Hodgson, 1988, Scott, 1995, Cohen et al, 2004), the links between individual understanding, actions and innovation are again in focus. When changes are institutionalised, the diversity of action and interdependences are not considered.

Yet, the diffusion of certain behaviour is not automatic. Once the leader has pointed out the way, "defined the reality", people need to follow. Think about this definition of a market. A market is i) a set of actual or potential customers, ii) for a given set of products or services,

iii) who have a common set of needs or wants, and iv) who reference each other when making a buying decision (Moore, 1991). There is a direct link between Kotter's (1996) eight-step change model and this view of a market. First, there has to be a need for the new thing. Then a group has to advocate the new thing and persuade others to buy it. Transferred to the concept of innovation, defined as "any process or product that sells and creates new value in a market" (Edquist, 2002), it is deeply dependent on a cognitive process where an idea roots itself and grow stronger as more and more people see it and value it. A successful innovation has gained a critical mass of buyers. A product offered by a company creates value to its buyer, otherwise there will be no repeat buying, nor any persuading elaborations to other potential buyers about the good qualities of the product.

The group that formulates the Trelleborg policy for patients' meals sets the stage for the base groups, which are now being formed and activated in the hospital. The base group activity is a conscious move to have the people reflect upon what they do and to discuss it with each other. In that way, under the supervision of the project members, there is a clear "guiding coalition" creating a sense of urgency. From this, the base group activities reduce the diversity of the staff's different conceptions of a hospital meal. This is the first ripple across the mental map of hospital food being redrawn.

5. BRIDGING REGIMES TO BOOST INNOVATION

The disparity among the concepts of market, institution, and culture may be perceived as significant. Nevertheless, adopting the aforementioned definitions as our perspective and considering Schein's (2010) interpretation of group culture elucidates the evident and intriguing connection among the three. The culture of a group Schein defines as *"a pattern of shared basic assumptions learned by a group as it solved its problems of external adaptation and internal integration, which has worked well enough to be considered valid and, therefore, to be taught to new members as the correct way to perceive, think, and feel in relation to those problems." (2010:36).* In the western business community, there is an adage that "culture eats strategy". This means that no matter what strategy companies formulate, the implementation will be determined by the culture of the organisation. Culture is based on past experiences, what has been working. Strategy, defined as a plan, is about future competition and profits. Culture is based on inductive emotional consequence, strategy is based on deductive rational choice. The statement could be tested on other cultural levels than the organisation.

Culture, Schein (ibid.) suggests, could observed on a macro level (national, ethnic), organisational level (private, public, nonprofit, government), on a sub level (occupational) and on a micro level (units, teams, groups within a larger organisation). Thus, if we extend the argument of culture eating strategy, it becomes imperative to take existing cultures into account if enduring change is wanted. In that sense, there is wind in our sails now, as the Swedish healthcare establishment increasingly directs attention to deteriorating hospital food. However, the real challenge of re-conceptualizing the hospital meals is found at the occupational and work group levels. These cultures need to be confronted and overtly discussed if they are to be altered. And to the extent they are, the innovation project will succeed.

Following Levinthal (1991) and Scott (1995) regimes have three dimensions: i) cognitive rules, related to belief systems, ii) normative rules expressed in missions, goals, and identity, and iii) strategies and strategic orientations towards the surrounding external socio- technical and politico–economic environment.

The Trelleborg case shows the three dimensions in relation to the re-conceptualization of a hospital meal. There were plenty of cognitive rules about how a hospital meal should be, that turned out to be an impediment for innovation, sometimes in direct conflict to the normative and strategic dimensions. Furthermore, the regimes of the public staff, the political level and the innovation facilitating actor (SFIN) were difficult to bridge, as shown in the initial conflicts about the hospital kitchen.

The pathway to effective innovation involved a coalition of individuals capable of serving as intermediates among the various operational regimes. Individuals employed by SFIN possessed reputation among the medical personnel, while the esteemed chef's expertise in food preparation was indisputable. The SFIN project leader possessed unrecognized abilities in mediating between the external organization and the senior administrative and political tiers. Significant efforts remain necessary prior to the successful implementation of the new paradigm of hospital meals in Trelleborg, particularly at the operational level. Innovation is not deemed successful unless the mission for change is ingrained at all levels..

FOOTNOTES

1. Here is an example of how innovations challenge existing norms. The Kotter book Power and Influence was read on an iPad, the new way of reading. In there, the pages are dynamically paginated. The traditional way to refer to a quote hence needs a paper copy - the old way is supported by norms and standards.

REFERENCES

1. Argyris, C, & Schön, D. (1978). Organizational Learning: A theory of action perspec tive. Addison-Wesley.
2. Christensen, C. M. (1997). The innovator's dilemma. When new technologies cause great firms to fail. Harvard Business School Press, Boston.
3. Christensen, C. M, Baumann, H, Ruggles, R, & Sadtler, T. M. (2006). Disruptive Inno vation for Social Change, Harvard Business Review, December, , 94-101.
4. Cohen, M. D, Burkhart, R, Dosi, G, Egidi, M, Marengo, L, Warglien, M, & Winter, S. (2004). Routines and other recurring action patterns of organizations: contemporary research issues, in Dosi et al (2004).
5. Dosi, G, Teece, D. J, & Chytry, J. (2004). Understanding Industrial and Corporate Change, Oxford University Press.
6. Drucker, P. F. (1994). The Theory of the Business, Harvard Business Review, September, , 95-105.
7. EdquistCharles and Hommen, Leif ((1999). Systems of innovation: theory and policy for the demand side. Technology In Society, 21, , 63-79.
8. Edquist, C. (2002). Innovation Policy for Sweden- Objectives, Rationales, Problems and Measures. The Swedish Agency for Innovation Systems, VFI 2002:2.
9. Ghoshal, S. (2005). Bad management theories are destroying good management practices. Academy of Management Learning & Education, 4, 1, , 75-91.
10. Govindarajan, V, & Trimble, C. (2010). The other side of innovation. Solving the execution challenge. Harvard Business Press, Boston.

11. Hodgson, G. (1988). Economics and Institutions, Polity Press.
12. Johnson, S. (2010). Where good ideas come from. The natural history of innovation, Penguin Group, New York.
13. Jönsson, H, Knutsson, H, & Frykfors, C-O. (2011). Facilitating Innovations in a Mature Industry-Learnings from the Skane Food Innovation Network, in Valdez (2011).
14. Lagnevik, M. (2006). Food innovation at interfaces: experience from the Öresund region', in Hulsink, W.; Dons, H. (Eds.) Pathways to High-Tech Valleys and Research Triangles. Wageningen: Wageningen UR Frontis Series, 24.
15. Levinthal, D. A. (1991). Organizational adaption and environmental selections-interrelated processes of change. Organization Science , 2, 140-45.
16. Lewin, K. (1946). Action research and minority problems. Journal of Social (2), 34-46.
17. Miles, R. E, & Snow, C. C. (1984). Fit, Failure And the Hall of Fame, California Management Review, 26, 3, , 10-29.
18. Mintzberg, H. (1978). Patterns in Strategy Formation, Management Science, 24, 9, ,934-948.
19. Moore, G. A. (1991). Crossing the chasm, HarperBusiness.
20. Normann, R. (1975). Skapande företagsledning, Aldus, Stockholm.
21. Normann, R. (2001). Reframing Business- When the Map Changes the Landscape, Wiley, London.
22. OECD ((2012). Territorial Reviews. Skåne, Sweden 2012. OECD Publishing.
23. Porter, M. E. (1985). Competitive Advantage: Creating and Sustaining Superior Performance, Free Press, New York.
24. Porter, M. E. (1990). The Competitive Advantage of Nations, Free Press, New York.
25. Porter, M. E. (1996). What is strategy?, Harvard Business Review, November-December, , 61-78.
26. Porter, M. E. (2008). The Five Competitive Forces That Shape Strategy, Harvard Business Review, January, , 79-93.
27. Rapaport, R. N. (1970). Three dilemmas of action research. Human Relations, 23, in Susman and Evered (1978)., 499-513.
28. Scott, W. R. (1995). Institutions and Organizations, Sage.
29. Selznick, P. (1957). Leadership in Administration, Harper & Row, New York.

30. Susman, G. I, & Evered, R. D. (1978). An Assessment of the Scientific Merits of Action Research, Administrative Science Quarterly, 23, 4, , 582-603.
31. Valdez, B. ed., ((2011). Scientific, Health and Social Aspects of the Food Industry, InTech.
32. Veblen, T. (1904). The Theory of Business Enterprise, Charles Scribner's Sons. New York.
33. Winter, S, & Nelson, R. R. (1982). An Evolutionary Theory of Economic Change, Harvard University Press.
34. Depree, M. (2004). Leadership is an art, Bantam Dell.
35. Schein, E. H. (2010). Organizational Culture and Leadership, 4th ed, Jossey-Bass.
36. Kotter, J. P. (1985). Power and influence,
37. Kotter, J. P. (1996). Leading change, Harvard Business School Press.
38. March, J. G. (1991). Exploration and Exploitation in Organizational Learning, Organization Science, 2, 1, , 71-87.
39. Collins, J. (2001). Good to Great. Why Some Companies Make the Leap... and Others Don't, HarperCollins.
40. Collins, J, & Hansen, M. T. (2011). Great by Choice: Uncertainty, Chaos, and Luck- Why some thrive despite them all. HarperCollins.

28
Chapter

COLLABORATIVE INNOVATION: EMPHASIS ON FOOD SMES

1. INTRODUCTION: AN OVERVIEW

1.1 Collaborative innovation: A set of agenda

In recent years, the concept of 'collaborative innovation' has emerged as the prevailing viewpoint in innovation literature, positing that inventions are efficiently cultivated through the interaction of various stakeholders from distinct businesses. (Steinle and Schiele, 2002; Trott and Hartmann, 2009; Nootebom 2004, Chesbrough, 2006).

Within the setting of SMEs, the establishment of robust networks for innovation is perceived as a need rather than an option (Goss, 1991; Pratten, 1991; Rothwell and Dodgson, 1993). A substantial amount of literature demonstrates that involvement in networks and collaboration in partnerships is crucial for SMEs, as these elements empower them to address new technology and business challenges and to adapt to the rapidly evolving environment. (Hanna and Walsh 2002; Van de Vrande et al. 2009).

Main advantages are the access to network competencies as well as the opportunities of engaging into supply chain innovation processes and of growing in collaborations with larger customers (Johnsen and Ford, 2006; 2007). At the same time, collaborative innovation creates challenges to SME resulting from the inability to nurture and maintain the necessary resources and capabilities for growing (Matthyssens, Vandenbempt, Berghman 2006), to build a competitive positioning (Colurcio and Russo Spena, 2009; Day and Nedugady, 1994) and to create value for the own company and for the network (Johnsen and Ford, 2006; Donada and Nogatchewsky, 2006; Ulaga, 2003).

Recently, the network perspective have triggered a fervent debate on the generation of knowledge and learning in inter-organizational and network collaborations (Dyer and Singh 1998; Nooteboom 1999, 2004, 2006; Lampela and Kärkäinen 2009, Hallikas et al. 2009, Lampela et al. 2008). Collaborative learning in innovation networks is said to stimulate the creation of new knowledge, processes, products and services as well as the motivation for networking itself (Araujo, 1998). Many contributions have demonstrated the role of learning in the context of innovation networks discussing it as particularly challenging but increasingly more important task for companies (Capaldo 2007, Lane and Lubatkin 1998; Inkpen and Tsang 2005, Dyer and Hatch 2006).

However, the integration of diverse knowledge sources and development of learning processes are carried out in relationships between a multiplicity of actors that may show different characteristics of asymmetry. The asymmetries become evident when the relationships involve actors with different positioning and power, as showed by Johnson and Ford (2007). Asymmetries in business relationships have been analyzed from their different perspectives and on their impact on innovation and network development (Mouzas & Ford, 2004 ; Johnsen and Ford 2001; 2006).

From other perspective a wide literature has identified enablers and barriers to network collaboration (Leonard, 1995; Szulanski, 1996; Knott, 2003). Among others Szulanski (1996) has recognized the main role of knowledge, motivation, trust and ambiguity while others researchers also considered the role of the context (the environment) (Nelson and Winter, 1982; Teece, 1986; Pihkala, Ylinenpaa and Vesalainen, 2002). Other studies have showed that the amount of social capital correlates significantly with the competitiveness of collaborative networks (Macke, Vallejos, Faccin, & Genari, 2010).

Notwithstanding these contributions, we observe that insufficient emphasis has been devoted to analyzing collaborative innovation within the framework of learning and asymmetric interactions. Our objective was to generate insights into the characteristics of relationships and to identify obstacles and facilitators to collaboration and learning within innovation networks from the viewpoint of SMEs.

The matters of collaborative innovation, asymmetry, and learning are crucial and insufficiently examined in elucidating the competitiveness and advancement of organizations.

2. ASYMMETRIC RELATIONSHIPS

The study examining the way in which a firm innovates through inter-organizational and network collaborations (Capaldo 2007, Lane and Lubatkin 1998; Inkpen and Tsang 2005, Dyer and Hatch 2006; Nooteboom 2004) has only a more recent tradition.

Many researchers focus on partner characteristics, such as size, interdependence, cultural compatibility and relative influence (Holmlund and Kock, 1996; Pittaway and Morrisay, 2005). They show that these factors may leverage resources to drive innovation but that they can also inhibit innovation. Other authors (Pfeffer and Salancik, 1978; Håkansson and Snehota, 1995; Axelrod, 1964) - although stressing that networking is often the only way to realize innovation within a small size context - emphasize the organizational (resource) dependencies that emerge from networking. Similarly, other contributions that have focused on power unbalance (Hakansson; 1987; Håkansson and Gadde, 1992) find that this relational characteristic may become a barrier to the build up of a fruitful relationship and it may inhibit the development of innovation capabilities for smaller partners. Power unbalance is a barrier since business relationships in such a situation are dominated very often by contracting more than by trust (Jeffrey and Reed, 2000; Blomqvist et al.2005); so they don't drive cooperation to innovation but simply consist of a contractual contribution of resources. Important studies are those by Ford and Rosson (1982) and Geser (1992) who have deepened insights about frequency, range and level of contacts between firms.

Although researchers focused on different characteristics which enable or inhibit the development of business relationships aimed at innovation, a common central theme emerged: the asymmetry of relationships. The concept of asymmetric relationships concerns different elements of the relationships such as size, power, resources, and competencies and strongly affects its innovation potential (Blomqvist, 2002; Whipple and Frankle, 2000; Bailey et al. 1998; Colurcio, 2009; Colurcio et al. 2012). On the topic of asymmetric relationships, a relevant contribution comes from Johnsen and Ford (2000; 2001, 2007) who have identified seven types of asymmetries in business relationships: mutuality, particularity, conflict, cooperation, interpersonal inconsistency, intensity, power and dependence.

All these characteristics express interdependency ties among partners involved in the relationship. These interdependencies unequally affect the partners of the relationships. Mutuality is conceived as an "attitudinal variable" since it concerns the willingness of an organization to advantage common goals in respect to its own goals; it requires a long-term perspective since a company may give up its own goals in order to advantage a counterpart. Over time this behaviour will create well-being for both (or all) the parts involved in the relationship. Particularity is the way to manage the relationship and concerns the interaction's level of standardization. It depends on the relational approach of the party and on its availability to customize the relationship and its contents. Conflicts are the essence of asymmetry since it conveys the inequality between the parties and conflicts are amplified by the level of the interdependence of the relationships. The interdependence strengthens the differences and in turn feeds relational pressure and conflicts. Co-operation concerns the willingness of the may concern specific goals or projects but it is mainly conceived as a way to work, as a perspective to manage the relationship and to extract value from it for all the involved parts. Interpersonal inconsistency relates to individual characteristics of subjects that interact in the relationship. Intensity stresses that the number of people who interact in a relationship affects the relationship and its value (cross-functional group; team working). Power and dependence hedges in different kinds of asymmetries and stresses difference in the partners' resource stock.

These characteristics may work both as facilitators and barriers to the development of innovation in a business relationship and affect the evolution of the relationship differently depending on value, culture and managerial system of the SME.

3. LEARNING PROCESS IN COLLABORATIVE INNOVATION

Recently knowledge and learning processes have become the main topics in the agenda of many scholars studying collaborative innovation (Dyer and Singh, 1998; Gemunden et. al, 1999). Among them, the inter-organizational and collaborative network learning perspectives have emerged as distinctive approaches providing a starting point for the analysis of development of innovations in collaborative relationships. The table 1 summaries the similarities and distinctions among these perspectives on the base of three main dimensions of knowledge and learning process in interaction i.e. firm's knowledge base, attributes of knowledge and characteristics of relationships.

Table 1. Learning in inter-organizational and collaborative perspective

Dimensions	Inter-organizational perspective	Collaborative network perspective
Firm's knowledge base and capacity	• Absorptive capacity • Relational, interaction and collaboration capability, • Relative absorptive capacity	• Collaborative competency • Cooperative competency • Network competence • Coordinator or orchestrator capacity • Positioning and visioning
Knowledge and processes characteristic	• Tacit/explicit knowledge • Similarity of knowledge • Specialized knowledge • Similarity and shared routine • Resource and knowledge appropriateness	
Relation characteristics	• Strong/weak ties • Commitment/opportunism • Trust, shared value and culture • Similarity of processes Shared inter-firm routine	• Multiple and connected relationships • Flexibility of network • Strong ties/loose ties • Managed/unmanaged network

Main references and studies	Dyer and Singh (1998); Lane and Lubatkin (1998); Adler, (2001); Johnson and Sohi (2003); Blomqvist and Levy (2004); Miles et al. (2000, 2004); Lampela and Kärkäinen (2006); Rasmussen (2007); (2002); Hurmelinna-Laukkanen et al (2007); Castaldo (2007);	Ford (1998); Holm et al (1999) Bloomqvist, (2006); Moller and Halinen (1999); Möller and Rajala (2007); Miles et al. (2000, 2004); Ritter et al. (2002); Möller and Rajala (2006); Dhanaraj and Parkhe (2006), Heikkinen and Tähtinen (2006)

The inter-organizational approach

Based on the foundation of organizational learning literature (Cohen and Lenvintal, 1990), the inter-organisational perspective (Johnson and Sohi 2003; Blomqvist and Levy 2004) stresses the firm's knowledge attributes of absorptive capacity (Levinthal, 1990; Koza and Lewin, 1998) as addressing the leveraging of firm's interaction (Lampela and Kärkäinen 2009, Hallikas et al 2009) in a dyadic relationship. Relational capability (Lorenzoni and Lipparini, 1999; Johnson and Sohi, 2003), collaboration capability (Blomqvist and Levy, 2004) and interaction capability (Johnson and Ford, 2006) are different terms used to refer to the ability of a firm to recognize the value of external partner's knowledge and to build and maintain high-quality and productive-firm relationships. These relationships have been recognized as important for accelerating the firm's knowledge access with an effect on growth and innovativeness (Lampela and Kärkäinen, 2006). Moreover, the understanding on the interactive process of knowledge flow has been deepened by Dyer and Singh (1998) who emphasized the similarity of the knowledge base and the level of knowledge specialization in learning between partners. Similarly, Lane and Lubatkin (1998) introduced the concept of relative absorptive capacity to take a wide set of characteristics of the partner into account. In addition, Dhanaraj et al. (2004) have showed that the strength of a relationship between firms positively influenced the transfer of both tacit and explicit knowledge, leading to higher performance in learning. Kachra et al. (2008) found that the presence of stronger social relationships and social capital correlates significantly with the innovativeness of learning based collaboration (see also Knight, 2002). Furthermore, scholars studied the role of motivation (Szulanski, 1996, 2002) and appropriability (Hurmelinna-Laukkanen et al, 2007) and identified trust as one of the most effective enablers to inter-firm knowledge and resources transfer because high trust decreases situational uncertainty (Simonin, 1999) and opportunism (Lubatkin et al., 2001) and encourages higher commitment to the relationship (Capaldo, 2007). At the same time, some authors have evidenced also the "dark side" of strong inter-organizational relationships as obstacles for innovation mainly of radical type. The main reason is that a strong relationship locks firms into a narrow network, making them dependent on the inspiration by only a small number of external sources of creativity (Capaldo, 2007) and reduces the likeliness to explore new knowledge paths (Collinson and Wilson, 2006).

The collaborative network learning approach

The focus on structural and social dimension of relationships has been further debated by collaborative learning perspective focused on “learning by a group of organizations as a group” (Knight, 2002). According to Hallikas et al. (2009) the network learning literature represents a novel topic for learning research as the innovation phenomenon becomes increasingly occurring with multiple participating organizations. Emphasizing the aspect of multiple and organized relations (Håkansson and Snehota (2006) and open and interconnected business relationships (Ford et al 2003; Ford and Håkansson 2006), these studies have shed new light on the mix of diversity, variety and continuity of relationships and the way in which they are conductive to learning and innovation. Among them, Möller and Rajala (2007) argued that in innovation networks, knowledge exploration through weak ties, i.e. sources external to well established relationships, is needed, and flexibility of network is essential. They furthermore recognized the role of network orchestration defined by the nodal position held by an actor in a flexible network as crucial from a knowledge transfer point of view, especially “because such an actor’s task is to connect multiple actors in the net”(Möller and Rajala, 2007; p 899).

The problem related to relational distance has also been discussed according to a cognitive perspective (Argyris and Schon, 1977). The lack of feedback for effective learning processes is seen as very likely when a relatively large number of agents interact with each other in various process steps. So it has been concluded that striving to learn more effectively in network means to enable trust-based mutual communication and continuous feedback as well as that the coordination and co-operation link between the organizations must be strong and kept active (Blomqvist, 2004; Miles et al., 2000; 2004). Miles et al. (2000, 2005) pointed out that the ability to collaborate in network is a meta-capability for innovation. Similarly, Sivadas and Dwyer (2000) discussed cooperative competency as “the ability of the partners to trust, communicate, and coordinate” (ibid, p 40). Moller and Halinen (1999) and Ritter et al. (2002) have developed a concept of network competence to understand the capacity of firm to drive innovation success through the effectively management of actors in the network. Many others authors furthered the role of coordination or orchestrator (Dhanaraj and Parkhe, 2006; Heikkinen and Tähtinen, 2006) and discussed these aspects in term of capacity to 1) support absorptive competences among the network actors, 2) foster articulation and codification of tacit knowledge when it is reasonable and possible, and 3) develop long-term inter-firm relationships and network vision and identities for members (Hurmelinna-Laukkanen and Natti, 2007).

However the ongoing debate on the nature and structure of the network and its impact on learning is far from a final resolution and there are many contributions that support different perspectives. Recent research has found that an open unformed network comprising of numerous and loose ties has to be particularly well-adapted to facilitate new knowledge creation, whereas the more closely and coordinated structure has been found to better facilitate the diffusion, implementation and exploitation of existing knowledge (Hallikas et al., 2009).

4. RESEARCH AIM

The above summarized literature studies the underlying principles of collaborative innovation at a rather abstract level or within the context of large companies (Chesbrough, 2003, 2006). Studies explicitly focusing on the SME network perspective look at advantages and opportunities for collaborations, whereas an in-depth debate about the asymmetric nature of relationships as well as the mechanisms that enable or hinder the development of effective collaboration and learning processes in SME innovation networks is yet missing. The main objective of our study is to contribute to bridge this gap.

More in detail the aim of this chapter is to provide a relationship approach to collaborative innovation in SMEs', specifically: 1) to investigate the dynamics of SMEs' relationships with partners different by size, power and resources and stock within a network 2) and exploring the barriers and facilitators to learning for Smes' innovation processes.

To address these efforts we chose to investigate in depth the food sector. We put two main points in the base of our business focus. First, food collaborative innovation has been analyzed so far mainly within the context of large, multinational firms (Fortuin & Omta 2008) stressing the role of this actors as transfer of formalized knowledge. Second, the topic of collaborative innovation of food sector SMS is particularly important as the food market is not very supportive to innovation. It is highly saturated (Sucher, 2007), consumers tend to be rather conservative concerning their food preferences (Rozin & Vollmecke, 1986), and the food industry is not extensively pushed by technical innovations (Moskowitz, 2008). Innovation is very much fraught with risk in the food sector; 60 – 80% of the new launched products fail (Grunert and Valli, 2001). To realize a successful food product innovation therefore a combined efforts of different network partners - like suppliers (often SMEs) and retailers – is needed for customizing the new product to the needs of the end-consumers (Gellynck, & Molnar, 2009).

Given the difficult market situation of food sector SME as well as the necessity to cooperate for being able to create and launch food innovations, there is a need for research that deepens understanding of how SMEs experience their relationships and configure modes of interaction with asymmetric partners. Also we want deepen the understanding of factors and barriers of food sector SME network learning, fostering integration and creation of new knowledge as antecedents and contribution to innovation and sustainable competitive advantage of all network.

4.1 Selection of industry and sample

For the study, we decided to investigate processing food SME that innovate in network collaborations. An aspect to study is how highly different partners like SME suppliers and large retailers can cooperate for innovation despite – or because of - large power asymmetries (Gellynck and Molnar, 2009; Colurcio et al. 2012). Beside product innovation, a network is also needed for the so-called "immaterial" organizational innovations like the adoption of quality standards and methods are of tremendous importance for food SMEs (Padel, 2001). Without network partners, these innovations would be out of their reach as they require inter alia scale economies in audit, control, certification activities.

Literature suggests that the opportunities for collaborative innovation depend on the market conditions (Chen and Chen, 2002).

Our sampling strategy followed three criteria: First, all selected companies had to be SME, i.e. have between 1 and 250 employees. Second, we based our sample on the "stylized model of agri-food vertical chain" originally defined by Raynaud et al. (2005:60) which is presented in table 2 below:

Table 2. Stages of the agri-food chain

Stage 1	Stage 2	Stage 3	Stage 4	Stage 5	Stage 6
Supplier (e.g. fertiliser)	Farmer	1st Processor (e.g. mills)	2nd Processor (e.g. semi-finished good, end products)	Wholesaler	Retailer

Within this model, we mainly focused on processors active in stages 3 and 4. We expected that these SMEs would have a high need to cooperate in innovation processes. Third, the accessibility of the company, i.e. the interest of the interviewee to participate in the study, has been crucial for the sample design (Merkens, 2007). The final sample consisted of interviewees from 167 Italian SME (148 small and 19 medium companies). All interviewees were indicated by the SME as a "person responsible for innovation".

4.2 Methods for data gathering and analysis

In our study, we were interested into the perspective of SME innovation managers on collaborative innovation. We particularly wanted to know what they perceive as barriers and enablers to inter-organizational collaboration and learning in innovation networks and gain detailed insights into attributes of network relations in innovation networks involving SME. We therefore decided to conduct semistandardized interviews (Flick, 2009; p. 156). The basic assumption behind the interview format that combines open, theory-driven and confrontational questions is that the interviewees hold a subjective theory on the topic of the study. This subjective theory consists of explicit knowledge which the interviewee can easily articulate as a response to an open question but also implicit knowledge where he or she needs methodological support in the articulation through other types of questions (Groeben, 1990). Our interview guideline therefore started with open questions on what the interviewee understands as innovation, what the major challenges of his company are in that respect and what he sees as networks. Then, interviewees where asked to describe the most important attributes of network relations in order to identified trigger and opportunities for collaboration and learning in innovation. Here, we added theory-driven sub-questions for being able to fully understand the attributes of the network relationships: If the interviewee did not mention it him- or herself, we asked for the cooperation duration and evolution, particularity, intensity, dependence, and mutuality of the relationship. These questions have been developed according to the theoretical framework proposed by Johnsen and Ford (2000; 2001, 2007) who identified different types of asymmetries as relevant for cooperation in business relationships.

The interviewee was asked both for the benefits and learning points from the cooperation but also for difficulties, conflicts, and barriers to collaboration and knowledge sharing and development. In this case, the questions have been drawn to deepen the firm's capacity, knowledge and relationships attributes as theoretically discussed at the base of learning processes in interaction.

Like this, the interviewee was confronted with the opposite of the own subjective theory regardless whether he or she was in favour or not in favour of cooperation in innovation networks.

Interviews were held via telephone. They took approximately half an hour and were then recoded and transcribed ad verbatim.

Data analysis and interpretation took several steps. We first separated the interview transcripts of small (up to 50 employees) and medium (between 51 and 250 employees) companies because we were interested whether size matters. We then created bodies of text related to the questions, for example all answers related to the question on "From your perspective, what is Innovation.

We then analyzed the question-specific data sets with a bottom-up approach and used techniques suggested by Miles & Huberman (1994). Between two and four researchers coded the material independently, then came together, noted themes and clustered them considering the relations and linkages between each other.

Table 3 provides an outlook on the typologies we have recourse to in order to analyze above mentioned characteristics of the relationship within the innovation network.

Table 3. A typology for the analysis of size asymmetry

Relationship Characteristic	Definition	Consequences Customer unilaterally influence/ Customer and Supplier bilaterally influence
Mutuality	Shared goals or common interests of firms.	• Formality/Informality in setting relationship goals • Use/purpose of writtren plans • Extent of interwining of goals and interest and experience
Particularity	Direction and uniqueness of interaction. Extent of standardisation of interaction.	• Standardization/adaptation of approach to interaction • Width of suppliers' customer portfolio • Building of confidence and emotional ties in interaction
Conflict	Differences between the parties over the direction of the relationship or over their respective contributions and benefits	• Conflict resolution efforts involving higher/lower -level actors • Development of approaches for coping withconflict
Cooperation	The extent to which companies work together to determne or implement a direction for the relationship Co-existence of co-operation and conflict	• Experience and inclusion of suppliers in co- operative projects • Development of personal expectations and contributions in interaction
Intensity	Level of contact and resource exchange between firms.	• Range, level and frequency of contact between customer and supplier • Extent of supplier involvement in contact and resource exchange
Power and dependance	Distinct types of power and dependance exist: Technical, Knowledge, Social, Logistic , Administrative.	• Strategic and operational aspects of relationship • Technical aspect of relationship and advisory roles • Decision-making process • Social/networking aspects of supplier's • relationships • Development of information and knowledge • Develeopment of values built on history

We analyzed data through a qualitative content analysis that is based on data from narratives and observations that requires understanding and co-operation between the researcher and the participants, such that texts based on interviews (Kondracki et al., 2002). Specifically, we analysed and categorized consequences of each characteristic of the relationship according a pattern of analysis based on abstraction (Graneheim, Lundman, 2003) since it emphasises descriptions and interpretations on a higher logical level including the creations of codes, categories and themes on varying levels. Then, we defined three different levels of intensity of the characteristic in the specific relationship: + = low level; ++ = medium level; +++ = high level.

5. FINDINGS

The findings below describe how and in what types of networks SMEs cooperate for developing innovation and what the interviewees identified as barriers and enables to inter- organizational collaboration and learning in innovation network.

5.1 The collaborative innovation of SMES

The understanding of innovation is different for interviewees from small and medium companies within the two data settings. When asked what they perceive as innovation, interviewees from medium companies often mentioned the development of new technologies for the improvement of production processes and technologies. The small firms instead associated process and technological innovation with the incremental improvement of existing products and processes. Generally they talked about product innovation as the improvement of existing products and the extension of the product range. All firms not or least associated the development of completely new products (See Table 4).

Table 4. What the firms understood as innovation

	Small	Medium
Development of completely new products		+
Innovative appearances at the market	+	++
New production processes and technologies	++	+++
Incremental improvement of existing product/ processes	++	++
New packaging solution	+	++

However, we observed a difference in the attributes of relationships depending on the size of companies. Table 5 provides a brief overview of these attributes. The general network of relationships comprises clients, suppliers, research centres, and other stakeholders. But looking specifically at innovation network it is understood by the large part of the interviewees as cooperation between processing companies and suppliers of raw material, equipment and services. Respect to the theoretical framework we adopted, in this table we didn't consider explicitly the attribute "conflict" since we investigated it through the variable "mutuality". However, although we didn't asked clearly about conflict, we will discuss it due to lateral information. The relationship with suppliers, especially with raw material and equipment suppliers looks very high cooperative to small companies are the suppliers were seen very often as source of innovation or as co-innovator partner. Suppliers act as development and implementation partners for new ideas; in some cases we even find common projects for developing innovations like new packaging solutions and new products. Suppliers also support small companies in designing and tailoring new production and logistic technology solutions for sustainable processes. Relationships to suppliers are characterized as long lasting, affordable and personal. In addition, small clients and experts or specialists are mentioned by small firm as important and strategic cooperative partners in some specific situation. In general collaborations judged to be important are those with specialized clients or firms that punctually provide services. Mentioned are market research companies or firms that have a high investment in R&D and don't want manufacturing or commercialize their new products or semi-finished goods by themselves. Interviewees from small companies reported least on universities and research centers and consumer collaboration. The cooperation with research centers is usually aimed at solving technical problems and has been described as project based with clear cooperation agreements.

Interviewees from medium companies defined network mainly as cooperation with suppliers of equipment and raw material and large distributors. According their view, a network is a strong cooperation within the supply chain, based on technology and market content, allowing to face competition and to guarantee the survival on the market. Suppliers are seen as important and they enable the medium companies to get access to update technology and complementary knowledge. The development of a win-win relationship is very important for the interviewees. Relationships with suppliers aret hus in most cases very stable (longer than ten years) and characterized by trust and reliability. Also relationships with large distributors have been recognized as very important to complement firm knowledge and demand a high management attention and regular contact. Particularly it is described as profiting when strategies are aligned, involve pear cooperation and partners can make reciprocal business deals. In that case, the company gains a profit because it can use the network of clients, take up new ideas, ask their clients to test the products with the end customer and in best case deliver products in exclusivity.

Only few companies named marketing agencies and consumers as important partners for the development of innovation.

Generally horizontal cooperation for innovation within the same branch is not so frequent within small and medium companies. Cooperation with competitors to reach common goals or to develop the whole branch for the profit is mentioned as generally scarce and without any effect when it comes to innovation or to a fruitful knowledge exchange.

Table 5. Attributes of Smes Network

<table>
<tr><th></th><th colspan="2">Cooperation</th><th colspan="2">Particularity</th><th colspan="2">Intensity</th><th colspan="2">Power and Dependence</th><th colspan="2">Mutuality</th></tr>
<tr><th></th><th>Small</th><th>Medium</th><th>Small</th><th>Medium</th><th>Small</th><th>Medium</th><th>Small</th><th>Medium</th><th>Small</th><th>Medium</th></tr>
<tr><td>Small clients</td><td>+++</td><td>+</td><td colspan="2">+++</td><td>+++</td><td>++</td><td colspan="2">++</td><td colspan="2">+++</td></tr>
<tr><td>Large distributors</td><td>++</td><td>+++</td><td colspan="2">++</td><td>+</td><td>++</td><td colspan="2">+++</td><td colspan="2">++</td></tr>
<tr><td>Suppliers</td><td colspan="2">+++*</td><td colspan="2">++</td><td>+++</td><td>++</td><td>++</td><td>+</td><td colspan="2">+++</td></tr>
<tr><td>Universities/ research</td><td>+</td><td>++</td><td colspan="2">+++</td><td>++</td><td>+++</td><td colspan="2">+</td><td colspan="2">+++</td></tr>
<tr><td>Competitors</td><td>+</td><td>+</td><td colspan="2">+</td><td colspan="2">+</td><td colspan="2">+</td><td>+++</td><td>++</td></tr>
<tr><td>Federal Agencies</td><td>+</td><td>++</td><td>++</td><td>+</td><td>++</td><td>+</td><td>+</td><td>+++</td><td>+++</td><td>++</td></tr>
<tr><td>Specialists and experts</td><td>+++</td><td>++</td><td>++</td><td>+</td><td>++</td><td>+</td><td colspan="2">+</td><td colspan="2">+++</td></tr>
<tr><td colspan="11">+ = low level
++ = medium level
+++ = high level
* very high</td></tr>
</table>

Regarding the particularity and evolution, interviewees reported that relationships develop and evolve over time built on increasing tied and personalised relationships. However, the particularity of the relationships of processing SMEs is described as high with small clients and research institute. With suppliers the particularity is a guarantee for the innovation since the benefit of innovation depends on the application and on the novelty of the equipment, especially for small companies. Processing SMEs however have innovation relationships with more than one supplier, so that from their side, the relationship with suppliers is less exclusive than the other way round. Relationships with large distributors are very often not particular; here, the mainly the bigger processing SMEs are usually just one innovation partners amongst many others for the distributers and often based on negotiated affair: technical know how is exchanged for marketing, knowledge on market trends for access to other markets, lower costs for product tests.

Asked about the intensity of the relationship, almost all interviewees mentioned the continuity and the frequency of exchange of information and knowledge with the main partners. Particularly, interviewees from small companies mentioned intensity mainly as team working and continuous knowledge and information sharing together with conjoint job training. Evidence shows the same dynamic in the intensity of relationship with the main partners also for medium companies. We however also find that the bigger the processing SME, the more intense cooperation with large distributors and the less intense cooperation with small clients, suppliers and Federal agencies are.

With respect to power dependence, only three few companies declared that the power is balanced and that there are no asymmetries in resources and power between the parts involved in the relationship. In tendency we find that the bigger the processing company, the more dependent are suppliers, and the less dependent is the company from large distributors. In addition, with the size of the company, its dependence from Federal agencies increases.

Mutuality of the relationships is generally mentioned at high level respect to all partners. Mainly small firms indicated as important to create a win-win situation that enables the (incremental) development of "new" products, services, production technologies and raw materials, and to solve common problems.

5.2 Benefits and difficulties in collaborative innovation

The reasons that companies provided on why cooperation for innovation is important include: 1) acquisition of know how on the market and trends, 2) the external view that helps to overcome blind spots, 3)the opportunity to enter new markets and to build up a market reputation, and 4) the possibility to join resources for radical new ideas.

Asked to medium firm for the benefits of cooperation in innovation networks, most interviewees mentioned the development of new products, new services, new production technologies and raw material. Very often, the interviewees mentioned that cooperation creates a win- win situation where they receive complementary know-how and information from each other, exchange concepts, solve together common problems without having to re-invent the wheel or get access to new markets. In some cases, interviewees stated that cooperation enables them to build up a completely new supply chain with producers of raw material in a developing country what makes the production and the products more sustainable innovation. Some medium firms additionally mentioned the opportunity to extend the firm's vision of network and to contribute to fostering the firm's influence on other relationships and partner's collaboration commitments. For the interviewees from small firms, especially the market effectiveness has been emphasized as important benefit during the interview, i.e. how the innovation allows them to improve the quality of their offering in a way that their products better fit the changing market requests.

Discussing on enablers and barriers of collaboration similar results emerged among small and medium firm. Cooperation, generally, has been reported ended due to missing reciprocity of efforts, changing interests or strategies of network partners or problems with the quality of the products. Small firms detailed this aspects in the missed promise of the partner or if he strives only for own benefit, the new product has now success on the market or can not be developed or the partners do not get along well. Rarely, cooperation ends because a contract ends and partners have new plans. Rarely, cooperation ends because a contract ends and partners has new plans. Small interviewees mentioned that cooperation with large distributers is difficult because of the small market power of their own company. For medium firms the challenges they faced are related to finding enough time for cooperation and clarifying expectations. A barrier to cooperation is also the necessity to confidentiality of new ideas. Some interviewees underlined that the number of partners involved into an innovation project should not be too big because otherwise, the process is too slow.

Looking at the difficulties, more than half of the interviewees declared that collaboration is not always easy. More into detail, some interviewees mentioned the difficulty to share information in real time; others referred to procedural and routine problems arising from the interaction between organizations with different process rules. Technological and technical difficulties have been mentioned mainly with reference to supply chain and R&D relationships where a larger distance has been perceived with respect to knowledge and experience background of partners. One interviewee well elicited this issue indicating that it is difficult (though not impossible) to work with specialized suppliers because of their different views on time, technology and ways of working. Similarly, to work with universities or R&D agencies needs the parties to get first used to each other. Within the perception of the most of interviewees these aspects represents the major challenge of collaboration for innovation. It has been widely outlined by the interviewees that the firms really innovate if they are able to conduct and participate in exploratory activity outside of their organizational boundaries absorbing novel practices from external. In addition, a difficulty other interviewees referred to is the lack of a shared vision with the partners and missing clarity in communication and the definition of goals and expectations.

When asked to expand on the reasons of such difficulties, a majority of respondents referred to the presence of cultural, organizational and resource-related barriers. Particularly trust issues, distances (geographical, dimensional, technological and commercial) between firms and strategic competences exchange increases the risk perceived by the interviewees to integrate into and participate in innovation networks.

The lack of trust and benefits understanding have also mentioned as the main reason of the failure and interruption of long standing collaborations. Furthermore, the i) imbalanced power between cooperating parties, ii) insufficient resources and highest changes required, iii) quality problems or iiii) better alternatives have been found as reasons to break up business relationships. Finally, to avoid dependency from other firms is also mentioned; as well as the time and money investment to innovate within a network was seen as unprofitable in certain circumstances. In addition it is also remarkable that more than the half of firms who disclaims problem or conflicts in collaboration are mainly the smaller ones. Surprisingly, there is a strong tendency among these managers to regard relationships in some general way as a "good thing" and there is also a common belief that all partners work towards closer and mutual relations.

A general agreement among all interviews regarded the belief that the cooperation for innovation needs openness as well as transparency on interests and on the own contribution. The partners should operate at the same eye-level and respect each other, develop together something new while building upon existing knowledge and both benefit from the cooperation. Both should be ready to invest time and efforts into cooperation. Long lasting, personal relationships and trust has been said to be essential for cooperation. Transparency and openness, collegiality and a good gut feeling have been seen by all interviewees as essential for cooperation at the same eye level and to avoid conflicts.

Again, the trust issue has been mentioned as a key driver of networking orientation of SMEs with the distinctive feature that trust is perceived to be built gradually through the continuity of cooperation among partners. A deep understanding of shared risks as well as of mutual benefits of all network partners have been identified as trigger to the openness of firms and to team working allowing to focus on common goals amongst all network members. Personal and face to face daily contacts - often within existing long standing relationships –have been seen as the preferred way to collaborate and to solve conflicts.

5.3 Barriers and enabler to inter-organisational learning

Discussing on barrier and enabler to inter-organisational learning the main finding which is noticeable is that about all firms (there is only two exception,one for each market) declared that innovation partnerships allow them to increase their competence and knowledge assets. First of all the complementary nature of knowledge and competence predominates innovation relationships. Acquisition and upgrading of technical and technological competences have been seen as the main results of learning collaboration by about the half of investigated firms. Interviewees from the other half of firms declared that their company mainly benefits from the development of market capability and subsequent increases of the market perception of firm offerings' value.

To improve the ability to learn is also mentioned as core element of innovation relationships by about two thirds of respondents. The interviewees stated that collaborating in innovation networks allows them to i) better define what they want and what is expected by partners, ii) develop a clearer vision of their relationships, iii) gain better insights on external knowledge through understanding its meaning for the own organization. As main inter-organizational learning results medium companies valued the better understanding of causes and consequences of their actions that allows for the detection and correction of errors. Some interviewees stressed the unlearning issue as an open approach to innovation. Innovation in networks is mentioned to promote the company to proactively question its older routines, assertions and beliefs and fostering the need of dismissing or replacing outdated knowledge.

Among those interviewees who confirmed the learning results of innovation network relationships, only a minor part identified some barriers or obstacles to their learning processes. The latter have been mainly identified as cultural and power distances between partners or low commitment of partners too. These difficulties have been often seen in relation to the different business and value chain position of partners. In addition, some interviewees stated to be not aware of competences and activities of their partners and this obviously leads to an inadequate understanding and knowledge of the competences available in the network. A common language is required to reduce the distance between partners and to reinforce social and cognitive dimension.

At the opposite site, the complementarities of the partners' knowledge have been seen as the main stimulus to the learning processes. Cultural issues instead feed a contrasting debate: whereas some interviewees named cultural barriers as critical obstacles, others perceived them as a trigger allowing a different view on problem solving issues. When supported by personal and social ties, the different culture of partners has been seen as great opportunity for a more effective knowledge integration allowing the firm to increase flexibility and therefore its ability to seize strategic opportunities.

6. DISCUSSION AND CONCISE SUMMARY

The objective of the works was to enhance the status of research on network innovation in SMEs by cultivating a profound comprehension of the obstacles and facilitators to collaboration and learning. The study focused on determining the networking attitudes, preferences, and behaviors of food processing SMEs pertinent to innovation creation in the Italian food sector.

The primary outcomes of our study indicate that food SMEs are inclined to engage with partners for innovation. Collaboration inside innovation networks appears to be significant to them, provided it yields reciprocal advantages and partners engage on an equal footing. Nonetheless, the receptivity to innovation is concentrated on select connections with a limited number of partners, typically inside the existing network of SMEs, where enduring relationships mitigate trust issues. The results underscore the significance of trust in innovative partnerships. The techniques for fostering collaborative innovation are little formalized and primarily reliant on personal and daily contacts.

We also ascertain that size is significant. We discerned variations among the partners in the collaborative innovation relationship based on the small and medium enterprises in our sample. Small organizations appear to collaborate more closely with clients that share a similar background and habit. Medium-sized enterprises frequently collaborate with universities and research institutions, thereby acquiring specialized technological expertise. The enhanced collaboration with the research sector elucidates a broader receptiveness to the advancement of new information and signifies that medium-sized enterprises conceptualize innovation within a long-term perspective. We did not observe the same trend among small food processing SMEs. This study underscores a significant issue for Italian small enterprises: the challenge of accessing certain types of relationships that are not project-specific but rather aimed at broader knowledge and development collaboration.

In our sample of SMEs, suppliers are the primary partners for innovation collaboration, while a pronounced innovation reliance defines the value chain interactions among imbalanced partners. Primarily, small first-stage processing SMEs tend to rely more on larger second-stage processors for innovation, while the latter are more dependent on major distributors. The reliance of partners in network innovation throughout the agri-food supply chain invariably encompasses the impact of innovation development expenses being transferred to the most vulnerable partner. Nonetheless, even in such circumstances, participation in the innovation network is mutually advantageous for both partners, as the larger partner provides marketing and technological contributions to enhance products. Remarkably, the issue of uneven electricity has not been identified as significant for small firms. The rationale may be that they typically function within a niche, fostering growth alongside small clients and suppliers through collaboration.

The foundation of innovation collaborations underscores the research on resource and knowledge complementarities (Lampela and Kärkäinen 2009).

The impact of SMEs on innovation (Holmlund, Kock, 1996) has been substantiated by our findings and expanded to encompass networking and learning orientation inside SMEs. Cognitive and relational advantages arise from knowledge acquired through "on-the-job" interactions: by collaboratively addressing common challenges, SMEs enhance their processes without unnecessary reinvention, while partners deepen their understanding of SMEs' potential and the necessity for increased engagement and interaction. The results indicate significant advantages stemming from the reference impact of the relationships (Ritter and Gemünden, 2004). For more knowledgeable SMEs, engaging in innovation facilitates access to an extensive network of interconnected relationships and enhances their positioning within value networks.

In summary, the disparities in resources and power appear to be regarded as a catalyst for network and learning interactions. This indicates that inequalities are not very crucial to creativity and learning network interactions. It is often considered that innovation partnerships provide significant benefits to the less powerful parties, particularly in terms of learning and knowledge acquisition.

REFERENCES

1. Araujo, L. (1998). Knowing and Learning as Networking", Management Learning 29 (3), 317-336.
2. Axelrod, R. (1964). The evolution of cooperation, Basis Book New York.
3. Bailey, W. J, Masson, R, & Raeside, R. (1998). Choosing successful technology development partners: a best-practice model", International Journal of Technology Management, , 15
4. Blomqvist, K, Hurmelinna, P, & Seppanen, R. (2005). Playing the collaboration game right-balancing trust and contracting", Technovation,. , 25, 497-504.
5. Blomqvist, K, & Levy, J. (2004). Collaboration capability - a focal concept in knowledge creation and collaborative innovation in networks", in Proceedings of the 20 the Annual Conference of Industrial Marketing and Purchasing (IMP) Group. Copenhagen.
6. Bougrain, F, & Haudeville, B. (2002). Innovation, collaboration and SMEs internal research capacities", Research Policy, 31(5), 735-747.
7. Capaldo, A. (2007). Network structure and innovation. The leveraging of a dual network as a distinctive relational capability", Strategic Management Journal, , 28, 585-608.
8. Chesbrough, H. (2003). Open Innovation: the new imperative for creating and profiting from technology, Harvard Business School Press.

9. Chesbrough, H. (2006). Open Business Models: How to Thrive in the New Innovation Landscape, Harvard Business School Press, Boston.
10. Cohen, M. D, & Levinthal, D. A. (1990). Absorptive capacity. A new perspective on learning innovation", Administrative Science Quarterly, 35 (1), 128-152.
11. Collinson, S. C, & Wilson, D. C. (2006). Inertia in Japanese Organizations: Knowledge Management Routines and Failure to Innovate,' Organization Studies, 27 (9), 1359-1387
12. Colurcio, M, Kocher, P, & Wolf, P. Russo Spena T. ((2012). Asymmetric relationships in networked food innovation processes",. British Food Journal, 114, (5) forhcoming.
13. Colurcio, M. (2009). Asymmetric business relationships and interaction capabilities: an exploratory study", Sinergie, Quaderno di Ricerche, December.
14. Dhanaraj, C, & Parkhe, A. (2006). Orchestrating Innovation Networks, Academy of Management Review, 31(3), 659-669.
15. Dhanaraj, C, & Lyles, A. (2004). Managing tacit and explicit knowledge transfer in IJVs: the role of relational embeddedness and the impact on performance," Journal of International Business Studies, 35(5), 428-442.
16. Dyer, J. H, & Singh, H. (1998). The Relational View: Cooperative Strategy and Sources of Interorganizational Competitive Advantage", Academy of Management Review, 23 (4), 660-679
17. Dyer, J. H, & Nobeoka, K. (2000). Creating and Managing a High Performance Knowledge Sharing Network: The Toyota Case," Strategic Management Journal, , 21, 345-367.
18. DyerJe. H., Hatch N. (2006). Relation-Specific Capabilities and Barriers to Knowledge Transfers: Creating Advantage through Network Relationships," Strategic Management Journal, , 27 Flick, U. (2009). An introduction to qualitative research, 4 edition. Sage, London.
19. Ford, D, & Rosson, P. J. (1982). The Relationships between Export Manufacturers and their Overseas Distributors, FORD ET AL. (eds) Export Management, Praeger, New York.
20. Ford, D, Gadde, L-E, Håkansson, H, & Snehota, I. (2003). Managing Business Relationships, Second Ed. John Wiley & Sons Ltd, England.
21. Ford, D, & Håkansson, H. (2006). IMP-Some Things Achieved, Much More To Do", European Journal of Marketing, 40 (3/4), 248-258
22. Gellynck, X, & Molnar, A. (2009). Chain governance structures: the European traditional food sector", British Food Journal, 111 (8), 762-775.

23. Geser, H. (1992). Towards an interaction theory of Organisational Actors", Organisation Studies, 13 (3), 1992.

24. Goss, D. (1991). Small Business, New Technology and Innovation. Routledge, London.

25. Graneheim, U. H, & Lundman, B. (2004). Qualitative content analysis in nursing research: concepts, procedures and measures to achieve Trustworthiness", Nurse Education Today , 24, 105-112.

26. Groeben, N. (1990). Subjective theories and the explanation of human action", in G.R. Semin and K.J. Gergen (Eds.), Everyday understanding: social and scientific implications (19-44). Sage, London.

27. Grunert, K, & Valli, C. (2001). Designer-made meat and dairy products: consumer-led product development", Livestock Production Science, , 72, 83-98.

28. Hadjimanolis, A. (1999). Barriers to innovation for SMEs in a small less developed country (Cyprus)" Technovation, , 19(9), 561-570.

29. Håkansson, H, & Gadde, L-E. (1992). Supplier Relations, Professional Purchasing, Routledge: London,

30. Håkansson, H, & Snehota, I. (1995). Developing Relationships in Business Networks, Routledge, London.

31. Håkansson, H. (1987). Industrial Technological development: a network approach, Croom Helm, London.

32. Håkansson, H, & Snehota, I. (2006). No business is an island" 17 years later", Scandinavian Journal of Management, , 22, 271-274.

33. Hallikas, J, Kärkkäinen, H, & Lampela, H. (2009). Learning in networks: an exploration from innovation perspective', Int. J. Technology Management,. 45 (3-4), 229-243

34. Hanna, V, & Walsh, K. (2002). Small Firm Networks: A Successful Approach to Innovation?, R&D Management, 32 (3), 201-207.

35. Heikkinen, M, & Tähtinen, J. (2006). Managed Formation Process of R&D Networks", International Journal of Innovation Management, 10 (3), 271-298.

36. Holm, D. B, Eriksson, K, & Johanson, J. (1999). Creating value through mutual commitment to business network relationships", Strategic Management Journal, 20 (5), 467- 486.

37. Holmlund, M, & Kock, S. (1996). Customer dominated relationships in a supply chain- A case study of four small-sized suppliers", International Small Business Journal,.33 (1), 19-37.

38. Hurmelinna-laukkanen, P, & Nätti, S. (2008). Network orchestrator and knowledge mobility in innovation networks" Proceedings. 24th IMP conference, September 2008, Uppsala, Sweden, 4-6.

39. Hurmelinna-laukkanen, P, Sainio, L, & Jauhiainen, M. T. ((2008). Appropriability regime for radical and incremental innovations", R&D Management 38 (3), 278-289.

40. Inkpen, A, & Tsang, E. (2005). Social capital, networks, and knowledge transfer", Academy of Management Review, 30 (1), 146-165.

41. Johnsen, R, & Ford, D. (2006). Interaction capability development of smaller suppliers in relationships with larger customers", Industrial Marketing Management, , 35, 1002-1015.

42. Johnsen, R, & Ford, D. (2007). Exploring the concept of asymmetry: A typology for analyzing customer-supplier relationships", Industrial Marketing Management, 37 (4), 471-483.

43. Johnsen, R, & Ford, D. (2001). Asymmetrical and symmetrical customer-supplier relationships: contrasts, evolution and strategy, Paper presented at IMP Conference, Oslo, September., 9-11.

44. Johnson, J. L, & Sohi, R. S. (2003). The development of interfirm partnering competence: platforms for learning, learning activities, and consequences of learning", Journal of Business Research, 56 (9), 757-766.

45. Kachra, A, & White, E. (2008). Know-how transfer: the role of social, economic/competitive, and firm boundary factors," Strategic Management Journal, 29 (4), 425-445. "Content Analysis: Review of Methods and Their Applications in Nutrition Education".

46. Knight, L, & Pye, A. A Learning and change in inter-organisational networks: the case for network learning and network change, Proceddings Third European Conference on Organisational Knowledge, Learning and Capabilities.

47. Kondracki, N. L, Wellman, N. S, & Amundson, D. R. (2002). Content analysis: review of methods and their applications in nutrition education", Journal of Nutrition Education and Behaviour 34 (4), 224-230.

48. Koza, M. P, & Lewin, A. Y. (1998). The co-evolution of strategic alliances," Organization Science , 9(3), 255-263.

49. Lampela, H, & Härkkäinen, H. (2009). Views and practices on inter-organisational learning in innovation networks," Int. J. Electronic Business, 7 (2), 130-148.

50. Lampela, H, Kuvaja, M, Kyrki, A, Härkkäinen, H, & Piirainen, K. (2008). Challenges and practices for learning within and by innovation networks Proceedings of the XIX ISPIM Conference, Tours, France- June., 15-18.

51. Lane, P. J, & Lubatkin, M. (1998). Relative absorptive capacity and inter-organisational learning', Strategic Management Journal, 19 (5), 461-477.

52. Lorenzoni, G, & Lipparini, A. (1999). The leveraging of interfirm relationships as a distinctive organizational capability: a longitudinal study", Strategic Management Journal, 20 (4), 317–338.

53. Lubatkin, M, Florin, J, & Lane, P. J. (2001). Learning together and apart: a model of reciprocal interfirm learning", Human Relations, 54 (10), 1353-1382.

54. Miles, M. B, & Huberman, A. M. (1994). Qualitative Data Analysis. A Source Book of New Methods. Sage, Beverly Hills.

55. Miles, R. E, Snow, C, & Miles, G. (2000). The future.org', Long Range Planning, 33, 300-321.

56. Miles, R. E, Snow, C, & Miles, G. (2005). Collaborative Entrepreneurship. How Groups of Networked Firms Use Continuous Innovation to Create Economic Wealth?, Stanford, CA: Stanford University Press.

57. Möller, K. K, & Halinen, A. (1999). Business relationships and Networks: Managerial Challenge of Network Era", Industrial Marketing Management, , 28, 413-427.

58. Möller, K. K, & Rajala, A. (2007). Rise of strategic nets- New modes of value creation", Industrial Marketing Management, , 36, 895-908.

59. Moskowitz, A, & Hartmann, J. (2008). Consumer Research: creating a solid base for innovative strategies", Trends in Food Science & Technology, , 19, 581-589.

60. Nooteboom, B. (1999). Innovation and Inter-Firm Linkages: New Implications for Policy", Research Policy, , 28(8), 793-805.

61. Nooteboom, B. (2004). Inter-Firm Collaboration, Learning and Networks, London, Routledge.

62. Nooteboom, B. (2006). Learning and innovation in inter-organizational relationships and networks, CentER Discussion Paper Series, Tilburg University, The Netherlands, 0924-7815Available at: http://ssrn.com/abstract=903754.(2006-39), 2006-39.

63. Padel, S. (2001). Conversion to organic farming: a typical example of the diffusion of an innovation", Sociologia Ruralis, 4 (1), 49-61.

64. Pfeffer, J, & Salancik, G. R. (1978). The External Control of Organisations: a Resource Dependence Perspective, Harper Row, New York.

65. Pittaway, L, & Morrisay, W. J. (2005). Buyer-Supplier Relationships in Small Firms: The Use of Social Factors to Manage Relationships", Discussion Paper n. 03, Sheffield University Management School.

66. Pratten, C. (1991). The Competitiveness of Small Firms. Cambridge University Press, London.

67. Raynaud, E, Sauvee, L, & Valceschini, E. (2005). Alignment between Quality Enforcement Devices and Governance Structures in the Agro-food Vertical Chains", Journal of Management and Governance, 9 (47), 47-77.

68. Ritter, T, & Gemünden, H. G., ((2004). The impact of a company's business strategy on ist technological competence, network competence and innovation success", Journal of Business Research,. 57, (5), 548-556.

69. Ritter, T, Wilkinson, I. F, & Johnston, W. J. (2002). Measuring Network Competence: Some International Evidence", Journal of Business & Industrial Marketing, 17 (2/3), 119-138.

70. Rothwell, R, & Dodgson, M. (1993). SMEs: their role in industrial and economic change", International Journal of Technology Management, (Special (8-22), 8-22.

71. Rozin, P, & Vollmecke, T. A. (1986). Food Likes and Dislikes", Annual Review of Nutrition, , 6, 433-456.

72. Simonin, B. (1999). Ambiguity and the process of knowledge transfer in strategic alliances", Strategic Management Journal, 20 (7), 595-624.

73. Sivadas, E, & Dwyer, F R. (2000). An examination of organizational factors influencing new product success in internal and alliance based process", Journal of Marketing, 64 (1), 31-49.

74. Steinle, C, & Schiele, H. (2002). When do industries cluster? Research Policy, , 31, 849-858.

75. Sucher, J. (2007). Die Milchmacht", Manager Magazin, , 8, 71-76.

76. Szulanski, G. (1996). Exploring internal stickiness: Impediments to the transfer of best practice within the firm," Strategic Management Journal, (17), 27-43.

77. Szulanski, G, & Winter, S. G. (2002). Getting it Right the Second Time," Harvard Business Review, , 80, 62-69.

78. Trott, P, & Hartmann, D. (2009). Why 'open innovation' is old wine in new bottles", International Journal of Innovation Management, 13(4), 715-736.

79. Trott, P. (1998). Innovation management and new product development, Pitman publishing, London.

80. Ulaga, W, & Eggert, A. (2006). Value-based differentiation in business Gaining and Sustaining Key Supplier Status", Journal of Marketing, 70 (1). , 119-136.

81. Ulaga, W, & Eggert, A. (2005). Relationship value in business market: the construct and the dimensions", Journal of Business-to-Business Marketing, 12 (1), 73-99.

82. van de Vrande Vde Jong J., Vanhaverbeke W., de Rochemont M, ((2009). Open innovation in SMEs: Trends, motives and management challenges", Technovation, , 29, 423-437.

83. Whipple, J.M, & Franke, . . (2000),. "Strategic alliance success factors", Journal of Supply Chain Management, 36, 21-28.1.

29 Chapter ENHANCING VALUE IN FOOD PRODUCTION

1. INTRODUCTION: AN OVERVIEW

This chapter examines the practice of value addition in food production from the perspective of a food manufacturer that has become increasingly integrated into the food industry in recent years: small-scale farming producers. Their construction and practice of enhancing 'authenticity' in culinary goods is specifically revealed.

This chapter has two objectives. The objective is to examine the process of value addition, focusing on the concept of 'authenticity' in food. This examination imparts fresh knowledge and offers future guidance to small-scale farming producers and the broader food business. The secondary purpose is to illustrate the significance of Applied Cultural Analysis in food development.

The behaviors of small-scale producers will be examined using an ethnographic method that integrates participant observation and interviews. This inquiry aims to elucidate the value contributed by small-scale producers, namely in the attributes they impart to their food commodities, the value inherent in their lifestyle, and the value embedded in their production practices. This study analyzes subtle, unconscious behaviors uncovered through ethnographic research and cultural theory, illustrating the practice of value-added 'authenticity' as a culturally produced phenomenon. These findings will be significant not just for producers in comprehending their procedures but also for the broader food business focused on enhancing authenticity in their products.

1.1 Adding value – A problem definition

First of all we have to establish what value is. When following, in a business context, the end- user of a product or service appreciates its use, he or she values the consumption. In return for the value that the business firm produces, the customer pays for the value. The economic value is determined by the subtraction of cost (effort) from the perceived benefit of using the product in question (Peteraf and Barney, 2003). Perceived benefits, then could be anything – status, self-confidence, health, satiety, excitement or whatever the user gets from the product.

A simplistic view of adding value to food brings with it straightforward notions of 'adding' to raw ingredients and processing them in a way that increases economic return. A useful supporting definition of this starting view to 'value added' is provided by the United States Department of Commerce and Industry, which outlines any processing of food as a value added act in and of itself:

Processed foods are "value-added" products, referring to the fact that a raw commodity or commodities are transformed into a processed product through use of materials, labour, and technology. Any product that requires some degree of processing is referred to as a processed product, regardless of whether the amount of processing is minor, such as for canned fruit, or more complex, such as for snack foods (2008; p. 3)

However, the processing of food is not as straightforward as gathering materials, labour and technology together equals automatically producing a value added product. In recent years, a increased emphasis on the addition of culture as a value is an evident phenomenon in the food market today that introduces an element of complexity and requires deeper investigation and further definition. Paul DuGay and Friedmann defines the addition of culture as a value that adds the "inscription of meanings and associations as they are produced and circulated in a conscious attempt to generate desire for them amongst end users" (DuGay & Friedmann, 2002, p. 7).

In close connection to the discussion of culture as value is the commercial value of authenticity in food products. Several large industry consultant groups refer to the importance of value adding, with statements such as:

Key trends to impact the food and beverage market through 2012 and beyond... relate to purity, authenticity and sustainability, as consumers continue to look for products with added value. (Innova Market Insights, 2011)

Added value still matters. Despite lingering economic uncertainty and mounting scrutiny of product health claims, consumers remain willing to spend a bit more on food that does, or stands for, 'something'. (Euromonitor International, 2010)

Value adding could be seen as the reinvention of product for economic gain. But what is the reinvention happening - what are the processes at play? Let's continue with the idea of processing as value adding in food. This processing of a raw commodity to product can also be described as a process of cooking. The french anthropologist Claude Levi-Strauss (1994), wrote in his classical article on the culinary triangle about the processes of cooking and described it as an "...act of mediation, where we transform raw materials into a cooked product... not only does cooking mark the transition from nature to culture, but through it and by means of it, the human state can be defined with all its attributes" (p. 164-5).

Focusing on the words of mediation, transformation and culture, cooking can be viewed as a process to mediate what and how is added in, or transited out; to transform into a final product is a practice to inscribe cultural meaning to the human state. It's the selection of these elements that positions the function of the cook (or food manufacturer) as a mediator at the conjunction of final product to the consumer and therefore places them in the position of cultural agent(s). Culture becomes an important part of the production in cooking - the combination of what cultural meaning and associations are selected and represented in this transformation is the challenge in 'cooking' and thus the 'value add'. Cultural processing and processes bind the selection of chosen elements that can be simplicity, rarity and tradition. Therefore to produce a 'value added' food product, an understanding of cultural processes is required – a complex task in which the food industry finds itself involved.

According to Mintel (2011), an astonishing 20,000 new "value added" products per month are launched in the global food industry. However, a demonstration of the complexity involved with cultural interpretation, and just how challenging creating a cultural product can be, is found in the exceptionally high failure rate, "exceeding 90 per cent for some categories, which suggests that firms have difficulty in developing products that appeal to enough people to warrant continued distribution" (Connor & Schiek, 1997).

Combined, this has created a food industry both focused on creating new added-value products while at the same time, working to minimise the potential risk of failure and subsequent economic loss. The sheer number of products also hints to the increasing complexity of adding value to products by the inscription of cultural values. These cultural values often take the form of ideologies, experiences, morals, mythology, magic and sometimes illusion infused within the product offer and branding, such as Coca Cola's inscription to transform a carbonated soft drink to the embodiment of the spirit and essence of America, summarised in their advertising plans:

Coke is real, authentic, honest, passionate and refreshing. It taught us to sing in perfect harmony. It even introduced us to Santa Claus. Coke is the kind of cultural fabric that unites all of us in some way. It has always encouraged us to share who we are and what we believe in (Identity Advertising, 2007, p. 2).

The creation of 'value added' inscribed in food is thus an effort to elicit emotion from consumers and entice them to spend more money in return for feeling this emotion. Value adding is a basis for economic gain via emotion, and it is emotion that becomes critical in the evaluation of how successful a value add food product is. It is also one of the more difficult to interpret – one cannot simply walk up to a consumer and ask what their emotions regarding a certain product are – they would not be able to tell you. It leaves mediators (cooks) charged with the responsibility to decode cultural meaning and charge their products with it by their practices. It also leaves the assessment of their labours as ephemeral and complex. It is information that needs to be teased out, tantalisingly revealed and then hidden, uncovered by other means - lending qualitative research methods such as ethnology and anthropology as critical in the understanding and evaluation of value added food.

For the food industry, the tension created between potential and seemingly boundless creative transformative opportunity available in developing 'new' value added products is balanced against the underlying understanding of risk in both the emotive assessment of the culture that is infused and also in economic terms has led to, as Belasco (1989, p. 210) states, "corporate conservatism", neatly outlining a response from the food industry to address both perceived risks: there has been increased focus to consumer-directed research, in some instances as broadly to elevate its importance into the heart of day to day business operations, while at the same time to reduce the costs in bringing a product to market. The emergence of qualitative research methods to uncover 'insight' in the food industry saw an increasing investment to understanding food consumption within a daily life context when developing value added products. This however, is a difficult task when we consider the unconsciousness of taste and how unaware we can be about the food we eat on a daily basis.

The environment in which we eat our meals also demonstrates the quiet absorption of food into the background of daily life. In a 2011 report, it was suggested that in an average person's day, 56% of the time spent eating (85 of 152 total minutes) is while engaged in another activity considered primary such as watching television, driving, preparing meals, or working (Hamrick, 2011, p. 5). We're not even paying full attention to over half the food that goes in our mouths, yet significant amounts of time and money are invested by large food manufacturers trying to get consumers to tell them why they eat what they do.

The complexity in dissecting everyday life within business constraints such as timelines / budgets in a climate of conservatism and cost reduction has conceivably led to an avoidance of radically innovative products.

The reduction of economic risk by bringing down costs in value adding is also demonstrated with the use of materials (such as synthetic / non-food additives, flavour enhancers, thickeners, colours, stabilizers), a change in business operations such as off shore manufacturing, combined actions of which have increased the length and complexity of the food chain and the distance – both geographical and mental - between the raw and the cooked. It is this distance created over the past 50 years that's had a hand in creating the demand for authentic products and places small-scale production in the position it finds itself today.

Given the ongoing involvement and importance of value adding authenticity in food by these large companies, it may seem surprising that small-scale food Producers were not run out of the industry long ago. Instead, these Producers find themselves in a strong and growing position within the food market for value added authenticity. For example, direct to consumer sales from small scale farming production have grown 104.7% from 1997 to 2007 (Martinez, 2010), and 60% of the Swedish population prefers to buy local produce (Ekelund, Fernqvist & Tjärnemo, 2007, p. 233).

2. UNDERSTANDING VALUE ADDING

2.1 Food and cultural studies traditions

There is a long tradition in cultural and sociological food studies regarding the consumption or meaning of food and does provide a basis for understanding a consumer perspective, and how demand is created. Existing food studies credit food consumption for a vast array of cultural process from the reproduction of a stable society (Lupton, 1996; Goody, 1982), the decoder of the unconscious attitudes of a society (Levi Strauss, 1994), of nationality (Barthes, 1967), as an indicator of class (Mennell, Murcott & van Otterloo, 1992; Bourdieu, 1984), social significance (what is important in a given society) (Douglas, 1982), to the relation of food in feminist studies and body shape (Adams, 1990). Taken together, these studies point to food as central to culture, a suitably broad implication that requires narrowing.

In narrowing to the central theme of this chapter - value adding and authenticity - the central ideas that emerge from a consumer perspective is that consumption of small-scale, authentic food is enjoyed largely by the elite upper classes in westernised societies – the diets of the poor have remained relatively unchanged: they are excluded from this possibility (Bourdieu, 1994). However, Alan Warde (1997) argues that the emphasis is shifting from class formation towards the formation of new groups in society that share 'lifestyle' rather than social class, and the consumption of 'authentic' food is a representation of chosen lifestyle group. A study by Anthony Giddens (1991) takes a wider view, outlining that the agency of consumers is emphasised over both the social and economic structures in which they find themselves, and this is a crucial means of establishing an individual's identity. The emergence of consumption of small-scale authentic food is also evidently associated with the mitigation of food risk and potential for disease as explored by Ulrich Beck (1992) and Mary Douglas (1996, p. 123).

Further, in specific regard to the term of authenticity, a concrete definition is a little difficult to locate. It has been described as a product of shared systems of signification (Ashley et al, 2004, p. 7) and is a modern phenomenon (Appadurai, 1986). What constitutes as 'authentic' changes over time (Peterson, 1997), an example of which is exemplified with the shift in lard heavy foods recipes - such as Cassoulet - where goose fat may be substituted for cheaper and lighter alternative fat sources but still acceptably remains an authentic French dish.

In some cases, authenticity has been deciphered as an act of micro-resistance to dominant forms of cultural production (De Certeau, M; Giard, L & Mayol. P, 1998) - where one makes an appeal for authenticity in response to their perception of inauthenticity in others. For example, you are what you don't eat. Josée Johnston & Shydon Baumann (2010, p. 70) provide a key view on authenticity with their claim that it "doesn't really exist" as it is socially constructed. The definitions of authenticity are difficult to define concretely and are deliberately vague in use here, as it is the cultural construction through practice that this thesis aims to reveal.

It's important to note that the term of 'consumer' can be applied to Producers in this chapter, also - after all, as we all are when it comes to food - the small-scale producers in this study are consumers, too. There is argument that Producers are developing products they personally consume.

2.2 A new food economy

The increasing intertwinement of "culture" and "economy" is a focal point in both cultural studies and economics during recent years (duGay & Pryke (2002), Pine & Gilmore (1993), Lash & Urry (2002)). The framework of a "New Economy" based on cultural values with labels such as Experience Economy or Dream Society has been launched during the last decades. We will discuss some of the major writings in order to define their usefulness for understanding an emerging "New Food Economy" with cultural production of added value as a common denominator. Firstly, the definition of New Economy in use here is reference to the increasing intertwinement of symbolic cultural messages and economic processes, the desired outcome of which is economic gain. The use of symbolic cultural messages is a key element in the New Economy, as it involves the responsibility for consumers to develop skills in 'reading' cultural values and, as Lash & Urry (2002) and duGay & Pryke (2002) point out, brings in references to the New Economy as being an 'economy of signs'. DuGay & Pryke's argument also outlines that there are an increasing number of goods and services that can be regarded as 'cultural' goods. Their use of 'cultural' goods refers to the inscription of meanings and associations to elicit consumer desire for their purchase - exactly what our cooks are trying to achieve - but from this view it is unclear if food is part of the basket of 'cultural' goods in the New Economy. They further describe that:

There is a growing aestheticization or 'fashioning' of seemingly banal products whereby they are marketed to consumers in terms of particular clusters of meaning, often linked to 'lifestyles' and this is taken as an indication of the importance of 'culture' to the production and circulation of a multitude of goods and services (as quoted in duGay & Prkye, 2002, p. 7)

The inclusion of 'banality' would seem to infer that food is indeed inclusive; as Jönsson (2005) outlines in his discussion of milk, he demonstrates an obvious connection in the New Economy due to rapid development pace and inclusion of cultural messages in added value products in the dairy counter. While the mention of rapid development would seem to indicate the there is something new happening with the 'fashioning' of meanings embedded in (banal) food products in the New Economy, Jönsson continues that this may not necessarily be the case. As he outlines, "the production of such "value-added products" can be seen as a part of a general conversion (to the New Economy), but it is suggested that it can just as well be that the economy is increasingly bent on trying to capture experiences that have always been conveyed via food" (2005). In other words, the meanings inscribed in value added food are an amplification of information that is already known, and communicated via experiences, bringing in an inclusion to the 'Experience Economy' as well. A concept discussed by Pine & Gilmore in 1993, they outline the construction of experience 'spaces' as key in the eliciting of emotion for economic gain. In this world, no one sells mere commodities ... they sell 'lifestyles'. That the spaces Producers develop to sell their products are in part to create emotion further embeds them in the experience economy.

It is here that I position small-scale Producers - within the New and Experience Economies, producing products inscribed with cultural values: but as will be shown, the inscriptions made utilise previously held cultural knowledge. The role of those who create cultural inscriptions is continued by Lash & Urry:

This process, they argue has been accompanied by the increased influence of what are often termed the 'cultural intermediary occupations' of advertising, designing and marketing; those practitioners who play a pivotal role in articulating production with consumption by attempting to associate goods and services with particular cultural meanings and to address those values to prospective buyers... these signifying practices in doing business is evident not only in the production, design and marketing of goods and services, but also the internal life of organisations as well (as quoted in duGay & Pryke, 2002, p. 7).

This further embeds the position of cooks as 'cultural intermediaries'. Another key supporting element of Lash & Urry's description is the increased importance accorded to signifying practices that create symbolic value in the production of goods and services, such as those of small-scale Producers under investigation here. It also calls into account for the wider food industry to understand that it's not only the communication messages that are interpreted here: it is the reading of the entire production process.

Accordingly, an investigation of understood cultural values and production processes over time provides foundational information in understanding the context of value adding in the New Economy. Searching the historical roots of the emergence of late modern value added food production, Warren Belasco stress the importance of the radical counter cuisine against mainstream food ways that began to emerge from 1966. Some of the more important elements were:

- A consumerist component offered survivalist advice and suggested what to avoid, especially processed 'plastic' food.
- New ways to make food more fun - e.g. through the delight of improvisation, craftsmanship, ethnic and regional cooking.
- The organic paradigm, which posited a radically decentralised infrastructure consisting of communal farms, cooperative groceries, and hip restaurants (Belasco, 1989, p. 4).

While the countercuisine initially was interpreted as a threat by the major food companies, the ideas soon became the main inspiration source for product development. The wide range of rural, authentic, hand made, old-fashioned, ethnic and light products launched the last decades owe a lot to the counterculture of the 1960s and supports the use of this theory as still applicable in the food industry today.

Further support that these elements are still in demand today is an appreciation that the decentralised infrastructure Belasco refers also includes Farmers Markets, Farm Shops and specialist stores - all of which are preferred sales channels of Producers in this study. The emergence of the Farmers markets as an alternative food channel became evident during this cross over period between the second and third food regimes. The first Farmers Markets were reported in the US during the 1970's, with a slow expansion over the subsequent twenty to thirty years - however increasing at a faster rate in more recent years. The market in observation for this study has been in operation since 2000. As shown on the chart below, the number of operating Farmers Markets in the US grew by seventeen per cent to 7,175 from 2010 to 2011 alone.

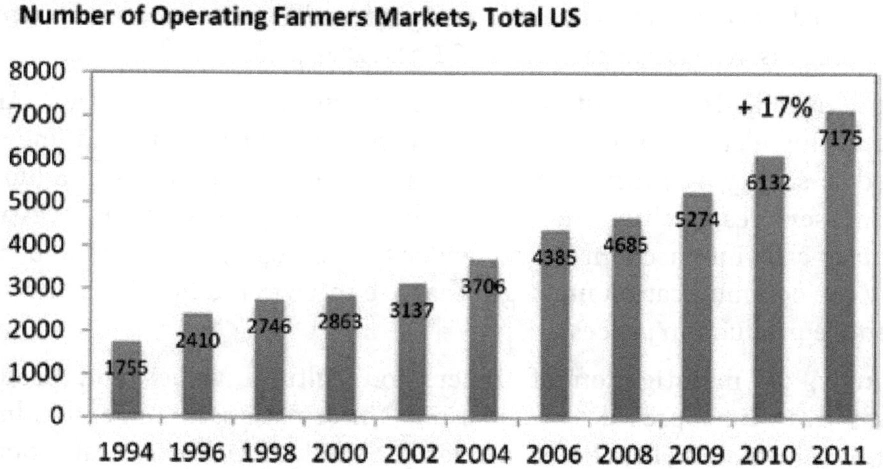

Chart 1. Number of Operating Farmers Markets, Total US Source: United States Department of Agriculture, 2011

Richard Wilk (2012), in his description of the contemporary food movement, also points to the growth of Farmers Markets, local food councils, regional food alliances, school food initiatives, community gardens and orchards, and popular interest in the quality of food over the past five years... and like Belasco, wonders if these will be exploited as a marketing opportunity or turned into an ideological foundation for change in the food landscape.

This counter also marked the emergence of the consumer as a voice of influence in the food industry, after the focus to internationalisation and mass industrialisation in the first and second regimes. A key point of Belasco's analysis is his argument that marketers who 'abused' the process turned this counter into mainstream belief. To what process does he refer? He describes the way in which marketers turned these beliefs into profitable products was by turning food messages to meet desired consumer lifestyle and identity values, and it was this focus to meeting consumer lifestyle needs that created their voice of influence. In order to communicate to identified values, marketers repositioned food added values. Belasco describes a deceptive 'picking and choosing' process where 'good' elements of products were chosen to anchor repositioning in lifestyle values, and the 'bad' elements were ignored or glossed over - for example, carbonated soft drink became 'low in fat' to embody consumer desire for a healthier lifestyle (Belasco, 1989, p.212).

The success evident in relevant consumer re-positioning, demonstrated below with organic retail sales - even with potential drawbacks of its use, particularly when deception or excess 'glossing over' is involved - has lead to virtually an entire industry shift to how food products are communicated to consumers.

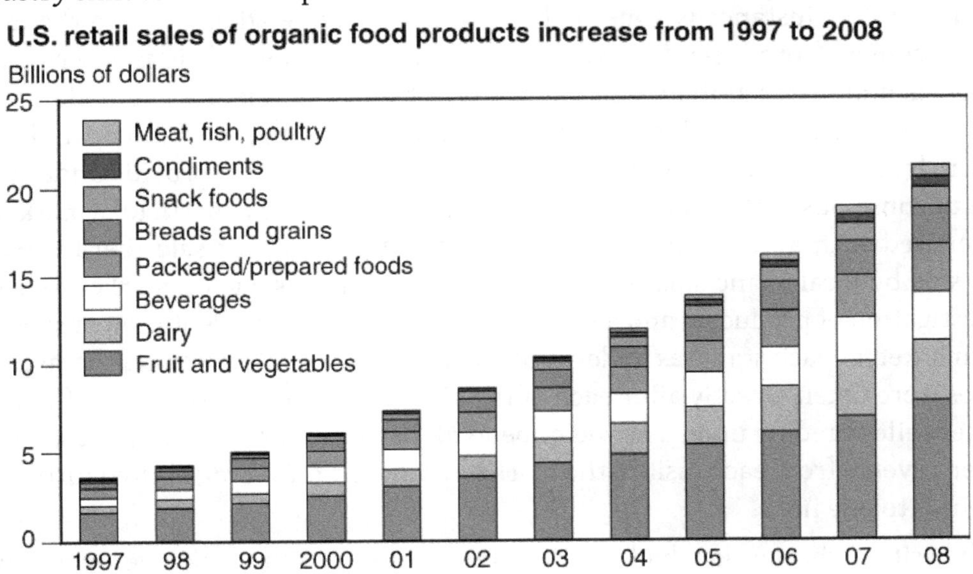

Chart 2. Retail Sales Growth of Organic Food Products. Source: Dimitri & Oberholtzer (2009, p. 13)

Products and their messages began to transform from functional communication style into symbolic lifestyle and experience messages - and it is this process of how consumers read these messages that underpins Belasco's argument in how the counter became mainstream, for without the culturally held knowledge to 'read' the message conveyed, for instance, using a picture of a tractor on a green field is a farm, is completely lost.

It is the process of developing symbolic representation and cultural construction of small scale Farmers through their production and selling methods that allows a simple understanding and competitive advantage in a confusing food landscape. This occurrence is made all the more interesting in a wider food industry of careful, researched product positioning and communication from larger scale manufacturers: to achieve this, the small scale Producers in this study are unconsciously just doing what they do.

I am a farmer's daughter, and my mother, she cooked jam and juice and all this that we also do, but she was never paid for it, but we are... and that is the difference between the 20th and 21st century! (Skåne Producer, November 2011).

3. METHOD AND MATERIALS

For this investigation a specific empirical study involving small-scale value-added producers in the Skåne region of Sweden was undertaken.1 Taking an ethnographic research approach, observation and participant consumer observations were undertaken at a periodic Farmer's Market held in Malmo, Southern Sweden, over six Saturday mornings during 27 August 2011 to 15 October 2011. Ethnographic enquiry in this instance is a method to investigate value adding due to the focus on consumer / producer interactions and the evaluation of signs inscribed in products and material items used in display. It is also a relevant methodology to 'move the conversation forward' regarding value adding, as common industry methods such as quantitative surveys have apparently failed to provide underlying and unconscious motivations in their subjects of study. This particular market is advertised with an express requirement that all produce for sale there is made and sold by local Skåne small- scale farmers. As such it is a good market to begin investigation of producers and also analyse the material representation of space at the market, as each stall was under control of the individual producers themselves. Notes were taken directly after each of the six visits to the market (it is difficult to write while standing up in a crowd of people), in a nearby café, with a focus to the salient events from each visit to the market. Each visit on average lasted forty- five minutes to one hour.

Each visit to the market had specific item to review. In week one, which was also the first market day of the market season, the organiser of the market was interviewed, and introductory contacts with all twenty one stall-holders present on that day. The attention on this day was drawn to the ways in which the organiser the perceived competitive advantage of the Farmer's Market, the type of consumers she believes shops at the market, expectations of the coming weeks' market sales and also the best way to market Producers' product to consumers, (particularly for newcomers to the market, of which there were two). The second week the investigation of the market were directed to the specific intention to review the material representation of space – display items outside the products themselves and signage that producers commonly use.

The third week, focus was to body language and behaviour between stallholder and consumer while tasting and/or buying products. Week four was focused to obvious advertising promotional or 'special' deals available at the market. Week five required another review of material representation (a repeat of week two), and the final week was spent looking for details missed on the previous weeks.

Additional ethnographic observations were made during farm visits, and hour long depth interviews were also conducted with four producers during October and November 2011, in addition to an online questionnaire circulated to farmers in attendance at the market, of which fifteen responded. The online questionnaire, which participants were able to respond anonymously is a recognised departure from ethnographic methods, however was circulated primarily as a way of locating participants for deeper investigation.

However, the understanding of why certain questions were answered as they did in the questionnaire revealed again that qualitative methods uncovered much more about Producer value adding practices than straightforward quantitative line of questioning, and adds further support for the use of a cultural analysis approach in the articulation of meaning. All of the depth interviews were conducted with the use of an interview guide allowing to focus the majority of attention to the subtle emotions the became evident during discussions. What subjects created the most emotion? These subjects became focus for later analysis. In addition, web ethnography was also undertaken of Producers at market to understand their representations of self and product online, and was also taken into account during analysis and depth interview guide preparation as required.

Of the four depth interviews conducted, three were held on the site of farming production – which usually also housed home, office and Farm Shop where their produce was sold. This gave an opportunity to view participants in the scene of their day-to-day lives, which often involved family.

The main field work was conducted at the Farmer's market in Malmö in the South of Sweden. As other Farmer's markets throughout the world, it was built around a specific regional provenance (in this case Skane, the most southern part of Sweden, which produces around 50 % of the Swedish food production) the number of stalls selling produce that had been mediated and transformed in some way (such jams, smallgoods, rapeseed oils, breads, cheeses), and the stallholders were also the producers of the products that they sold.

Johnston & Baumann (2010, p. 70) have tried to decode the elements of Authenticity that consumers demand in high-value food. They claim that geographic specificity, simplicity, personal connection, tradition, or 'ethnic' connections are the manin elements of authenticity. They report that from a consumer perspective, a 'good' authentic product does not need to have all these elements, perhaps one or two... but a really authentic product from a consumer perspective has elements of all five. This market communicated most, if not all of these elements.

While watching Producers carefully setting up their stalls to display their product in a way they deemed the most appealing, and engaged in passionate descriptions about the product they sold with consumers, it became obvious that there was a very close connection they made with their product. Their products were tended with almost maternal care. Jars were carefully and individually placed, labels facing forward... not too cluttered, each jar having a right to stand on it's own. Esteemed flavours stood centrally, close to where the stallholder could easily access for offering tastings and making serving recommendations. It was explained that this set up and staging was taken very seriously, with the utmost 'attention that must be paid to the small details in a display' – right down to the pattern on a table cloth used to conceal a rickety, aged wooden table. The care and thought placed into the display interestingly revealed a pattern of placing the stallholders themselves as co-stars and leads to the first Producer practice in adding value.

Observing the interaction between consumers and producers at the market, it became evident that exchanges were based not only on direct economic issues such a price. There was much more under discussion underway. Even though the elements of authenticitiy described by Johnston & Baumann were found at the market, the elements as such does not give full credit to the practices involved. And they certainly do not answer the question why certain producers and products seem much more authentic than others, although they share the same elements.

The study of the actual practices of value making made some new themes emerge. They are summarized under four headings; Co-creation, The farmer myth, Art and Magic.

3.1 Co-creation

The online questionnaire circulated to locate participants for this research included a question that asked: 'of your current distribution channels, which do you prefer? Please select one from below'. The majority of respondents (66.6%) indicated Farm Shop or Farmers Market.

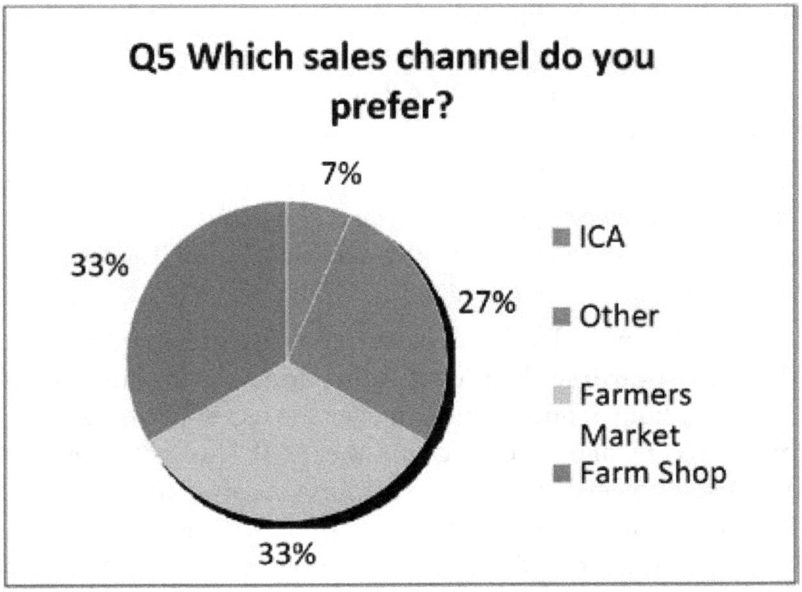

Chart 3. Preferred Sales Channel, Producers in Skåne, Sweden. Source: Online Questionnaire, August 2011. (Note: 'Other' is predominately 'Restaurants')

This piece of information on its own is a little bit 'nice to know' as it fits with the preferred channels of the counter movement outlined previously. When asked about why Farmers Market or Farm Shop is preferred, most participants indicated it was personal selling and consumer contact that they like the most about these channels. There was a consensus that selling in this way was intensive work, but it was 'worth it' - there was significant return for labour. Which begs the question: What is it that they are getting in return? Is it purely economic gain?

Evident from the exchanges observed at the Farmer's market and discussed in later depth interviews indicated this return contained multiple components. A key component was one of education. Producers and consumers educate each other about the meaning infused in the food (consumers commonly rewarded the product for being hand-made), new ways of consumption and suggested occasions (try this with blue cheese, perfect for a summer party!) - important in meeting the consumer and Producer need in identity building and class positioning through cultural capital. As Producers spoke about their product, they educated consumers about varieties of produce included in the food (i.e.; The tomato used is Roma, best suited for this type of relish), and also, importantly, Producers gained ideas for new product development through suggestions from consumers. The drawing of new products into the discussion may at first seem a little offensive, but later discussions with Producers revealed that this type of communication is invaluable information for new product generation and development, and has been credited with sparking ideas for product and also material displays of space. Two examples of this involved a recent award-winning product with the unusual inclusion of a savoury, exotic spice to a sweet condiment – tapping into an emergent consumer trend for food exoticism. The second example regarded the co-creation of one producer's use of the ubiquitous 'chalkboard' sign:

When I started, I wrote by hand (the signage) and I do not know how many times I bowed and curtsied for customers and apologised, how it was awful and all of that. And then a lady comes and she says to me like this: 'But, it is proof it is handmade'. Huh! (Laughs)... Sigh! It was, of course! (Skåne Food Producer, November 2011).

It was also revealed during depth interviews that disagreements with consumers may occur (usually regarding their 'confusion', or as one Producer referred 'blasé and spoilt', when talking of one discussion over the use of plastic wrapping around a herb plant, the consumer thought this meant it was not grown at site and objected to its use), however Producer position was always made in reference to their relation as the solution to confusion rather than the cause of it. It was also revealed that Producers rarely discuss new product ideas with other producers, as 'the market in small in Sweden - you have to learn to stand on your own', indicating that there are limits to the farming community in Skåne. Exceptions to this were evident, however only when a philosophical connection and strong friendship had been formed over a number of years.

These exchanges with consumers are very much a two-way communications process, and, to reference to Prahalad & Ramasway, an exchange "between the firm (Producer) and the consumer (in) becoming the locus of value creation and value extraction" (2004, p. 6) and is an invitational act of co-creation. The co-creation is the centre of value extraction and knowledge for Producers to build their producer offer and the way they sell their product, and as such is held at a high level of importance.

While Prahalad & Ramasway's (2004) investigation examples primarily involve larger corporations, such as car manufacturing giant General Motors – what stuck with the discourse with consumers was that small-scale Producers are routinely in a position that larger business pay large amounts of research money for... in direct contact with individual consumers who are eagerly sampling their products, freely offering feedback, sharing tips from other sources, with the added bonus that the consumer is speaking with a person that has a level of influence within the business structure to initiate change. Small-scale producers are very close to the consumer via their preferred distribution methods, and in a position where they can respond relatively quickly to changing consumer demand. However, as Prahalad & Ramasway articulate, this is achieved not by only just 'showing up' at the Farmer's Market or opening a Farm Shop. To do so, Producers must invite feedback. Do they? If so, how?

Exploring the practices that Producers engage to begin the co-creation process at market started to tease out more information. The approach to addressing this question firstly lay not by observing Producers that appeared to be engaging in a co-creation with consumers, but by observing Producers that were not. There were two stalls at the market during fieldwork, both of which consistently had uncrowded stalls while the remainder of the market teemed with activity (one sold shiitake mushrooms, the other corn).

There were two common denominators between these stalls: 1. Their product offer was not immediately apparent it was 'value added' and 2. Their stalls featured little to no elaboration in their stall displays – no or very limited signage, no extra items to add visual appeal, and symbolically represent a return to the farm / authenticity such as the previously mentioned flower arrangements or wooden crates. There was no elaboration in the space to communicate their values, which as discussed earlier, are important in the social construction of authenticity.

The absence of continuous consumer - producer dialogue at these stalls would seem to indicate that consumer feedback and discourse is in support of Prahalad & Ramasaway's theory in that it is invitational. This occurrence implies that there is prerequisites involved the staging of experience in order to co-create with consumers. The invitation extended by a Producer must materially demonstrate their authenticity by the use of symbolically loaded material objects, and further offer an experience such as tasting plates before consumers will engage in discourse. It would also seem to indicate that non-value added products are perhaps not necessarily seen as something in which to discuss in greater detail than necessary to complete a financial transaction - again bringing to mind Levi Strauss with his cooking as a mark of culture.

3.2 The farmer myth

The symbolic meaning communicated by the material display items in use is a key point here. As the symbolic loading in these items is culturally held, unconscious knowledge, further analysis is required. After all, the person that builds the displays first imagined it in their mind - perhaps they didn't realise how those images got there. The displays heavily utilise cues for 'authenticity', and for discussion of this I'll elaborate on what I am calling the Farmer 'myth' utilising Barthes.

One of Barthes "lasting contributions... was the identification and interpretation of certain 'mythologies' that he drew from everyday life in France" (as quoted in Atkins & Bowler, 2001, p.6). Barthes utilises semiology as a tool to understand the basis for myth creation, and relates the expression of myths as inherent in food. As Barthes stated: "for who would claim in France that wine is only wine?" (1997, p.20) relating wine as a symbol of France and to the myth of what it means to be 'French'. Likewise, in the current food regime of counter-values to mass industrialisation, the myth of farming is utilised in small-scale food production to embody a 'return to the farm' in support of their authenticity.

The Producers are met with a challenge, though: this use of these 'return to farm' signs is not a unique in the food industry today, as Belasco previously outlined in discussion of the emergence of the food counter culture. Most of the items in use here to symbolise the return to farm can also be found represented on numerous food packaging, and in virtually every food distribution channel, including mass supermarkets and speciality stores. Items such as chalkboards, wooden crates, wicker baskets, striped awnings, are in common use in supermarkets and speciality stores around the world.

But what of our Producers in this, using the same material items, yet successfully inviting co-creation with consumers? With this proliferation of material displays in use in the market today, why are these displays not approached with cynicism, disbelief, or just plain dismissal? Here Baudrillard makes a contribution with the outline of two distinctive features of mythology to be drawn as relevant: "firstly the nostalgia for origins, and secondly the obsession with authenticity" (1996, p. 76). Baudrillard outlines that the obsession with authenticity is reflected with an obsession with certainty - specifically, certainty as to the origins, date, author and signature of a work, continuing:

The fascination of handicraft derives form having passed through the hands of someone the markers of whose labour are still inscribed thereupon: we are fascinated by what has been created, and is therefore unique, because that moment cannot be reproduced (1996, p. 76).

In co-creation the feature of certainty in myth here is a distinct element in invitation: the certainty that in a confusing food landscape and the new economy of signs, where authenticity is hinted at almost every turn, that the Producers are farmers. There is certainty created in regards to the origins of the food, as the author and signature to the work, often accompanied by undeniable physical presence. What could be more certain than that? This invitational certainty is in direct contrast to Prahalad & Ramasway's theory where they propose that 'trust' (or certainty) is an outcome of co-creation (2004, p. 13). Here we propose it is a central element in use at the beginning of the co-creation process. It's the certainty that Producers can truthfully and safely take consumers from myth and into reality.

In the next section, the displays used at the Farmer's Market is contrasted to the spaces constructed at Farm Shops - with the consumer on the farm, the invitational space to co-create becomes something else entirely.

3.3 Art

When discussing five-year plans, most Producers indicated a desire to increase business through their on-site Farm Shops - all of the depth interviewees had a

Farm Shop, and thirteen of the fifteen respondents to the online questionnaire did also. These shops are under their full control, and thus ideal to explore the more permanent environment (compared to the Farmers Market) in which they have constructed to sell their product. All the producers referred in depth interviews to a desire to elicit positive emotion from consumers from their Farm Shops, something they felt that large supermarkets could not do[2].

One Producer spoke about their delight with their shop layout creating happiness, astonishment and surprise emotional reactions from consumers:

In my farm shop I don't stock these types of everyday products (sold at ICA, a large supermarket chain in Sweden). No, what I have here has nothing to do with everyday commerce. There is no milk... I do not have those types of products, this is really speciality items... I really take it to myself, how I have been thinking when I planned the store. "What should it look like in the shop to make it inviting to the customer, inviting to enter and to have a look?" Often I say... (Pauses: body language changes, pulling away a little bit - it's like a self reprimand for becoming a little too pleased with oneself, quickly decides on an acceptable rephrase, leans forward again) ...the customers usually stand in the door and say. "OOOOH what a lot of, AHHHH how nice!" Then they are of course happy, I think! (Laughing). Yes indeed! Most of them are perhaps a little astonished and surprised, because they don't expect anything more. (Skåne Food Producer, November 2011).

Arguably this is an effort to create a unique positioning and distance to the dominant food sales channel today (supermarkets), however the exclusion of what is perceived to be everyday products raises some important questions regarding the 'specialness' infused their products and reveals beliefs about the work they do. Why is an everyday food item such as milk not considered 'special', particularly when considering that a number of products for sale in this store could be considered as 'everyday' use items such as jams and mustards? Consideration of the space implies an effort to raise their product above the everyday and into a form of art, and in doing reveals Producer cultural capital and class aspirations.

To communicate their perception to the specialness, non-everyday, uniqueness of their products, Producers reveal a tacit, cultural understanding to their construction of space, and have turned their Farm Shops into spaces resembling Art Galleries. The Producer quoted earlier in this chapter created a large space, beautiful to look at, with a significant amount of non-food items, and set up in a maze of stalls to encourage exploration and discovery. Another Producer housed several buildings, one of which featured a café with soaring cathedral ceilings, artwork on the walls, and with their produce barely visible in - again - a beautifully constructed space. The farm shop was located in the same building where the produce was 'cooked', separate to the main house that was a short distance away, however all cooking space was off limits and invisible to regular custom in the shop, as were all of the cooking spaces in the farm shops visited as part of this study.

In guided tours with Producers at their farm shops, there was always a respectable distance kept from the Product in store, to allow the viewer to take the entire scene in. The lighting scheme shined directly on produce, to illuminate its form. The display of goods was uncluttered, so as not to catch a wandering eye's attention from the main event. Discussions commenced about why the products on display were chosen with slightly hushed tones, even if we were the only people in the store. These actions struck as being very similar to behaviour that would be found in art galleries - here, a discussion was ongoing with an Artist. An artist creates. What are these spaces creating?

The construction of these spaces to resemble art galleries attracts the use of the term 'Experience Economy' as introduced in theoretical framework to elaborate and further embeds Producers as active participants in the Experience Economy. The art gallery spaces are carefully staged for consumption and construction - consumption of the producer, consumption of myth, consumption of art...and hopefully for our Producers, what follows is the consumption of authentic 'special' food products.

Tom O'Dell & Peter Billing (2010) outline a useful interpretation regarding spaces that are staged for consumption using a theoretical framework that combines the conceptual tools of Arjun Appadurai's (1996) 'scapes' and Henri Lefebvre's (1991) production of space. Herein they have developed an understanding that spaces where experiences are manufactured for consumption can be interpreted as "places in which the global and the local are entwined and where power relations played out, political interests are material interests are materialised, cultural identities are contested and dreams are redefined" (O'Dell & Billing, 2010, p.18). The building of producer's identity through practice is a key function of the space created. The environment feels very much staged for artistic expression and exploration... a space has been created that Pine & Gilmore (1993) could interpret as a carefully staged production to create experiences such as those available at Disney or the Ritz Carlton.

A main criticism of Pine & Gilmore's description of the 'Experience Economy' is that the spaces as provided as examples (such as Disneyland and the Ritz Carlton) are created with the assumption of a passive receptivity – consumers have no hand in creating the space and limited opportunity to personally shape it. However, as is known from the previous research finding, co-creation is a key element to Producer practices and the impact of consumer feedback on the construction of this space cannot be underestimated - regardless if the Producer is conscious of this or not. The inclusion of art indicates a desire for the accumulation of cultural capital in identity construction, not only for Producers, but for their consumers, also. The construction of this space seeks to confirm Producer position within upper 'cultural capital' class, and the investment of time, money and planning into the ongoing operation of their galleries demonstrates a long-term desire to retain this class positioning.

3.4 Magic

Magic, the art of influencing through supernatural means, could infer secret, hidden and disappeared acts in production practices without clarification of its use. Here, magic is described and applied using two references from the social sciences - Sir James George Frazer (Social Anthropology) and Marcel Mauss (Sociology).

In '*The Golden Bough*' first published in 1890, Frazer notes magic generally falls into two categories: the Law of Similarities and the Law of Contact or contagious magic. The former applies to the belief that like produces like - an example used was "the then current belief in among peasants in the Balkans that swallowing gold could relieve the symptoms of jaundice" (Barnard & Spencer, 1996, p. 341). The Law of Contact or contagious magic was causal idea of things that had once been in contact continued to influence on each other at a distance.

In describing themselves in their products, Producers emphasise 'them' as integral to the product offer. This introduces a belief that they hold continued influence in and over their product before, during and after consumer purchase, to continue the communication of their authenticity, thus asserting their influence at a distance via magical means. In discussions with Producers, the appeared to be little separation between Producers and their products ... the way their products were referred was often in the same way that you would refer a person. The extension of themselves as represented in their products created a desire for their products to be treated as how they would personally like to be treated. For example, it encouraged a slight concern about selling in delicatessens where there is a perception consumers are particularly discerning and demanding, which means the chance of product rejection (and thus, themselves) is higher. Producers also held a strong resistance to sell in supermarkets where their product could become the 'little' with scant consumer attention (a bit like going to a crowded party and standing alone in the corner), or face the potential of having their product abused in store by careless staff.

The Contagious Magic in use by Producers is a practice to infuse a competitive advantage they believe to be what no one else has into their products: themselves. Producers are the magicians in their kitchen, infusing themselves in their products. Two quotes from research summarise this belief in infusion of self very well. One Producer mentioned that their soul must be in the product:

I have nothing to sell if there is no soul in my product... then I find it difficult to justify why the customer should buy my (product), then he can buy another, because what's unique in mine is that I cultivate it. It should be something that no one else has. (Skåne Food Producer, November 2011).

While another in responding to a question asked in the online questionnaire: 'If your favourite product could talk, what would it say to consumers?' The response neatly stating: 'I am special, unique and wonderful. And you?'

Additional theory to view producer practices using Magic is provided by Mauss in 'A General Theory of Magic' (2001), which is that of a social phenomenon containing belief and rites.There must be a magician to produce an act of magic, and an audience to believe in this act. Rites are developed to enact and perform the magic. Magic is not a one-time occurrence; magical performances are repeated patterns of behaviour and when at it's most successful, refer to "those things that society as a whole considers magical" (Mauss, 2001, p. 22). As O'Dell (2010) describes, the inclusion of belief in this theory provides 'special problems' that will be revealed as particular relevant for this discussion:

Magic is an institution only in the most weak sense, it is a kind of totality of actions and beliefs, poorly defined, poorly organised even as far as those who practice it... It's existence necessitated two different forms of belief that he (Mauss) called a 'will to believe' and 'actual belief' (p. 57)

The problem being, for acts of magic (or influence at distance) to be not seen as trickery, they require collective belief by all actors involved, with each actor having a role to play. The magician in this act is the Producers and it is here where belief does exist - but it is not without tension. Present and seemingly unwelcome is a level of uncertainty about their own skill and knowledge in their practice, that in conversation betrayed an oscillation between a 'will to believe' and 'actual belief'. This tension was managed by the development of magical rites, and as far as they could possibly manage, a self-imposed 'no- cheating' policy on these rites.

These rites specifically managed the symbolic values they use to describe their products as authentic - themselves in the product made it hand-made, home-made, crafted. The translation to practice is rather literal: to be home made, the production practice must be either in their home or in a nearby building as part of their faming complex, never in a removed and distant facility. To be hand made, their own hand must be involved in making the product, and they must perceive a level of craftsmanship involved in the transformational cooking process. One Producer described the importance of seeing and inspecting each individual raw main ingredient with his or her own eyes. Another Producer describes the importance of the cook (themselves) to be directly involved in the cooking, cooking in low batch volumes and never on a stove that could be considered 'industrial':

I stir the pot myself. I never use a pot that does not sit on the stove. I never cook in volumes that are larger than 25 litres. No, no, no. Not ever. (Skåne Food Producer, November 2011).

...And also describes readying themselves psychologically as to not infuse the products with stress, which might negatively taint the flavour:

It is an experience, every day, every batch that is made; you do not always get the same batch or the next batch exactly the same with the same products. It depends on your peace of mind, on how much you put in your measuring cup. It has to do with... well, if you are under stress... all those things matter (to the final product). (Skåne Food Producer, November 2011).

However, these assertions were made at the same time limitations and exceptions were made obvious, and showed clearly the edges between a 'will to believe', 'actual belief' and deceit. It and also showed differences of opinion and poor definition where these edges actually lay between the Producers involved in this study - there was no common approach. Individuals had determined what fit with their magical act. For example, it's understandably difficult to grow all of the required raw ingredients for a product on the one farm, so while it is important that they farmed the 'star' ingredient in their products, it appears that consideration to all origins of the ingredients in use are not considered as heavily - one Producer mentioned popping into the local ICA store whenever an ingredient was needed, and had never checked the provenance of the product bought there for that purpose; while another mentioned checking provenance of all the ingredients they use. Another mentioned the recipe for one product was something they looked up on the Internet, hardly conjuring images of crafted, while a different Producer claimed that all recipes they used were the result of experimentation in their home kitchen. Another mentioned that they had employed Polish workers to pick their raw ingredients as a cost saving, and wondered if this were doing the 'right thing' in terms of authenticity. In this instance they looked but did not find an exception to the rules of hand and home made, as they were selecting the produce to cook from the yields themselves - but this Producer conveyed feelings of unease and insecurity about their decision. It appeared to be a compromise in their hand made claims.

As these edges are indistinct, blurry, and left to the Producers themselves to navigate, the inclusion of magic rites to infuse themselves in their products brings a host of anxieties in addition to an immense sense of pride. It has also turned their attention to and creates a high level of reliance on consumer approval. The consumer is seen as the ultimate judge of magical efforts, and also aligns with Mauss's outline of magic as an act that "society considers magical" (2001, p. 23) - everybody has to believe in the act of magic to be true. In this case, consumers are representations of society. When Producers are asked why they engaged in the business that they do, this is commonly the first aspect mentioned - it's a checking of consumer belief in their magic, expressed in a desire to please:

'(I make this) for the customer to enjoy!'

'I do this to delight my clients'

'I can't sell anything unless the customer is satisfied directly'

The importance of bringing consumer satisfaction - and by extension, belief - through their produce underpins and adds meaning and reinforces their belief in their rites of magic. Feedback from consumers is actively gained from Producers when possible, and it is this information that is actively sought and helps guide further practices, as discussed 'Co-Creation and Certainty of Myth' finding.

The execution of these magical rites also brings small-scale food production to more closely resemble a creative processes (such as art or fashion) rather than a production process (such as Fordist production methods), and again brings food production to in closer alignment within the New Economy. In order to execute their magical rites, Producers must manage themselves, their lives and their work in a way that optimises their creativity and mood to enact their magical acts, which blurs the line between 'work' and 'play'. Here, Löfgren and Willim (2005) refer to the use of magic as central to emergent people and self-management methods in the New Economy, where the introduction of 'fun' has become a central concern.

This management of self is referred to as 'reflexivity' by Lash and Urry (2002) and described by McRobbie (2002) "as a form of self disciplining where subjects... are increasingly called upon to inspect themselves and their practices, in the absence of structures of support... reflexivity marks the space of self responsibility, self blame" (p. 522). Producers are highly cognizant in the requirement for individual responsibility in their rites, as I was informed 'you are held accountable for what you do'. Giddens' and Becks description of the concept as undertaking self monitoring activities' (McRobbie 2002, p. 522), is relevant in describing this accountability here and is further articulated by Lash and Urry:

Social agents are increasingly 'set free' from heteronomous control or monitoring of social structures in order to be self-monitoring or self-reflexive. This accelerating individualisation process is a process in which agency is set free from structure; a process which, further, it is structural change itself in modernisation that so to speak forces agency to take on powers that heretofore lay in social structures themselves. Hence for example structural change in the economy forces individuals to be freed from the structural rigidity of the Fordist labour process. That is, it is increasingly a pre-requisite... that the labour force becomes increasingly self-monitoring as well as develops an even greater reflexivity with respect to the rules and resources of the workplace" (2002, p. 5)

In this instance, the use of magic to add symbolic value of Authenticity in production methods has transferred Producer association away from the Fordist economic labour process, but embedded them more closely with the counter-structure that has developed in resistance to it within the New Food Economy. Furthermore, their intense self-monitoring has also developed skills of greater reflexivity with respect to the rules of their workplace that extends across their day to day lives, and creates a unique set of challenges inherent in operating in this way, which is discussed in the Research Finding Summary. As McRobbie describes the fashion industry in London (2002), this may mean holding down additional jobs in order to make ends meet - one Producer interviewed shovels snow in winter.

4. CONCISE SUMMARY

4.1 Summary of findings

This paper presents new insights on value-adding strategies to ensure authenticity in food items. The viewpoint of small-scale farming producers was considered, emphasizing co-creation, myth, art, and magic as essential techniques for enhancing product authenticity.

Value addition is a prevalent strategy in the food industry, increasingly involving the incorporation of lifestyle and cultural capital messages into food, such as health, knowledge, and social class, frequently by asserting a form of "authenticity." The thesis posits that authenticity is a social construct, characterized by allusions to handcrafted, homemade, and produced messaging as perceived by producers in this study. These messages are comprehended through the deconstruction of cultural codes. An illustration of this is the societal idea of 'farming' as a mythic market that the food industry has extensively leveraged in marketing communications. The misleading communication strategies employed by large food corporations generated consumer uncertainty regarding the true origins and production methods of mass-produced food, thereby allowing small-scale producers to offer consumers a sense of certainty through their products.

To enhance their notion of 'authenticity' in things, they engage in rituals of magic, co-creation, the dissemination of certain myths, and the establishment of artistic spaces. These techniques are distinct and have been formulated in opposition to the prevailing food system to provide a competitive edge. Producers disclose culturally created views regarding their relationship with their product, comprehension of symbolic meanings in spatial representation, and their aspired identity through cultural capital. It also demonstrates a lack of understanding of the foundational cultural framework and historical context of food that has resulted in their current status in the food industry.

The use of New Economy theory indicates that the entire production process is increasingly considered when interpreting food communications, granting small-scale producers a competitive advantage in the marketplace. Nonetheless, the producers' ignorance on the origins of their fame has engendered anxiety about their current and future behavior. In response, producers have established enchanting rituals and a profound self-reflexivity concerning their authenticity methods, alongside a significant dependence on co-creation with consumers to enhance their products.

Product range and material depiction of space. The architecture of this room instills a sense of assurance in consumers regarding their engagement with farmers, while also serving as an inviting environment for dialogue. In developing spaces and products that consumers desire, producers merely replicate existing market dynamics and rely solely on the information provided by consumers; paradoxically, they seek to astonish consumers with insights that the consumers themselves disclosed. This is where a significant advantage of Cultural Analysis is demonstrated. The food industry is delineated through a comprehensive theoretical and historical perspective, revealing motivations and aspirations that remain unexpressed by both consumers and producers.

This paper discovered that, as producers and cultural mediators, the quest for identity and purpose of producers is embedded in their products. To invoke the Law of Contact or contagious magic and affect their products from a distance, they enhance and augment their products by assuming the role of magicians and performing magical ceremonies in their kitchens. These rites, by imbuing their products with the sought-after virtues of authenticity associated with handcrafted, homemade, and crafted items, have imposed constraints on production methods and underscore the tenuous distinction between enchantment and deceit. Producers, endowed with cultural capital manifested as knowledge, recognize and document this capacity, so transforming the irreproducible into their products: their own identities. By meticulously observing consumer behavior, producers have implicitly recognized that their products signify a quest for cultural capital and social status, prompting them to transform their farm stores into distinctive, extraordinary environments akin to art galleries. Through their actions, they transition from cooks to artists, aligning food with the Experience Economy, while simultaneously disclosing and reinforcing their interpretation of 'culture' as high art, in contrast to the quotidian. These actions necessitate a reevaluation of the actual economic domain of these Producers: such practices indicate a greater alignment with the service and artistic sectors than with the food industry.

Nonetheless, the conviction in art, myth, and self as value-added determinants of authenticity has engendered practices that constrain income streams, resulting in certain producers facing economic precariousness. This also potentially introduces risk to their business strategy if their product offering loses consumer favor as inexplicably as it gained it.

4.2 Conclusions, Applications, and Recommendations for Future Research

The understanding of value-adding activities such as co-creation, the farmer's myth, art, and magic employed by small-scale producers has numerous applications for food producers. Initially, it necessitates an examination of the extent and rationale of corporate engagement. In this instance, small-scale producers have integrated components of the food and creative sectors as a manifestation of cultural capital for both themselves and their clients.The 'blurring' of industry edges in the New Economy is in itself not necessarily a brand new phenomenon in the food industry as Jönsson (2012) has outlined in 'The Gastronomic Revolution'. However, in this context it is evolving the question of why Producers are engaged in the business that they are, where more inspiration for application can be found. As Simon Sinek, cultural anthropologist and management and leadership author proposes in 'Start with Why' (2009), inspired leaders such as Steve Jobs, Martin Luther King, Orville Wright looked for their why in places based in altruism, common good and social betterment (which he suggests is another construction of authenticity), which in turn motivated and inspired others, and in doing turned their ideas into social movements. In other words, their why is the basis for construction of consumer desire. Sinek suggests that it is the articulation of why as the reason that consumers ultimately buy products or services. A further description in Sinek's book is in support of a key message in this article - that consumers will be unable to tell you what this why is. There may be some concerns regarding the why Producers in this study have employed, but it is currently a compelling and motivating proposition for them nonetheless.

A less individualistic, more collective and cooperative approach within the food industry, to educate and empower themselves and consumers, develop sustainable, kind and ethical business practices, and authentically communicate about how their food is really made could be a more compelling and sustainable why than what is currently employed. However, the thrust of this application remains for the wider food industry: using a cultural analytical approach, define what industries your business is involved and describe why it's important.

Additionally, it is the certainty that Producers can create through their value adding practices that constitutes part of their competitive advantage. The wider food industry, and mass production companies in particular, have heavily utilised and communicated the 'return to the farm' and 'farmer' myth. In so doing, they have shown just how powerful mythology is in our culture and how easily it can be communicated in marketing messages. Unfortunately, these messages in past use were often concealing large-scale production methods and laboratories that bear little resemblance to what is commonly perceived as a farm (or farmer). As this concealment was exposed over the years consumer distrust and cynicism crept in. What Producers are able to do in this climate is remain present in their farmhouse kitchens, and their Farm Shops, as consumers to come to them, and in doing, start to create a new view for consumers to build their perceptions regarding the myth of farming. The certainty that is created, and ability to begin changing the myth points to an opportunity not only for small-scale Producers to start using a heavier hand for the evolution of this myth, but for the wider food industry in developing certainty in myth-based marketing claims - by initially providing increasing transparency to business practices and building trust.

This knowledge also illustrates the magic in infusing authenticity - where a key element is the audience belief in the magical act. To develop rites without an audience means there is no magic; and conversely if you have an audience and there is no belief in these rites, again there is also no magic. There exists a thin line between magic and deceit. Magic is alike to authenticity in this regard: all actors involved have to believe it to be true. It is also a key element that provides competitive advantage, and Producers have selected a positioning that cannot be replicated: themselves. The authenticity in this study is predominately the Producers themselves; they have magically infused their sense of self as unique representations in homemade, hand-made and crafted production methods as part of their product offer. Which leads to ask the question to the wider food industry: What does authenticity mean to your business? Is it believable? Can it be replicated?

A warning arrives in the form of the risk involved when an over-reliance on consumer feedback is evident in the production of practice. Some research methods, such as quantitative surveys, will merely reinforce current modes of behaviour and reveal only what lies at the surface, arguably providing a perfect template for continuation of business as it is today. This article illustrates that consumer feedback can be too much of a good thing if taken out of a wider cultural context. To build business with a large reliance on this practice is problematic - as mentioned in the Introduction, one cannot simply walk up to a consumer and ask them why they like a certain food product. They can't tell you. Co-creation discussions with consumers reveal more about the widespread consumer ignorance regarding the food industry rather than revealing how they could be surprised and astonished. There is nothing game changing about approaching a consumer in this way - it doesn't challenge the status quo, and will serve only to reinforce what already is. How can one be surprised if they can articulate what constitutes a surprise? A deeper and wider view provided through methods such as Cultural analysis theory and ethnology as a method will reveal culturally unconscious held beliefs, the structures that surround society that influence behaviour and uncover what the underlying motivations really are in value making and value adding.

Future research recommendations pay attention to assisting the development of knowledge of practices identified in this thesis, and in doing proposes a greater involvement and collaboration of actors across the food industry. The examination of value adding by small-scale food Producers in the New and Experience Economies is a relatively new area of study within the social sciences and warrants further attention.

Continuing the development of knowledge in regards to co-creation, a point of interest is made with reference to Douglas Holt (2004). Like Prahalad & Ramasaway, his approach is positioned at the crossroad of marketing, brand management and cultural studies, and in his publication 'How Brands Become Icons' invokes the use of co-creation (what he terms as 'co-authoring') and mythology as a key to success in branding strategy. However he does not refer to the element of certainty or trust in myth. It appears to be a given fact that this commonly held cultural knowledge can and will be used by larger corporations in their branding and communication strategies, and begs the future study within these cross disciplines from a cultural analysis perspective as to the importance of certainty (or trust) as required before a marketing or communication messages is developed, particularly when mythology is in use.

The value shown in the Cultural Analysis of this thesis is relevant not just to small-scale producers in Skåne, Sweden, but also to other participants in the food chain. There exists an opportunity for reciprocal engagement and collaboration to enhance the value of food, with an increased emphasis on education and participation in shaping consumer preferences.

The integration of experience and art into food signifies a necessity for a deeper comprehension of these interrelated cultural production sectors within this framework. To enhance this discussion, it is advisable for Producers to transcend their established community boundaries and engage with industry organizations, such as the Skåne Food Innovation Network, to foster collaborative opportunities through networks and connections. Education forums that adopt a Cultural Analysis perspective on value addition will aid individuals in the manufacturing and production of consumer goods within the food sector in comprehending their position and role as cultural mediators.

In conclusion, the role of small-scale food producers can be enhanced through their collaboration, so fostering an appreciation of their distinctive position and activities within the food business. This will instill confidence in small-scale producers to enhance their methods for improved economic returns and long-term business stability as consumer demands in the food industry progress.

FOOTNOTES

1. The field work was undertaken by Rebecca Dare, for her Master's thesis in Applied Cultural Analysis at Lund University. This article is a revised version of the main parts of the thesis.
2. In the online questionnaire, Producers were asked to nominate the emotion they felt consumers experienced in different sales channels. 80% of participants believe customers feel 'negative' emotions such as neutral, bored or confused in Supermarkets.

REFERENCES

1. Adams, C (1990). The sexual politics of meat: a feminist vegetarian critical theory. New York, Continuum Publishing.
2. Atkins, P & Bowler, I (2001). Food in Society. Oxford University Press.
3. Appadurai, A (1986). The Social Life of Things. Cambridge University Press.
4. Ashley, B et al, (2004). Food and Cultural Studies. Routledge.
5. Barnard, A & Spencer, J (1996). Encyclopaedia of Social and Cultural Anthropology. Routledge.
6. Barthes, R (1997). Toward a Psychology of Contemporary Food Consumption. In Counhihan, C & van Esterik, P (1997). Food and Culture: A reader. Routledge.
7. Barthes, R (1967). Elements of semiology. London: Cape.
8. Baudrillard, J (1996). The system of objects. London: Verso.
9. Beck, U. (1992). Risk Society: Towards a New Modernity. London: Sage.
10. Belasco, W (1989). Appetite for Change: How the Counter Culture Took on the Food Industry. New York: Pantheon.
11. Belasco, W & Scranton, P (2002). Food Nations: Selling Taste in Consumer Societies. Routledge.
12. Bourdieu, P (1984). Distinction: a social critique of the judgement of taste. London: Routledge.
13. Connor, J & Sheik, W. (1997). Food Processing: An Industrial Powerhouse in transition. New York, Wiley.
14. De Certeau, M; Giard, L & Mayol. P. (1998). The Practice of Everyday Life: Volume 2 living and cooking. Translated by Tomasik, T. University of California Press.
15. Dimitri, C & Oberholtzer, L (2009). Marketing U.S. Organic Foods: Recent Trends From Farms to Consumers. ERR-58 United States Department of Agriculture
16. Douglas, M (1996). Thought Styles. London: Sage.
17. Douglas, M (1982). Food as a system of Communication in Idem: in the active voice, London, Routeledge, 81-118.
18. DuGay, P & Pryke. P (2002). Cultural Economy. Sage.
19. Ekelund, L; Fernqvist, F & Tjärnemo, H. (2007). Consumer preferences for domestic and organically labelled vegetables in Sweden. In Food Economics Acta Agricult Scand C, 2007; 4: 229-236, Taylor and Francis.
20. Euromonitor (2009). Retrieved at: http://www.euromonitor.com/packaged-food-2010-part-2-global-market-opportunities/report

21. Frazer, J.G (1995). The Golden Bough. Wordsworth Editions Ltd.
22. Friedmann, H., & McMichael, P. (1989). Agriculture and the state system. Sociologia Ruralis 29(2): 93–117.
23. Giddens, A (1991). Modernity and Self-Identity. Self and Society in the Late Modern Age. Cambridge: Polity.
24. Goody, J (1982). Cooking, Cuisine and Class: a study in comparative sociology. Cambridge University Press.
25. Hamrick, K; Andrews, M; Guthrie, J; Hopkins, D; & McClelland, K. (2011). How Much Time Do Americans Spend on Food? EIB-86 United States Department of Agriculture, Economic Research Service
26. Holt, D (2004). How Brands Become Icons: The principles of cultural branding. Harvard Business Press.
27. Identity Advertising (2007), 'Final Coca Cola Plans Book', Retrieved at: http://www.scribd.com/Sportzchick2287/d/2473813-Final-CocaCola-Plans-Book
28. Innova Market Insights (2011). 'Top 10 Trends for 2012 and Beyond'. Retrieved at: http://www.food and beverage-network.com/blog/food_innovation?category=6
29. Jönsson, H (2012). Den Gastronomiska Revolutionen. Stockholm: Carlsson Bokförlag.
30. Jönsson, H (2005). A Land of Milk and Honey: The Dairy Counter in an Economy of Added Values. In Magic, Culture and the New Economy. Edited by Löfgren, O & Willim, R. Berg.
31. Johnston, J & Baumann, S (2010). Foodies: Democracy and Distinction in the Gourmet Landscape. Routledge.
32. Lash, S & Urry, J (2002). Economies of Signs and Space. Sage.
33. Le Heron, R (1993). Globalised Agriculture, Political Choice. Oxford: Pergamon Press.
34. Levi Strauss, C (1994). The raw and the cooked. Pimlico.
35. Lupton, D (1996). Food, the body and the Self. London: Sage.
36. Mc Robbie, A (2002). Clubs to Companies: Notes on the decline of Political Culture in Speeded Up Creative Worlds. In Cultural Studies, 16(4), 516-531.
37. Mauss, M, (2001). Mauss: A General Theory of Magic. London: Routledge
38. Mennell, S., Murcott, A. & van Otterloo, A.H (1992). The sociology of Food: eating, diet and culture. London: Sage.

39. Mintel, (2012). Retrieved at:http://www.gnpd.com/sinatra/gnpd/front page/? cc=1&utm _ source = food and drink. mintel.com&utm_medium=menu
40. O'Dell, T (2010). Spas: The Cultural Economy of Hospitality, Magic and the Senses. Nordic Academic Press.
41. O'Dell, T & Billing, P (2010). Experiencescapes: Tourism, Culture, and Economy. Co-penhagen Business School Press.
42. Peteraf, M.A. and Barney, J. (2003), Unraveling The Resource-Based Tangle, Managerial and decision economics, 24, pp. 309-323.
43. Peterson, R (1997). The Rise and Fall of Highbrow Snobbery as a Status Marker. In Poetics: Journal of Empirical Research on Culture, the Media and the Arts. 25: 75 - 92.
44. Pine, J & Gilmore, G (1993). Welcome to the Experience Economy. Harvard University Press.
45. Prahalad, C.K & Ramaswamy, V (2004). Co-creation experiences: the Next Practice in Value Creation. Journal of Interactive Marketing 18(3), 5-14.
46. Reed-Danahay, D (2005). Locating Bourdieu. Indiana University Press: Bloomington and Indianapolis.
47. Sinek, S (2009). It all starts with why: How great leaders inspire everyone to take action. Penguin, London.
48. United States Department of Agriculture (2011). 7,175 Ways to Know Your Farmer, Know Your Food: Visit a Farmers Market. Agriculture Deputy Secretary Kathleen Merrigan. Retrieved at: http://blogs.usda.gov/2011/08/05/7175-ways-to-know-your- farmer-know-your-food-visit-a-farmers-market/
49. United States Department of Commerce (2008). Industry Report Food Manufacturing NAICS 311. Retrieved at: http://www.trade.gov/td/ocg/report08_processedfoods.pdf, pp 1- 13.
50. Warde, A (1997). Consumption, Food and Taste: culinary antinomies and commodity culture. London, Sage.
51. Wilk, R (2012). Lecture, Lund University, Sweden.

SECTION 6

PRODUCTS MANAGEMENT OF FOOD INDUSTRY

30

Chapter

MANAGEMENT OF WATER AND WASTEWATER AND BIOMASS-TO-ENERGY CONVERSION IN A BRAZILIAN MEAT PROCESSING FACILITY – A CASE STUDY

1. INTRODUCTION: AN OVERVIEW

A dedication to sustainability and comprehension of "cleaner production" principles are essential for attaining eco-friendly industrial operations. These concepts encourage the reduction of freshwater usage, the decrease of wastewater generation, and the recycling of waste materials. Consequently, in a world characterized by water scarcity and climate change, it is imperative for all industries, regardless of type or size, to implement measures that safeguard freshwater resources and augment renewable energy capability. Concerning solid waste, social and environmental responsibility extends beyond legal requirements and necessitates comprehensive scientific research to formulate a rigorous environmental management program.

Globally, meat processing facilities utilize around 62 million cubic meters of water annually. A mere fraction of this quantity constitutes a component of the final product. The residual portion transforms into wastewater characterized by elevated biological and chemical oxygen demands, significant fat content, and high levels of dry residue, silt, total suspended solids, nitrogen, and chloride compounds. (Sroka et al., 2004). Of the components usually found in these effluents, blood can be considered as the most problematic due to its capacity to inhibit floc formation during physicochemical wastewater treatment and its high biochemical (BOD5, biochemical oxygen demand during decomposition over a 5-day period) and chemical oxygen demand (COD). In fact, even with correct handling during meat processing, this activity generates 2.0 and 0.5 liters of blood as effluent for each bovine animal and pig, respectively (Tritt & Schuchardt, 1992). The treatment of both the solid wastes and the wastewater from the meat processing industry represents one of the greatest concerns associated with the agro-industrial sector globally, mainly due to the restrictions that international trade regulations have imposed over their use and the related environmental issues.

In order to meet this challenge, one of the largest meat processing companies in Brazilian initiated a series of investments in scientific research to improve its environmental performance. Biomass as an energy source, air pollution control, and water and wastewater management were the main issues addressed in research projects carried out from 2003 to 2010.

The Brazilian agro-industrial sector consumes large amounts of fresh water and produces large amounts of residues and by-products, which can potentially be used as energy sources. The Brazilian legislation itself admits the need for water management in industrial plants to implement cleaner production techniques, which include the conscious uses of water. There are several legal documents that promote the recognition of water as public property and a finite resource with economic value. These legal norms and legislation are gathered in a single official document called "Set of legal regulations: water resources" (Brazil, 2011) and promote: (1) the rationalization of water use and its conservation, reconditioning and sustainable management; (2) investment in pollution control, reuse, protection and conservation as well as the use of clean technologies to protect water resources; (3) the practice of water reuse to reduce discharges of pollutants into receiving waters, conserving water resources for public supply and other uses which demand high quality water; (4) the practice of water reuse to reduce the costs associated with pollution, contributing to the protection of the environment and public health; and (5) the creation of guidelines to regulate and encourage the practice of direct reuse of non-potable water. Official Brazilian reports highlight that the costs of water treatment have been raised by the contamination of water resources and water shortages (aspects of quality and quantity) in certain regions of the country. Consequently, they emphasize that high quality water should not be used in activities that tolerate water of lower quality (Brazil, 2011).

Regarding the solid waste materials generated in agro-industries, these are commonly generated during the processing of crops, but are also produced by all sectors of the food industry including everything from meat production to confectionery, such as peelings and scraps from fruit and vegetables, food that does not meet quality control standards, pulp and fiber from sugar and starch extraction, sludge from physicochemical and biological wastewater treatment and filter sludge. The co-digestion of energy crops and a variety of residual biomasses may be a good integrated solution for energy recovery from such waste materials, particularly with wastes that are unsuitable for direct disposal on land, as proposed by Schievano et al. (2009). These authors evaluated the suitability and the costs associated with many substitutes for energy crops in biogas production such as: swine manure, municipal solid waste, olive oil sludge, glycerine from biodiesel production and other agro-industrial by-products and residues. They concluded that farms could implant biogas plants to treat their own biomass generated and other urban and agro-industrial organic wastes, providing power for the neighborhood and improvements in the agrarian economy.

The use of the biosolids originating from the physicochemical treatment of meat processing wastewater can reduce the costs associated with its disposal (which has been prohibited in many locations by strict regulations) as it can directly and significantly reduce the mass and volume of such wastes, allowing energy recovery and generally lower toxic gas emissions when compared to fossil fuels. As long as emissions are below the specified legislative limits, changing energy policies lend support to the use of this type of biomass as a fuel source, as part of a move towards achieving low carbon economies.

The EIA Annual Energy Outlook 2011 reported that the global marketed energy consumption is expected to rise by nearly 50 percent from 2009 through 2035 (US EIA, 2011). Unless the world energy matrix is altered, fossil fuels will account for 90% of this increase.

The requirement to reduce carbon dioxide emissions has sparked interest in the use of many types of biomass as alternative energy sources. Since biomass is produced by the photosynthetic reduction of carbon dioxide, its utilization as biofuel can essentially be carbon neutral with respect to the build-up of atmospheric greenhouse gases, increasing both the demand for the characterization of alternative fuels and encouraging the proliferation of scientific papers concerned with this subject (Demirbas, 2004, 2005; de Sena et al., 2008, 2009; Floriani et al., 2010; Obernberger et al., 2006; Virmond et al., 2010, 2011 2012a, 2012b; Werther et al., 2000).

Brazil is currently implementing advanced programs aimed at the use of biomass energy, and several experimental and commercial projects are being implemented, such as those presented by Lora and Andrade (2009), to provide important information in order to overcome the technical and commercial barriers which inhibit the extensive implementation of bioenergy. The solid wastes produced by the meat industry have been applied mostly to the production of animal feed, which include the slaughter wastes and the wastewater treatment solids as main ingredients (Johns, 1995; Tritt and Schuchardt, 1992). However, diseases such as BSE (Bovine Spongiform Encephalopathy) have led to restrictions over the use of these wastes for feed production.

The first actions taken by the case study meat processing company, between the years 2003 and 2004, as shown by de Sena et al. (2008), were related to the in-depth investigation of the physicochemical treatment carried out at the wastewater plant with regard to its solids removal, mainly to achieve an increase in the chlorine-free biomass obtainment with a view to its utilization as a biomass fuel for steam generation. The data obtained indicate that the raw wastewater has a high organic load comprised basically of blood and organic materials that cause the red color, the greater part of the turbidity, the high concentration of total solids, oils and greases, the BOD5 and the COD. The combustion of these wastes, especially the sludge from the wastewater treatment plants, might be a nobler utilization for economic reasons, however, many parameters related to the combustion must be monitored due to the formation of pollutants such as polychlorinated dibenzodioxins (PCDD), polychlorinated dibenzofurans (PCDF), volatile organic compounds (VOCs), NOx, and SO2. The authors showed that the physicochemical treatment carried out at the meat processing wastewater plant provides around 20% of sludge, an organic solid residue, using the chlorine-free coagulant ferric sulfate (instead of aluminum or ferric chloride). In order to avoid discharge and subsequent environmental problems, the authors performed a preliminary combustion test with a mixture of biosolids and sawdust in a mass ratio of 4:1. The results suggested that the use of the biosolids as an alternative energy source would offer a favorable solution, reducing disposal and processing costs, as well as avoiding environmental and health problems for staff and the community close to these processing plants, thus establishing a cheaper and cleaner energy source for the meat industry segment (de Sena et al., 2008).

Another point related to sustainability in the meat processing industry is the associated farm residues, like pig and chicken manure. A reasonable solution for these wastes is their anaerobic digestion to produce biogas and/or fertilizers. Many farms in Brazil are implementing biodigestors in order to obtain biogas to produce electrical or thermal energy. Boersma et al. (1981), for example, studied the energy recovery from biogas produced from pig waste and verified an economy of 86%, showing a very good potential for this kind of solution. Also, carbon credits could be sold when the biodigestion process together with the energy recovery are applied to the farms.

Biogas is composed basically of methane, carbon dioxide, hydrogen sulfide, and other components in lower concentration. The gas production and the proportion of each compound are dependent on the biodigestor parameters and the chemical composition of the substrates (Lucas Jr., 1994).

A standard composition of biogas consists of 55-65% CH4, 35-45% CO2, and a minimal proportion of H2S. The presence of H2S can lead to corrosion issues when biogas is utilized as fuel, and its atmospheric emission has a greenhouse potential 21 times greater than that of CO2. A high concentration of methane is advantageous, as its presence enhances the calorific value of the gas, rendering it more appealing for energy production.

This chapter aims to elucidate complementary research initiatives conducted between 2005 and 2010, focusing on strategies to diminish freshwater consumption, advocate for water recycling and reuse, and explore the utilization of biomass residues as energy sources and their gaseous emissions during combustion processes.

2. CASE STUDY – MEAT PROCESSING PLANT

The industrial plant which formed the basis of this case study is located in the west of Santa Catarina State (southern Brazil), where water pollution and overexploitation, the uneven distribution of rainfall over the seasons and long periods of drought, especially in summer, have become a significant problem. The activities of the meat processing plant of this case study include the slaughtering and processing of poultry and swine, while the poultry hatchery plant includes all activities involved in poultry growth: breeding, hatching, rearing, food production and waste handling.

Figure 1 shows the wastewater treatment process of the case study meat processing plant.

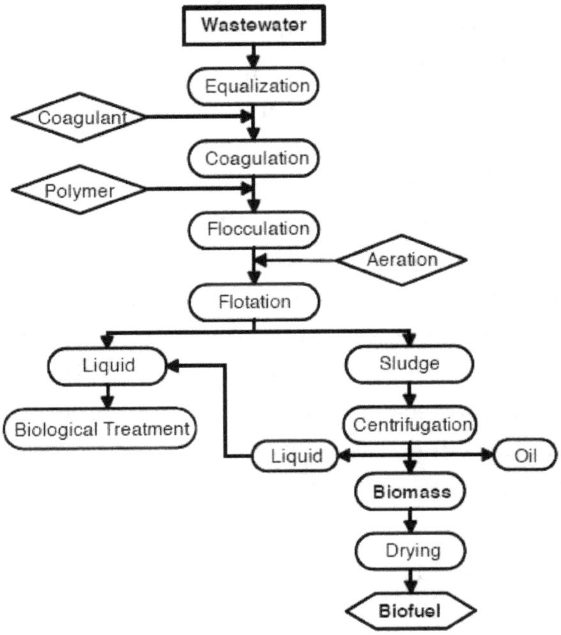

Figure 1. Processes involved in the wastewater treatment plant and obtainment of biofuel (de Sena et al., 2008)

The meat processing plant has its own drinking water treatment plant (DWTP) and wastewater treatment plant (WWTP). The major water resource of this unit is from a river called Rio do Peixe. Its DWTP produces around 8,600 m3 d-1 of drinking water, and the WWTP treats around 7,900 m3 d-1 of wastewater. As described by de Sena et al. (2008), after the flotation process with a continuous capacity of 350 m3 h-1, the treated effluent undergoes a biological treatment, while the biosolids are transported by pumps to a three-phase centrifugal system, which separates oil, water and solid parts (biomass). Afterwards, the biomass is dried in an industrial rotating granulator drier with an operating capacity of 400 kg h-1 (model Bruthus, Albrecht, Brazil), where the moisture content was reduced from approximately from 80 wt% to 10-20 wt% in order to make the burning process feasible.

The wastewater treatment plants of meat processing units in Brazil usually undergo the same type of treatment process, where a flotation system is the most commonly used solid- liquid separation step, due to the natural characteristics of these effluents, which possess high oil and grease contents. To increase the flotation performance the use of coagulants and coagulation aids are mandatory. Dissolved air flotation (DAF) has become an attractive separation process because of its well-known higher efficiency in terms of organic matter abatement, although the increase in costs associated with the production of micro-bubbles and system maintenance must be considered. Flotation processes are preferred in relation to sedimentation considering their faster solids separation, the lower moisture content of the sludge produced and the smaller area requirements. Coagulants themselves are very efficient for floc formation during the coagulation process, but the use of coagulation aids (e.g., anionic polyacrylamide polymers) after the rapid mixing of the coagulant-wastewater to disperse the coagulant, have been shown to increase of floc size and to provide higher floc stability and high solid separation rates. Since the coagulation and flocculation (addition of coagulation aids during gentle dispersion) are successive steps applied to neutralize the suspended particles and achieve strong flocs, the addition of these reagents must be carefully and precisely controlled to enhance solid-liquid separation. If some variables related to the process efficiency are not properly controlled, such as pH, reagent concentrations, mixing speed and contact time, the whole process will be unsuccessful. During the increase in floc size air bubbles of different diameters are incorporated into the flocs and this is responsible for the flotation phenomenon. Flotation efficiencies may vary from 60 to 95% of organic matter removal, according to the technology applied.

When the flotation of the solids is complete, the froth on the surface is separated from the water and skimmed off. It is collected in chambers and is pumped to a three-phase centrifuge where another polymer, a cationic polyacrylamide, is added to improve the oil-water- sludge separation. The water undergoes biological treatment and the oil is collected and sold as a raw material for the soap and detergent manufacturing industries. The remaining solid fraction is the sludge which was formerly used as an ingredient for animal food and feeds, especially the pet segment. However, due to the above-mentioned restrictions regarding its use in feeds, there are currently two other available options for the correct discharge of this so-called waste: combustion/incineration or land disposal. The combustion of the sludge for steam generation was the option chosen in this case study due to both economic and environmental aspects, since the use of an existing waste as part of the fuel content will decrease the fuel costs for internal energy supply, and the amount of sludge added to the fuel used (wood chips) could be properly controlled with regard to the gaseous emissions. On the other hand, land disposal might bring extra costs associated with transportation and long-term storage. All of the results obtained, as well as their pros and cons, are discussed in detail in the following sections.

The importance of the Brazilian poultry industry can be verified by its strong presence in the rural regions, mainly in the southern and south-eastern states. In many cities, poultry production is the main economic activity. The poultry hatchery unit of this case study, as in the case of the meat processing unit, also has its own WWTP. The wastewater originated from the processes of this unit is characterized by a high organic content, with the presence of nutrients such as nitrogen and phosphorus, as well as persistent organic compounds such as the residues of sanitizing products (e.g. pesticides) and veterinary drugs (Genena, 2009). The treatment system for the poultry hatchery wastewater comprises a screening stage (primary treatment), followed by equalization and finally biological treatment (secondary treatment: stabilization ponds). The treated wastewater is then discharged into a river (surface water).

3. WATER AND WASTEWATER MANAGEMENT

The water and wastewater management (W2M) proposed for the pilot plant aimed to minimize the water consumption and evaluate the possibilities for water and wastewater reuse in the food industry. The W2M, described in a previous publication (Luiz et al., 2012a), proposed strategies for water management in slaughterhouses considering the restrictions imposed by Brazilian legislation and hygiene concerns particular to the food industry. The objective was to present alternatives for the minimization of water consumption and wastewater production.

The proposed W2M is a practical model of industrial water management, which consists of seven stages:

1. Collection and analysis of documents;
2. Measurement of water consumption and wastewater production (water balance);

3. Verification of the points of greatest water consumption;
4. Minimization of water consumption with emphasis on the points of greatest water consumption;
5. Evaluation of the potential for water reuse and recycling without reconditioning;
6. Evaluation of the potential for water reuse and recycling after reconditioning; and
7. Maintenance of water management.

The points identified as being associated with major water consumption were: (1) pre-cooling of giblets, (2) washing of poultry carcass before pre-chilling, (3) transportation of giblets, poultry necks and feet, and (4) washing of swine carcass after buckling. The potential for reducing the fresh water consumption in-line with the current Brazilian legislation in these four process steps was approximately 806 m3 d-1 (Luiz et al., 2012a).

After the minimization of water use, the most important action is the evaluation of direct recycling and reuse of wastewater without reconditioning or treatment (direct reuse). The direct reuse could be "in processes without direct contact with food products, that is, in non-potable uses (e.g., as cooling water, for flushing toilets or as irrigation around the plant), thus saving fresh potable water" (Luiz et al., 2012a). Hence, according to the water balance carried out, "the wastewater with the possibility for direct or indirect recycling or reuse was evaluated physically, chemically and microbiologically to verify if and where it could be recycled and reused" (Luiz et al., 2012a). The four types of wastewaters which offered the possibility of reuse originated from: (1) the defrosting of refrigerating and freezing chambers, (2) the purging of condensers, (3) the cooling of smoke fumigator chimneys, and (4) the sealing and cooling of vacuum pumps. These residues had similar water quality parameters; hence they could be mixed before reuse, totaling approximately 1,383 m3 d-1 of wastewater. Depending on the final use, this mixed wastewater could be reused without major treatment or following simple filtration; thus, this approach can be considered as direct wastewater reuse.

The theoretical reduction in water consumption, after applying the principles of water minimization and wastewater reuse, was 25.6%, representing a financial saving of around $434,000 per year (Table 1). However, new regulations need to be elaborated together with national environmental, sanitary and water supply agencies, processing industries and research institutions aiming at the legalization and promotion of water reuse in the food industry.

Table 1. Water and financial savings

Condition	Water flow (m3 day-1)	Water saving (%)	Annual Costs1 ($)
Production in 2007	8616.0	-	1,539,353
Theoretical production after water minimization	7810.0	9.4	1,366,000
Theoretical production after wastewater reuse	7216.8	16.0	1,256,996
Theoretical production after water minimization and wastewater reuse	6410.0	25.4	1,104,731

[1]Considering costs in 2007: $0.10 and $0.42 per m3 to treat water (DWTP) and wastewater (WWTP), respectively, in the case study meat processing plant. Data reproduced from Luiz et al., 2012a.

3.1 Tertiary and advanced treatment for indirect wastewater reuse

Additionally, tertiary treatments are a good alternative to produce high quality indirect reuse water, reducing the percent of fresh water consumption. Tertiary treatments can be applied to recondition secondary effluents (i.e., after secondary activated-sludge treatment), further increasing the possibilities for indirect wastewater reuse inside or outside the building. For example, an industrial wastewater treatment plant can produce high quality tertiary wastewater to be used as reuse water in its processes that do not involve contact with the food product, without the risk of adverse effects in terms of the product quality and human health. Alternatively, it can provide this high quality reuse water for another industrial activity, which does not require fresh potable water for all of its processes.

To improve the quality of the wastewater to be reused, it is necessary to disinfect it and to decrease or eliminate the concentration of biologically persistent organic compounds. The inefficient removal of these organic compounds from the wastewater before reuse or discharge into natural watercourse is promoting their accumulation in fresh water bodies and causing environmental and human health problems and, especially, harming the aquatic animals (Esplugas et al, 2007; Liu et al., 2009; Luiz et al., 2009, 2010, 2011; Oller et al., 2011).

Biologically resistant pollutants or persistent organic pollutants (POPs) are compounds which are not eliminated through the metabolic activity of living organisms (mainly bacteria and fungi) in natural waters and soils (Oller et al., 2011). Thus, conventional primary (removal of suspended compounds) and secondary (such as activated sludge) wastewater treatments are inefficient in removing these pollutants (Luiz et al., 2009). These compounds are present in municipal wastewaters primarily as pharmaceuticals and personal care products (PPCPs) (Esplugas et al., 2007), and also in industrial wastewaters which contain a large number of synthetic and toxic compounds, mainly polar and non-polar hazardous compounds, pharmaceuticals, phenols, pesticides, endocrine disruptor compounds (EDCs), and non-biodegradable and toxic chlorinated solvents (Esplugas et al., 2007; Liu et al., 2009; Luiz et al., 2009, 2010, 2011, 2012a; Petrović et al., 2003).

Due to the large variety of recalcitrant organic contaminants, the tertiary treatment applied to produce water for reuse must be exceptionally efficient. The advanced oxidation processes (AOPs) are an excellent alternative. During AOPs, highly reactive oxidizing radicals are formed, mainly hydroxyl radicals (•OH) (Koning et al., 2008). These radicals are non-selective, promoting the oxidation of all organic and inorganic contaminants and, in the presence of a sufficient amount of oxidant and optimized reaction conditions, complete mineralization can be reached, the final products being CO2, H2O and inorganic anions. Thus, AOPs are applied to totally or partially remove recalcitrant organic compounds, increasing the biodegradability of wastewater (Rizzo, 2011).

3.1.1 Proposed tertiary treatments of slaughterhouse secondary wastewater

In previous studies carried out by our group we evaluated the different options of tertiary treatments to produce high quality reuse water to be used in processes without contact with food products, that is, non-potable uses (e.g. as cooling water, boiler feed water, toilet flushing water or for irrigation around the plant) (Cornel et al., 2011; Luiz et al., 2009, 2011, 2012a). However, since Brazilian legislation only allows the use of fresh potable water in the food industry, our research, using real wastewater and aiming to obtain reclaimed water with drinking water quality which adhered to Brazilian legislation, was carried out in bench-scale and pilot-scale.

The tertiary treatments evaluated for the slaughterhouse secondary wastewater included: UV, H2O2, O3, and AOPs (H2O2/UV, O3/UV; O3/H2O2/UV; TiO2/UV; and H2O2/TiO2/UV). Additionally, for the best combinations, the kinetics of the photo-induced degradation of color, UV254, total organic carbon (TOC) and/or total coliforms were evaluated (Luiz et al., 2009, 2010, 2011, 2012a,b).

Two main problems were encountered during this research. The first issue was the variation in the quality of the target slaughterhouse wastewater over time, which affected the treatment efficiency and the determination of the best treatment (Luiz et al., 2011). The second issue was the high concentration of nitrate and nitrite: 45.9(±17.7) mg NO --N L-1 and 3.74-3.77 mg NO --N L-1, respectively (Luiz et al., 2012). The Brazilian drinking water standard (Brazilian Ministry of Health Administrative Ruling 518/2004) allows 10 mg NO --N L-1 and 1 mg NO --N L-1, respectively.

In order to remove the recalcitrant organic compounds and the nitrate/nitrite, to reduce the color and turbidity, and to disinfect the secondary wastewater in a single treatment, micro-filtration followed by an AOP employing H2O2/TiO2/UV was identified as the best combination evaluated (Figure 2).

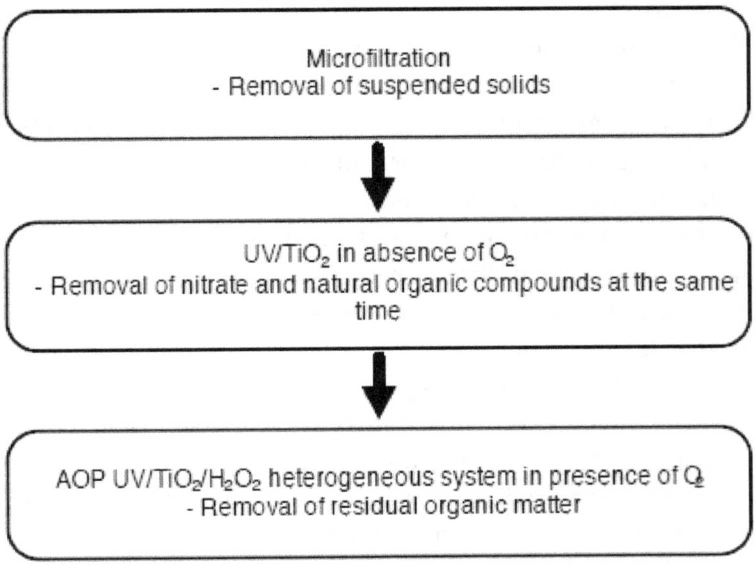

Figure 2. Process proposed for the treatment of secondary wastewaters with high concentration of nitrate/nitrite and recalcitrant organic compounds.

The photocatalytic removal of nitrate/nitrite is more effective in the absence of dissolved oxygen, because if O2 is present this oxidant agent will be a better final electron acceptor than nitrate or nitrite. The catalyst is activated by the absorption of high energy photons, promoting the excitation of the electrons from the valence band (VB) to the conduction band (CB), and consequently an electron (e-) and a positive hole (h+) are formed in the CB and VB, respectively (Luiz et al., 2012b). The electron reduces the oxidizing agent adsorbed on the catalyst, and the hole oxidizes the organic compound or H2O. In the latter case, the oxidation of H2O produces •OH radicals, which will also oxidize organic matter (Ahmed et al., 2010).

Therefore, during the photocatalytic removal of nitrate/nitrite by UV/TiO2, nitrate and nitrite ions will be the final electron acceptor and they will be reduced to N2 gas. The natural organic compounds or residual, biologically persistent, organic pollutants (in the case of industrial wastewaters) will be the hole scavengers (electron donors, reducing agent). In cases where the natural concentration of organic compounds in the aquatic medium is not sufficient to promote the reduction of nitrate/nitrite to below the desired concentration, a carbon source should be added. Formic acid is a good alternative since its residue can be completely decomposed into the harmless compounds CO_2 and H_2O (Luiz et al., 2012b; Rengaraj and Li, 2007; Sá et al., 2009; Wehbe et al., 2009; Zhang et al, 2005). Finally, the heterogeneous AOP system UV/TiO_2/H_2O_2 was applied in the presence of O_2 to remove residual organic matter and achieve the required standard of drinking water quality (Luiz et al., 2012b).

The proposed treatment was also successful in removing recalcitrant organic compounds present in the secondary treated slaughterhouse wastewater, which include antibiotics, pharmaceuticals and personal care products which are commonly found in industrial, but predominantly in sanitary and domestic, wastewater (Luiz et al., 2009). One such compound found was the macrolide antibiotic erythromycin A and its removal and degradation products resulting from direct ozone attack and hydroxyl radical attack (AOPs O_3/UV, O_3/H_2O_2 and UV/H_2O_2) were evaluated. However, the research indicated that the degradation of organic micropollutants, such as erythromycin, in the AOP may be faster than under ozone treatment, because the hydroxyl radical attack (AOP treatments) is not selective and is usually diffusion-controlled. On the other hand, the direct attack of ozone is selective and is typically targeted toward functional groups with a lone valence electron pair where the electrophilic addition of ozone occurs (unsaturated compounds with carbon-carbon double or triple bonds - π bonds, aromatic rings, amines and sulfides) (Luiz et al., 2010, 2012a).

3.1.2 Poultry hatchery wastewater treatment

Industrial wastewater is generally comprised of various effluent streams generated at different points in a particular process. Its physicochemical characteristics can present considerable variation over time due to, for instance, changes in operating procedures and cleaning activities. Therefore, the complexity and variation of its composition are typical attributes of industrial wastewater (Genena, 2009).

The poultry hatchery wastewater of this case study was collected and passed through the stages of screening and equalization. The wastewater variability was investigated over a period of 48 h and its quality was evaluated by chemical oxygen demand (COD) analysis, which is an overall pollution indicator and represents the amount of organic matter present in the sample. The COD values ranged from 218±2 to 997±5 mg O_2 L^{-1}, which confirms the high variability in the nature of the poultry hatchery wastewater (Genena, 2009).

Wastewater in the poultry hatcheries originates mainly from the washing of equipment and utensils. Therefore, a series of diverse compounds may be present, such as veterinary drugs administered to the animals through feed and excreted by them in urine, and sanitizer agents and pesticides used in the cleaning and disinfection of the work environment (Genena, 2009). These compounds, which are persistent compounds, are very harmful to the environment, presenting high toxicity and bioaccumulation (Almeida et al., 2004). They are complex and often difficult to degrade in the biological treatment systems commonly present in industrial wastewater treatment plants.

The food industry is constantly seeking ways to improve the quality of its wastewater through changes in treatment systems. The growing concern regarding emerging and persistent compounds has resulted in researchers focusing their attention on alternative methods of wastewater treatment to minimize or avoid the discharge of these pollutants into water resources, since the biological treatment processes typically used by the food industry are not able to destroy these types of compounds (Genena et al., 2011). The application of oxidative elimination methods, e.g., direct oxidation with ozone or hydrogen peroxide and AOP have been highlighted as strong alternatives for the treatment of wastewater containing compounds which do not degrade easily (de Sena et al., 2009; Genena, 2009, Genena et al., 2011; Luiz et al., 2009, 2010; Tambosi et al., 2009).

The proposal for the use of physicochemical processes for the treatment of the poultry hatchery wastewater of this case study was based on the value of 4.6 for the COD/BOD5 ratio (low biodegradability) and the presence of persistent compounds. Therefore, the application of different AOPs (H2O2/Fe2+ – Fenton, H2O2/Fe2+/UV – photo-Fenton and H2O2/UV) for the poultry hatchery wastewater treatment was investigated. The wastewater treatment process by photo-Fenton reaction was found to be the most appropriate, resulting in better organic matter removal efficiency (approximately 91.9% of COD and 66.3% of TOC). Additionally, the COD/BOD5 ratio obtained for the treated wastewater indicates that all physicochemical treatments applied improved the biodegradability, i.e., there was an increase in the amount of material susceptible to degradation by biological processes, reaching a value of 1.5 in the photo-Fenton process. Thus, the biological process can be considered as a post- treatment stage, which would reduce the total costs of the wastewater treatment process (Genena, 2009).

An important consideration in the degradation processes is the potential for the generation of toxic intermediates or compounds which are even more toxic than their parent molecule, and thus it is necessary to monitor the process using toxicity assays (Bila et al., 2005). The Daphnia magna acute toxicity evaluation showed that all treatments promoted a significant reduction in the wastewater toxicity effects, and a 94% reduction was reached in photo-Fenton process (Genena, 2009).

Photo-Fenton and Fenton processes result in the formation of a sludge, which is usually deposited in landfills. Thus, better alternatives are being proposed and among them is sludge combustion for power generation. However, in this case study the amount of sludge obtained was insufficient for the determination of its calorific power (Genena, 2009).

The poultry hatchery wastewater was submitted to analysis by liquid chromatography coupled to mass spectrometry (LC/MS) with the objective of investigating the presence of persistent compounds. The presence of imazalil (pesticide) was confirmed among the investigated compounds. Imazalil is an organochloride compound used as a fungicide in the industry for sanitization (Genena, 2009, Genena et al., 2011). Organochlorine pesticides are typical persistent organic pollutants and are the subject of worldwide concern due to their persistence, bioaccumulation and potential negative impacts on humans and animals (Guan et al., 2009; Zhang et al., 2007). The biological treatment of wastewater containing micropollutants, like pesticides, is often very complicated or even impossible, because many pesticides are highly toxic to wastewater biocoenosis (Genena et al., 2011).

The treatment of ultrapure water to remove imazalil has been investigated applying the photo-Fenton (AOP) and ozonation processes. Tert-butanol (t-BuOH) was used in the ozonation process as an •OH scavenger to ensure that the study was focused only on the direct attack of imazalil by molecular ozone. For both processes the detection and identification of by-products were carried out, applying sophisticated analytical techniques such as LC/MS and LC/MSn (liquid chromatography coupled to mass or multiple tandem mass spectrometry). The toxicity induced by these by-products was also investigated. For each process of oxidative treatment, four degradation products not yet known were detected and their structures were elucidated. The toxicity analysis (Daphnia magna assays) revealed a decrease in toxicity over time for both treatments, indicating that the by-products were not more toxic than their parent molecules (Genena, 2009, Genena et al., 2011).

4. BIOMASS-TO-ENERGY ACTIONS

The biosolids originating from the wastewater treatment system of the meat processing plant, sawdust and their mixture in a mass ratio of 1:9 (w/w) were characterized as fuels. The correlations between the fuel properties, the operating parameters for the combustion and the gaseous emissions were then investigated in order to evaluate the feasibility of applying this organic residue as a substitute fuel for thermal energy generation.

4.1 Biomass properties

The fuel properties often form the basis for the selection of the most appropriate technology for the biomass-to-energy conversion process. Depending on these properties, a biomass fuel may not be suitable for specific conversion options, partially for technical and sometimes for environmental reasons. The characteristics of the biomass are influenced by its origin and also by the entire processing system preceding any conversion step. Biomass presents a wide variation in its physical and chemical properties. Many publications have investigated the effects of the biomass properties on thermal conversion processes (Demirbas, 2004; Jenkins et al., 1998; Obernberger et al., 2006; van Paasen et al., 2006; Werther et al., 2000; Werther, 2007). The use of biomass as a fuel in combustion processes is frequently desirable in the agro-industry sector because the residues, such as wastewater sludge, usually present high calorific value. However, burning biomass containing different mineral matter compositions may create various problems which can affect the boiler operation or make the firing of the biomass in conventional combustion systems unprofitable.

Wood and wood-based materials are extensively used as fuel for thermal energy generation particularly in the Brazilian food industry, which requires large amounts of steam.

In order to evaluate the potential for the utilization of the biosolids originating from the case study plant for co-combustion with sawdust, a sample of the biosolids obtained from the physicochemical treatment (LFP) was chemically and physically characterized, and its composition was compared to that of sawdust (SD), taken as a reference fuel. Additionally, a sample of a mixture of LFP and SD in a mass ratio of 1:9 w/w (LFPSD1:9) was also characterized and the results compared to SD and LFP properties.

The methodology applied for the biomass characterization and the results obtained for LFP, SD and LFPSD1:9 were reported by Floriani et al. (2010) and Virmond et al. (2008, 2011), and are summarized in Table 2.

Carbon (C), hydrogen (H) and oxygen (O) are the main components of solid biofuels. Carbon and hydrogen contribute positively to the HHV (higher heating value). The content of hydrogen also influences the LHV (lower heating value) due to the formation of water. The content of greases was also measured in the LFP composition (34.39 wt%, raw) and it contributes considerably to the high energy content of the LFP (LHV of 25.77 MJ kg-1, daf). The presence of chlorine in the biomass (0.18 wt%) occurs due to the utilization of chlorine-based products for hygiene purposes at the plant and is incorporated into the wastewater as well as into the remaining biosolids (LFP). The nitrogen content of the fuel mixture LFPSD1:9 (1.36 wt%), even though much lower than the concentration found in LFP, can still cause problems in terms of NOx emission during its combustion. The variability of components present in the biomass is mainly due to the chemical compounds used as ingredients during meat processing operations, especially salts and additives. The sulfur content in LFP is mainly due to the conversion of sulfur-containing proteins, but some may remain from the precipitation agent used in the wastewater treatment (ferric sulfate).

Table 2. Biomass properties

	Units	SD[1]	LFP[1]	LFPSD1:9[2]
Proximate analysis[3]				
Ash	(wt%, db)	0.43	12.30	1.75
Moisture	(wt%, raw)	19.97	15.00	50.23
Volatiles	(wt%, daf)	79.78	85.29	83.08
Fixed carbon	(wt%, daf)	20.22	9.58	17.01
Ultimate analysis[3]				
Carbon	(wt%, daf)	55.30	58.04	51.06
Hydrogen	(wt%, daf)	7.14	9.01	6.64
Nitrogen	(wt%, daf)	0.21	9.24	1.36
Sulfur	(wt%, daf)	< 0.01	0.34	0.03*
Oxygen	(wt%, daf)	37.34	22.68	40.94
Chlorine	(wt%, daf)	< 0.01	0.18	< 0.01
Fluorine	(wt%, daf)	< 0.20	< 0.20	n.d.
Phosphorus	(wt%, daf)	0.01	1.03	n.d.
Lower Heating Value[3]				
LHV	(MJ kg^{-1}, daf)	16.62	25.77	20.31
LHV	(MJ kg^{-1}, raw)	16.55	22.60	19.76

[1]Data reproduced from Virmond et al. (2011); [2]Data reproduced from Floriani et al. (2010); [3]Maximum experimental uncertainties equal to 0.30%; db is on a Dry Basis; daf is on a Dry and Ash Free basis; *Value previously presented by Floriani et al. (2010) corrected; n.d. is Not Determined; LHV is Lower Heating Value

In previous publications (Floriani et al., 2010; Virmond et al., 2008, 2011), the authors have addressed the effects of the LFP ash composition on the fouling and slagging tendency in the combustion systems, showing that the occurrence of this problem can be reduced when burning a mixture of the biosolids with wood residues such as SD compared to LFP alone.

As shown in Table 3, the ash melting temperatures of LFP are much lower than the values estimated for LFPSD1:9 through mass balance analysis considering a homogeneous mixture, and its utilization is recommended in low blending proportions.

Table 3. Biomass ash properties

	Units	SD[1]	LFP[1]	LFPSD1:9[2]
Ash composition[3]				
Fe_2O_3	(wt%, db)	4.44	32.40	9.34
CaO	(wt%, db)	31.27	17.40	22.77
MgO	(wt%, db)	11.64	1.30	4.59
Na_2O	(wt%, db)	1.67	1.70	1.14
K_2O	(wt%, db)	10.44	1.70	8.77
SiO_2	(wt%, db)	15.69	4.90	17.84
Al_2O_3	(wt%, db)	12.30	1.70	8.22
TiO_2	(wt%, db)	3.94	0.00	3.07
P_2O_5	(wt%, db)	2.74	36.30	8.50
MnO	(wt%, db)	2.06	n.d.	n.d.
SO_4	(wt%, db)	3.02	n.d.	n.d.
Ash melting temperatures				
Deformation temperature	(°C)	>1150*	750	1335
Softening temperature	(°C)	>1170*	990	1359
Hemispherical temperature	(°C)	>1190*	1010	1361
Fluid temperature	(°C)	>1230*	1040	1364

1Data reproduced from Virmond et al. (2011); 2Data reproduced from Floriani et al. (2010); 3Maximum experimental uncertainties equal to 0.30%; db is on a dry basis; n.d. is not determined; 4Data reproduced from Llorente & García (2005) for eucalyptus sample

It was observed that the main element found in the sludge ash was phosphorus, followed by iron. This is considered a problem because P forms compounds with lower melting temperature, which may have influenced the results presented in Table 3. As expected, the mixture of biomasses maintained a relatively high ash melting temperature, which is a desirable aspect when considering the combustion of solid fuels. Additionally, the design of the equipment and the definition of the operating conditions are extremely important to control, or even avoid, the occurrence of such problems.

Table 4. Results for the determination of trace metals and nutrients in the sludge samples

Concentration (mg kg-1, db)				
	Limit1	**BSFlot**	**BSCent**	**BSBiol**
Trace metal				
Hg	5	< 0.50	< 0.50	< 0.50
Cd	5	< 0.50	< 0.50	0.64
Cr	800	6.7	28.4	26.7
Cu*	800	16.2	29.8	182.1
Ni	200	1.9	9.9	22.0
Pb	500	1.3	3.4	6.1
Zn*	2000	88.2	183.8	1090.3
As	75	0.57	< 0.50	< 0.50
Mo*	75	0.50	1.7	4.4
Co*	5	< 0.50	< 0.50	4.1
Micronutrient				
K	-	427	599	6903
Fe	-	9360	25600	20900
Al	-	1750	498	3420
P	-	6350	15900	28400
Secondary nutrients				
Ca	-	1520	5080	18600
Mg	-	148	259	7185
S	-	3140	6630	9810

db is Dry Basis; 1Upper limit of pollution for the disposal of sewage sludge in the environment (EU, 2000); * Trace elements, biologically essential in small quantities. Reproduced from de Sena et al. (2009)

Besides the determination of the biomass properties for biofuel applications, the organic and inorganic contents of the sludge generated from the WWTP were characterized, since it is necessary to assure that sludge containing high pollutant loads is not applied as fertilizer, in order to avoid contamination of agricultural soil and cultivated plants, i.e., to avoid the transfer of contaminants into the food chain. Inorganic and organic pollutants not removed during physicochemical wastewater treatment processes are either bio-chemically degraded or adsorbed by the sludge. The characterization of the sludge was reported by de Sena et al. (2009), where the trace metal, PAH, PCB and PCDD/PCDF concentrations in the WWTP sludge were determined. Trace metals might end up in the effluent from the meat processing plant through sources like equipment, sanitizers and cleaning agents, as well as equipment and pumps used in the wastewater treatment plant itself. Also, some metals such as arsenic, copper and zinc are occasionally added to animal feed as mineral food supplements and/or as growth promoters (US EPA, 2004). The group of PAHs, generated undesirably mostly during manifold incomplete incineration processes, includes numerous compounds with three or more condensed aromatic rings. PCBs synthesized for specific applications as non-inflammable insulators, hydrolic liquids and plasticizers are pollutants which today are ubiquitously found in the environment, although they were phased out from production worldwide at the end of the 1970s. PCDDs/PCDFs are not intentionally produced by humans but they are released into the atmosphere as sub-products of incineration and combustion processes, both domestic and industrial, when carbon, hydrogen, chlorine and oxygen together with copper as a catalyst are present. The incineration of municipal or clinical wastes, iron ore sinter plants and non-ferrous metal industries (Quass et al., 2000) as well as the chemical synthesis of chlorophenols and electrolysis of sodium chloride are sources of PCDDs/PCDFs. These compounds accumulate in sludge due to their extreme lipophilicity, and it is very difficult to assess the various sources of these compounds and their pathways into the environment and into the food chain (Klöpffer, 1996). Tables 4, 5 and 6 show the results of the biomass characterization, including 3 (three) types of sludge (BS) collected at different points in the WWTP. BSFlot refers to the sludge remaining after the flotation process, BSCent to the sludge remaining after the three-phase centrifugation, and BSBiol to the sludge collected from the activated sludge bioreactor (not primarily intended for combustion purposes).

Table 5. Results for the determination of PAHs and PCBs in the sludge samples

Compounds	Concentration (μg kg-1, db)				
	TEF1	Limit2	BSFlot	BSCent	BSBiol
Polycyclic Aromatic Hydrocarbons (PAHs)					
Naphthalene (Nap)	-		110.0	< 40.0	< 30.0
Acenaphthylene (Acy)	-		84.0	< 30.0	< 30.0
Acenaphtene (Ace)	-		< 7.0	< 7.0	< 6.0
Fluorene (Flu)	0.001		20.0	< 7.0	< 6.0
Phenanthrene (Phe)	0.001		84.0	< 8.0	< 7.0
Anthracene (Ant)	-		< 2.0	8.0	8.0
Fluoranthene (Fla)	0.001		320.0	510.0	< 10.0

Pyrene (Pyr)	0.001		97.0	72.0	11.0
Chrysene (Cry)3	0.01		92.0	< 1.0	5.0
Σ PAHLMW	-		816.0	616.0	113.0
Benzo[a]anthracene (BaA)3	0.1		24.0	< 1.0	4.0
Benzo[b]fluoranthene (BbF)3	0.1		38.0	25.0	< 3.0
Benzo[k]fluoranthene (BkF)3	0.1		11.0	< 1.0	< 1.0
Benzo[a]pyrene (BaP)3,4	1.0		2.0	< 1.0	5.0
Dibenzo[a,h]anthracene (DbA)3,4	5.0		< 4.0	< 4.0	< 4.0
Benzo[g,h,i]perylene (BgP)	0.1		12.0	< 6.0	< 6.0
Indeno[1,2,3-cd]pyrene (InD)3	0.1		< 10.0	< 10.0	< 10.0
Σ PAHHMW	-		101.0	48.0	33.0
Σ PAH		≤ 6000	917.0	664.0	146.0
Σ TEFPAH	-		11.0	3.0	6.0
Polychlorinated Biphenyls (PCB)					
PCB non-ortho					
PCB 77, 81, 126*, 169	< 0.1		< LOQ	< LOQ	< LOQ
PCB mono-ortho					
PCB 105, 114, 118, 123, 156, 167, 189	< 0.005		< LOQ	< LOQ	< LOQ
Σ PCB	-	≤ 800	< LOQ	< LOQ	< LOQ

db is Dry Basis; 1TEF for dioxin and dioxin-like compounds (Nisbet et al., 1992; van den Berg et al., 2006); 2Limit for solid disposal onto soil (EU, 2001); 3Carcinogenic isomers; 4Isomers of PAH with TEF comparable to the most toxic polychlorinated dibenzo-dioxins and -furans; *Isomer of PCB with highest TEF; LOQ is Limit of Quantification. Reproduced from de Sena et al. (2009).

Table 6. TEF and concentrations of PCDDs/PCDFs in the sludge samples

Concentration (ng kg-1, db)				
	TEF1	BSFlot	BSCent	BSBiol
Polychlorinated dibenzo-p-dioxins (PCDD)				
2,3,7,8 – TCDD2	1.0	0.1	0.4	0.4

1,2,3,7,8 - PeCDD	1.0	0.6	1.3	2.2
1,2,3,4,7,8 - HxCDD	0.1	0.3	1.5	1.7
1,2,3,7,8,9 - HxCDD	0.1	0.2	1.9	3.3
1,2,3,6,7,8 - HxCDD	0.1	0.7	1.3	1.8
1,2,3,4,6,7,8 - HpCDD	0.01	0.8	4.2	4.6
OCDD	0.0003	2.1	21.1	23.1
Σ PCDD	-	4.8	31.7	37.1
Polychlorinated dibenzofurans (PCDF)				
2,3,7,8 – TCDF	0.1	0.2	0.3	1.3
2,3,4,7,8 – PeCDF	0.3	0.5	2.1	2.1
1,2,3,7,8 – PeCDF	0.03	0.7	1.5	3.8
1,2,3,4,7,8 – HxCDF	0.1	0.6	1.2	2.0
1,2,3,6,7,8 – HxCDF	0.1	0.6	1.6	2.3
1,2,3,7,8,9 – HxCDF	0.1	1.7	2.9	3.1
2,3,4,6,7,8 - HxCDF	0.1	0.7	1.9	3.8
1,2,3,4,6,7,8 - HpCDF	0.01	2.6	4.2	5.3
1,2,3,4,7,8,9 – HpCDF	0.01	1.4	3.1	3.6
OCDF	0.0003	2.1	4.4	10.2
Σ PCDF	-	11.1	23.2	37.3
PCDD:PCDF Ratio	-	0.43	1.37	0.98
Σ TEF PCDD/PCDF	100*	1.4	3.8	5.4

db is Dry Basis; 1TEF for dioxins and dioxin-like compounds (van den Berg et al., 2006; WHO, 2005); 2Isomer with highest acute toxicity; *TEF (ng kg-1) limit for solid disposal onto soil (EU, 2001). Reproduced from de Sena et al. (2009)

A study by de Sena et al. (2009) verified low pollution loads for the sludge (BS) originating from the WWTP, with respect to the most relevant inorganic and organic priority pollutants as monitored by the US EPA, at the case study meat processing plant located in the south of Brazil. Although other pollutants such as veterinary drugs, pesticides and surfactants were not investigated in this first analytical approach, they are of high concern. However, this study was a preliminary report for future monitoring of other food processing segments located in different regions of Brazil.

4.2 Biomass combustion

Co-combustion of agro-industrial residues in thermal power plants is not necessarily a low cost alternative for the thermal treatment of wastes. There is the possibility of interaction between the components and the main fuels in such a way that either the operating behavior of the conversion system is improved or the emissions are reduced (Werther, 2007). The emission of pollutants generated during combustion is strongly related to the biomass properties. Pollutant formation mechanisms and many other parameters related to the combustion process must be monitored due to the formation of highly problematic compounds such as NOx, SO2, benzene, toluene, ethyl-benzene, and (o-,m-,p-)xylenes (BTEX), polycyclic aromatic hydrocarbons (PAH), polychlorinated dibenzo-p-dioxins and polychlorinated dibenzofurans (PCDD/PCDF) (Chagger et al., 1998; Kumar et al., 2002; Mckay, 2002; Stanmore, 2004; Watanabe et al., 2004), and have to be controlled in order to comply with the stringent limits set by recent environmental legislation. Chlorine-associated, high-temperature corrosion and the potential corrosion problems associated with burning biomass fuels have been previously discussed (Nielsen et al., 2000). Fuel nitrogen can be a problem in terms of NOx emissions. The conversion of nitrogen in systems fired by solid fuels (mainly coal, but also biomass) has been reviewed in detail, as well as the combustion characteristics of different biomass fuels, the potential applications of renewable energy sources as the prime energy sources in various countries, and the problems associated with biomass combustion in boiler systems (Demirbas, 2004; Werther, 2000).

The pollutant emissions due to incomplete biomass combustion can be effectively controlled by an optimized combustion process, i.e., enhanced mixing, sufficient residence time at high temperatures (>850 °C), and low total excess air (Demirbas, 2005), as well as the appropriate choice of the combustion device.

Given the higher energy value of the biosolids (LHV equal to 22.60 MJ kg-1) compared to sawdust (LHV equal to 16.55 MJ kg-1), as shown in Table 2, the substitution of 10 wt% of sawdust with this residue can increase the thermal energy production by approximately 4% compared to the combustion of sawdust alone, leading to a sawdust saving of 1950 ton per year, besides the economic benefits related to reduced landfill disposal. However, the gaseous emissions have to be monitored so as not to infringe current legislations.

The evaluation of the feasibility of the utilization of the biosolids originating from the WWTP studied was performed through combustion testing of the biosolids as the sole fuel in a pilot-scale cyclone combustor (model Drako, Albrecht, Brazil) with a burning capacity of 100 kg h-1, as described by Virmond et al. (2011), and in a furnace equipped with a recip rocating-grate coupled to a boiler at the meat processing plant with the capacity to process 12000 kg h-1, as reported by Virmond (2007).

Grate firing is one of the main technologies that are currently used for biomass combustion aiming at heat and power production. Grate-fired boilers can fire a wide range of fuels with varying moisture content and show great potential in biomass combustion (Goërner, 2003). The plant-scale furnace was operated at a fuel feed rate of 2604 kg h-1 (moisture content of approximately 50 wt%) at 900 °C, with 59% excess air and without gas recirculation. The combustion test lasted for approximately 2 h after the system had reached steady state.

In the pilot-scale plant the combustion of SD was carried out at a fuel feed rate of 32 kg h-1 (moisture content of 9.16 wt%), with gas recirculation of 20% at an average temperature of 642 °C in a cyclone combustor (Drako, Albrecht, Brazil), which has been described by Virmond et al. (2011). The average temperature at the front stage of the combustor was 1052 °C and at the outlet it was 1024 °C, and an air-to-fuel (A/F) ratio of 9.08 (theoretical A/F ratio of 7.23 based on the fuel composition) was used. For the LFP combustion test, the conditions applied to the cyclone combustor were: fuel feed rate of 43 kg h-1 (moisture content of 11.44 wt%), with gas recirculation of 20% at an average temperature of 800 °C. The average temperature at the front stage of the combustor was 868 °C, at the outlet of the combustor it was 1080 °C, and the A/F ratio was 11.82 (theoretical A/F ratio of 7.76 based on the fuel composition). CO, O2, SO2, CxHy (measured as CH4), NO, and NO2 emissions were measured using a Greenline MK2 (Eurotron) analyzer and BTEX emission measurements were based on the adsorption and desorption of gases which were analyzed by gas chromatography. Emissions of BTEX were expressed as Total Organic Carbon (TOC). The detailed methodology for emissions sampling and analysis during the combustion tests, as well as the complete set of results, have been previously reported by Floriani et al. (2010) and Virmond et al. (2007, 2008) for LFPSD1:9 combustion, and by Virmond et al. (2011) for SD and LFP. Thus, only the main points are highlighted herein. The gaseous emissions observed in the combustion tests and the respective regulation limits were corrected to the reference oxygen content (O2ref) of 7% and are given in Table 7.

Since no data on the combustion of meat processing or slaughterhouse wastes had been previously published, no comparison was possible. However, the results for the contaminants reported were compared with the emission limits established by national and international environmental agencies, such as the Brazilian guidelines issued by The National Council of the Environment (CONAMA, 2002) for gaseous emissions in the thermal treatment of wastes; the American guidelines issued by the US Environmental Protection Agency (US EPA, 2002) for emissions from commercial and industrial solid waste incineration units; and the German Guidelines 17.BlmSchV (17.BlmSchV, 2003) for emissions from biomass combustion and from biomass co-combustion. Concerning the combustion tests performed in the pilot-scale cyclone combustor with SD and LFP as fuels, the emissions of CO, CO2,

CxHy and TOC were well controlled and their concentrations remained below the regulation limits considered for both biomasses, except CO in the SD combustion test in relation to the German guidelines, which refer to waste incineration and are stricter than the other regulations.

Table 7. Gaseous emissions from the combustion tests at O2ref=7%

	CO (mg Nm-³)	CO2 (%)	CxHy (mg Nm-³)	NOx (mg Nm-³)	SO2 (mg Nm-³)	TOC (mg Nm-³)
SD biomass2	93.58	10.32	0.00	241.58	0.00	1.27
	±21.97	±0.03	±0.00	±26.31	±0.00	±0.57
LFP biomass2	63.33	10.33	0.00	1727.43	363.54	1.23
	±10.30	±0.05	±0.00	±229.93	±90.80	±0.12
LFPSD1:9 biomass3	734.83	10.39	554.44	497.94	128.69	1.72
	±12.39	±0.02	±7.91	±19.04	±4.31	±0.83
CONAMA 4	124.88	n.a.	n.a.	560.00	280.00	n.a.
US EPA 5	196.16	n.a.	n.a.	796.02	57.09	n.a.
17.BlmSchV (24 h; <50 MW) 6	140.00	n.a.	n.a.	373.33	186.67	9.33
17.BlmSchV (24 h)7	70.00	n.a.	n.a.	280.00	70.00	14.00

[1]NOx expressed in terms of NO2; TOC is Total Organic Carbon; 2Data from Virmond et al. (2011); 3Data from Floriani et al. (2010) and Virmond et al. (2007, 2008); 4CONAMA 316/02, thermal treatment of wastes (CONAMA, 2002); n.a. is Not Applicable; 5US EPA, solid waste incineration (US EPA, 2000); 617.BlmSchV 24 h, <50 MW, co-combustion of wastes (17. BlmSchV, 2003); 717.BlmSchV 24 h, direct combustion of wastes (17.BlmSchV, 2003)

The effect of the biomass composition on gaseous emissions was clearly observed, especially considering the N and S fuel contents in LFP, which led to concentrations of these pollutants being higher than the established limits.

The use of the biosolids originating from the meat processing plant investigated in this study as a fuel in the pilot cyclone combustor was shown to be feasible; however, further research is required concerning the control of SO2 and NOx emissions to avoid exceeding the very strict emission limits as well as the occurrence of fouling and slagging.

In relation to the combustion test performed with the mixture of SD and LFP (LFPSD1:9), the high levels of CxHy and CO emitted indicate incomplete combustion. This can be attributed to the high moisture content of the biomass (50.23 wt%, as shown in Table 2), the lower combustion temperature (approximately 900 °C) compared to the other two tests performed in the pilot-scale combustor (approximately 1000 °C) and the absence of gas recirculation. Additionally, the control of the operating conditions of the large-scale plant is more difficult to achieve, due to the restricted testing time or minimal variation from the normal operation with the SD. Firing a biomass with low moisture content and flue gas recirculation could provide better oxidation conditions. Due to the lower nitrogen concentration found in LFPS1:9 compared to LFP, as well as to the low operating combustion temperature, NOx emissions remained below the limits established by CONAMA and US EPA.

The co-combustion of LFP and SD with lower-N fuel content reduced the NOx in the gaseous emissions compared to the burning of LFP alone. In fact, this option is the most feasible in Brazil considering the relatively high NOx emissions related to both the fuel nitrogen and to the fact that wood and wood-based materials are extensively used as fuel for thermal energy generation in the Brazilian food industry. Chlorine, PAHs and PCBs are among the elements or compounds that must be studied in greater depth as they are precursors to the formation of dioxins and partly of furans.

Industrial solid wastes must be disposed of safely, and co-firing them with SD has been shown in other studies by the authors, which are currently underway, to be profitable using the same pilot-scale cyclone combustor and different biomasses. The advantages are both a reduction in the consumption of primary fuels and the recovering of energy from wastes inside the plant, which would normally be disposed of in landfills, potentially causing environmental problems.

4.3 Biogas production and combustion

During this study, only preliminary tests were performed, such as the determination of the chemical composition of the biogas and the exhaust gases when the biogas was burned. Measurements were performed at several partner pig farms close to the case study plant that rear pigs from 45 to 110 days of life. At these farms there were relatively small horizontal-type biodigestors (approximately 450 m3) in which the input consisted only in pig wastes without previous treatment. The hydraulic retention time was approximately 30 days and the entrance flow 8 m3 d-1.

Table 8 shows the average chemical composition and calorific values of the biogas. To analyze the composition of biogas, a biogas analysis kit was used (Alfakit, Brazil) which is based on colorimetric methods.

Table 8. Average chemical composition of biogas

	Average chemical composition	Reference values1
CH4 (v/v%)	45-65	57
CO2 (v/v%)	35-55	34

H_2S (v/v%)	>1020	-
CH_4/CO_2	0.82-1.86	1.7
HHV (kJ Nm^{-3})	17996-25995	-
LHV(kJ Nm^{-3})	16200-23400	-
[1]Data reproduced from Silva et al. (2005); HHV is Higher Heating Value; LHV is Lower Heating Value		

At the time this study was completed the farms were burning only biogas to avoid harmful emissions and to provide a better disposal/reuse options for the waste. At all farms there were simple flares to burn the gas, and thus the gaseous emissions were evaluated. CO, SO_2, NO, NO_2, O_2, and CxHy were measured using a Greenline MK2 (Eurotron) analyzer and the sampling point was located at the top of the chimney. Results are not presented in this document because each burner presented different burning efficiencies and they were still in the adjustment phase. However, by the end of the project, all gaseous emissions were below the limits imposed by Brazilian Legislation. The SO_2 requires greater caution due to the presence of H_2S and therefore a pre-treatment has to be considered before the process can be considered adequate. H_2S can easily react with iron oxides and hydroxides, requiring the presence of water, and this can thus be considered a good method to remove the H_2S from biogas (Zichari, 2003).

5. CONCISE SUMMARY

The water balance analysis conducted at the case study plant revealed that reducing fresh water use at the four primary consumption points might result in a savings of around 806 m^3 per day. Regarding wastewater reuse, the four streams identified as viable for reuse amounted to around 1383 m^3 per day. These require basic reconditioning prior to use in operations that do not involve direct contact with food goods, namely in non-potable applications (e.g., cooling water, toilet flushing water, or watering surrounding the facility), hence conserving new potable water. The projected reduction in freshwater consumption following water minimization and wastewater reclamation was 25.4%, resulting in an annual financial savings of approximately $434,622.00.

Furthermore, to further diminish the proportion of freshwater usage, indirect wastewater reuse may be implemented post-reconditioning through tertiary treatments, such as advanced oxidation processes (AOPs), to process the secondary effluent following secondary activated sludge treatment. The tertiary-treated water effluent may then be utilized in various procedures devoid of interaction with food goods. The proposed tertiary treatment model is a heterogeneous advanced oxidation process (AOP) system ($UV/TiO_2/H_2O_2$) applicable to urban, rural, or industrial effluents. The factors impeding their reuse as potable water include the presence of suspended solids (even at low concentrations), dissolved organic matter, recalcitrant micro-pollutants (trace compounds), and elevated concentrations of nitrate and nitrite. Laboratory studies must be conducted using actual wastewater to assess the efficacy of each process stage.

Given the low biodegradability of the effluent from the poultry hatchery case study (COD/BOD5 = 4.6) and the existence of persistent chemicals resistant to biological treatment, a chemical or physicochemical approach would be the most effective solution. All assessed treatments, especially the photo-Fenton reaction, led to enhanced biodegradability of the effluent, signifying an increase in the fraction of the material amenable to breakdown by biological processes. Consequently, a biological process should be incorporated as a final step in the effluent treatment, mostly to eliminate the previously-formed more biodegradable chemicals (with lower molar mass) and the nutrients that remain after the physicochemical process, such as nitrate. A comparative analysis of the advantages and disadvantages, particularly regarding costs and efficiency, of the two treatments: (a) photo-Fenton combined with basic biological treatment methods (such as stabilizing ponds) and (b) UV.

The efficacy of TiO2/H2O2 in the treatment of secondary effluent from slaughterhouses could be evaluated. Therefore, the primary consideration to assess is which treatment can guarantee compliance with legislative standards and limits, thereby preventing adverse environmental impacts (such as the release of persistent organic pollutants into waterways) while also delivering economic advantages.

The replacement of 10 wt% of sawdust with biosolids from physicochemical wastewater treatment can enhance thermal energy production by roughly 4% compared to the combustion of sawdust alone, resulting in an annual savings of 1950 tons of sawdust and reducing landfill disposal costs. Furthermore, co-combustion is the most viable method for energy recovery from this waste in Brazil, enabling the regulation of the combustion process, the prevention of fouling and slagging, and compliance with the emission standards set forth in pertinent law. Given that wood and wood-derived materials are widely utilized as fuels for thermal energy production in the Brazilian food sector, the incorporation of a minor mass fraction of this solid waste with sawdust should not necessitate substantial alterations to the existing operational parameters.

Industrial solid wastes require safe disposal, and recent tests conducted by our research group demonstrate that co-firing these wastes with sawdust is economically viable when utilizing a pilot-scale cyclone combustor. Biogas generated from swine waste possesses significant potential to emerge as a vital bioenergy alternative for the Brazilian agroindustrial sector. Moreover, anaerobic digestion offers additional environmental advantages beyond generating a renewable energy source, such as nutrient recycling and trash volume reduction. Nonetheless, research is required to examine the impact of input fluctuations on a biodigester, how waste composition affects the overall stability of the process and output quality, and the alternatives for biogas utilization.

The thorough technical and scientific analyses of water, wastewater, and solid waste management actions conducted at the case study meat processing plant demonstrated that environmentally, financially, and socially sustainable practices can be effectively implemented in any food processing facility, regardless of type or size.

REFERENCES

1. 17th Ordinance on the Implementation of the Federal Emission Control Act – Ordinance for Combustion and Co-combustion of Wastes of August 14th (Siebzehnte Verordnung zur Durchführung des Bundes-Immissionsschutzgesetzes. Verordnung über die Verbrennung und die Mitverbrennung von Abfällen) 17.BlmSchV, Bundes- anzeiger, Bonn/ Germany, 2003 (in German).
2. Ahmed, S., Rasul, M.G.; Martens, W.N., Brown, R., & Hashib, M.A. (2010). Heterogeneous photocatalytic degradation of phenols in wastewater: A review on current status and developments. Desalination, Vol. 261, No. 1-2, (Oct 2010), pp. 3-18, ISSN 0011-9164.
3. Almeida, E., Assalin, M.R., Rosa, M.A., & Duran, N. (2004). Tratamento de efluentes industriais por processos oxidativos na presença de ozônio. Química Nova, Vol. 27, No. 5, (Sept/Oct 2004), pp. 818–824, ISSN 0100-4042.
4. Bila, D.M., Montalvão, F., Silva, A.C., &Dezotti, M. (2005). Ozonation of a landfill leachate: evaluation of toxicity removal and biodegradability improvement. Journal of Hazardous Materials, Vol. 117, No. 2-3, (Jan 2005), pp. 235–242, ISSN 0304-3894.
5. Boersma, L., Gasper, E., Oldfield, J.E. & Cheeke, P.R. (1981) Methods for the recovery of nutrients and energy from swine manure. 1. Biogas. Netherlands Journal of Agricultural Science, Vol. 29, pp. 3–14, ISSN. 0028-2928.
6. Brazil, 2011. Set of legal rules: water resources, Ministry of Environment, Department of Water Resources and Urban Environment. 7 . ed., Brasília, DF, Brazil, 640 p. (in Portuguese). Retrieved from: http://www.cnrh.gov.br/sitio/index.php?option=com_docman&task=doc_download&gid=822.
7. Chagger, H. K., Kendall, A., McDonald, A., Pourkashanian, M., & Williams, A. (1998). Formation of dioxins and other semi-volatile organic compounds in

biomass combustion. Applied Energy, Vol. 60, No. 2, (Jun 1998), pp. 101-114, ISSN 0306-2619.

8. CONAMA - National Council of the Environment (29 Oct 2002). Resolution 316, In: Ministry of the Environment (MMA), Brasília, DF, Brazil (in Portuguese), Date of access: 11 Jan 2009, Available from: <http://www.mma.gov.br/port/conama/legiabre.cfm? codlegi=338>.

9. Cornel, P., Meda, A., & Bieker, S. (2011). Wastewater as a Source of Energy, Nutrients, and Service Water, In: Treatise on Water Science, Peter Wilderer (ed.), Vol. 4,pp. 337-375, Oxford: Academic Press, ISBN: 9780444531933.

10. de Sena, R.F., Claudino, A., Moretti, K., Bonfanti, I.C.P., Moreira, R.F.P.M., & José, H.J. (2008). Biofuel application of biomass obtained from a meat industry wastewater plant through the flotation process - A case study. Resources, Conservation and Recycling Vol. 52, No. 3 (Jan 2008), pp. 557-569, ISSN 0921-3449.

11. de Sena, R.F., Tambosi, J.L., Genena, A.K., Moreira, R.F.P.M., Schröder, H.Fr., & José, H.J. (2009). Treatment of meat industry wastewater using dissolved air flotation and advanced oxidation processes monitored by GC–MS and LC–MS. Chemical Engineering Journal, Vol. 152, No. 1 (Oct 2009), pp. 151–157, ISSN 1385-8947.

12. de Sena, R.F., Tambosi, J.L., Floriani, S.L., Virmond, E., Schröder, H.Fr., Moreira, R.F.P.M., & José, H.J. (2009). Determination of inorganic and organic priority pollutants in biosolids from meat processing industry. Waste Management, Vol. 29, No. 9 (Sep 2009), pp. 2574-2581, ISSN 0956-053X.

13. Demirbas, A. (2004). Combustion characteristics of different biomass fuels. Progress in Energy and Combustion Science, Vol. 30, No. 2 (2004), pp. 219-230, ISSN 0360-1285.

14. Demirbas, A. (2005). Potential applications of renewable energy sources, biomass combustion problems in boiler power systems and combustion related environmental issues. Progress in Energy and Combustion Science, Vol. 31, No. 2 (2005), pp. 171-192, ISSN 0360-1285.

15. Esplugas, S., Bila, D.M., Krause, L.G., & Dezotti, M. (2007). Ozonation and advanced oxidation technologies to remove endocrine disrupting chemicals (EDCs) and pharmaceuticals and personal care products (PPCPs) in water effluents. Journal of Hazardous Materials, Vol. 149, No. 3 (19 Nov 2007), pp. 631-642, ISSN 0304-3894.

16. EU – European Union Commission. (2000). Working document on sludge, 3rd Draft. ENV.E.3/LM. Brussels, Belgium.

17. EU – European Union Commission. (2001). European Commission Directive, Joint Research Centre/Institute for Environment and Sustainability/Soil and Waste Unit, 2001. Organic contaminants in sewage sludge for agricultural use. Brussels, Belgium.

18. Floriani, S.L., Virmond, E., Althoff, C.A., Luiz, D.B., José, H.J., & Moreira, R.F.P.M. (2010). Potential of industrial solid wastes as an energy source and gaseous emissions evaluation in a pilot scale burner (ES2008-54355). Journal of Energy Resources Technology, Vol. 132, No. 1 (Mar 2010), pp. 011003-011010, ISSN 0195-0738.

19. Genena, A.K. (2009). Treatment of agroindustrial wastewater containing persistent compounds by coagulation-flocculation, Fenton, photo-Fenton, photo-peroxidation and ozonation (in Portuguese). Doctoral Thesis, Federal University of Santa Catarina, Florianópolis, Brazil. Date of access: 01 Aug 2012, Available from: http:// www2.enq.ufsc.br/teses/d069.pdf (in Portuguese).

20. Genena, A.K., Luiz, D.B., Gebhardt, W., Moreira, R.F.P.M., José, H.J. & Schröder, H.Fr. (2011). Imazalil degradation upon applying ozone—transformation products, kinetics, and toxicity of treated aqueous solutions. Ozone: Science & Engineering, Vol. 33, No. 4 (Jul-Aug 2011), pp. 308-328, ISSN 0191-9512.

21. Goërner, K. (2003). Waste incineration: European state-of-the-art and new development. IFRF Combustion Journal, Vol. 32, Article No. 200303 (Jul 2003), 32 p., ISSN 1562-479X.

22. Guan, Y.-F., Wang, J.-Z., Ni, H.-G., & Zeng, E.Y. (2009). Organochlorine pesticides and polychlorinated biphenyls in riverine runoff of the Pearl River Delta, China: Assessment of mass loading, input source and environmental fate. Environmental Pollution, Vol. 157, No. 2 (Feb 2009), pp. 618–624, ISSN 0269-7491.

23. Jenkins, B.M., Baxter, L.L., Miles Jr, T.R. & Miles, T.R. (1998). Combustion properties of biomass. Fuel Processing Technology, Vol. 54, No. 1-3 (Mar 1998), pp. 17-46, ISSN 0378-3820.

24. Johns, M.R. (1995). Developments in wastewater treatment in the meat processing industry: a review. Bioresource Technology, Vol. 54, No. 3 (1995), pp. 203-216, ISSN 0960-8524.

25. Klöpffer, W. (1996). Environmental hazard assessment of chemicals and products. Part V. Anthropogenic chemicals in sewage sludge. Chemosphere, Vol. 33, No. 6, (Sep 1996), pp. 1067-1081, ISSN 0045-6535.

26. Koning, J., Bixio, D., Karabelas, A., Salgot, M., & Schäfer, A. (2008). Characterisation and assessment of water treatment technologies for reuse. Desalination, Vol.218, No. 1-3 (5 Jan 2008), pp. 92-104, ISSN 0011-9164.

27. Kumar, A., Purohit, P., Rana, S., & Kandpal, T.C. (2002). An approach to the estimation of the value of agricultural residues used as biofuels. Biomass and Bioenergy, Vol. 22, No. 3 (Mar 2002), pp. 195-203, ISSN 0961-9534.

28. Liu, Z.H., Kanjo, Y., & Mizutani, S. (2009). Removal mechanisms for endocrine disrupting compounds (EDCs) in wastewater treatment - physical means, biodegradation, and chemical advanced oxidation: A review. Science of the Total Environment, Vol. 407, No. 2 (1 Jan 2009), pp. 731-748, ISSN 0048-9697.

29. Llorente, M.J.F., & García, J.E.C. (2005). Comparing methods for predicting the sintering of biomass ash in combustion. Fuel, Vol. 84, No. 14-15 (Oct 2005), pp. 1893-1900, ISSN 0016-2361.

30. Lora, E.S., & Andrade, R.V. (2009). Biomass as energy source in Brazil. Renewable and Sustainable Energy Reviews, Vol. 13, No. 4 (May 2009), pp. 777-788, ISSN 1364-0321.

31. Lucas Júnior, J. (1994). Considerations on the use of swine manure as substrate for three anaerobic biodigestion systems in Portuguese). Doctoral Thesis, - Faculdade de Ciências Agrárias e Veterinárias, Universidade Estadual Paulista, Jaboticabal, 137 p.

32. Luiz, D.B., Genena, A.K., José, H.J., Moreira, R.F.P.M., & Schröder, H.Fr. (2009). Tertiary treatment of slaughterhouse effluent: degradation kinetics applying UV radiation or H2O2/UV. Water Science and Technology, Vol. 60, No. 7, pp. 1869-1874, ISSN 0273-1223.

33. Luiz, D.B., Genena, A.K., Virmond, E., José, H.J., Moreira, R.F.P.M., Gebhardt, W., & Schröder, H.Fr. (2010). Identification of degradation products of erythromycin A arising from ozone and AOP treatment. Water Environment Research, Vol. 82, pp. 797-805, ISSN 1061-4303.

34. Luiz, D.B., Silva, G.S., Vaz, E.A.C., José, H.J., & Moreira, R.F.P.M. (2011). Evaluation of hybrid treatments to produce high quality reuse water. Water Science and Technology, Vol. 63, No. 9, pp. 2046-2051, ISSN 0273-1223.

35. Luiz, D.B., José, H.J., & Moreira, R.F.P.M. (2012a). A discussion paper on challenges and proposals for advanced treatments for potabilization of wastewater in the food industry. In: Valdez, B., Zlatev, R., Schorr, M. (Org.). Scientific, Health and Social Aspects of the Food Industry. : InTech, p. 3-24, ISBN 978-953-307-916-5.

36. Luiz, D.B., Andersen, S.L.F., Berger, C., José, H.J., & Moreira, R.F.P.M. (2012b). Photocatalytic Reduction of Nitrate Ions in Water over Metal-Modified TIO2. Journal of Photochemistry & Photobiology, A: Chemistry. In press. ISSN: 1010-6030.

37. Mckay, G. (2002). Dioxin characterization, formation and minimization during municipal solid waste (MSW) incineration: review. Chemical Engineering Journal, Vol. 86, No. 3 (28 Apr 2002), pp. 343-368, ISSN 1385-8947.

38. Nielsen, H. P, Frandsen, F. J, Dam-Johansen, K., & Baxter, L.L. (2000). The implications of chlorine-associated corrosion on the operation of biomass-fired boilers. Progress in Energy and Combustion Science, Vol. 26, No. 3 (Jun 2000), pp. 283-298, ISSN 0360-1285.

39. Nisbet, I.C.T., & La Goy, P.K. (1992). Toxic equivalency factors (TEFs) for polycyclic aromatic hydrocarbons (PAHs). Regulatory Toxicology and Pharmacology, Vol. 16, No. 3 (Dec 1992), pp. 290–300, ISSN 0273-2300.

40. Obernberger, I., Brunner, T., & Bärnthaler, G. (2006). Chemical properties of solid biofuels - significance and impact. Biomass and Bioenergy, Vol. 30, No. 11 (Nov 2006), pp. 973-982, ISSN 0961-9534.

41. Oller, I., Malato, S., & Sánchez-Pérez, J.A. (2011). Combination of advanced oxidation processes and biological treatments for wastewater decontamination - A review. Science of the Total Environment, Vol. 409, No. 20 (15 Sep 2011), pp. 4141-4166, ISSN 0048-9697.

42. Petrović, M., Gonzalez, S., & Barceló, D. (2003). Analysis and removal of emerging contaminants in wastewater and drinking water. TrAC Trends in Analytical Chemistry, Vol. 22, No. 10 (Nov 2003), pp. 685-696, ISSN 0165-9936.

43. Rengaraj, S., & Li, X.Z. (2007). Enhanced photocatalytic reduction reaction over Bi3+– TiO2 nanoparticles in presence of formic acid as a hole scavenger. Chemosphere, Vol. 66, No. 5 (Jan 2007), pp. 930-938, ISSN 0045-6535.

44. Quaß, U., Fermann, M.W., & Bröker, G. (2000). Steps towards a European dioxin emission inventory. Chemosphere, Vol. 40, No. 9-11 (May-Jun 2000), pp. 1125–1129, ISSN 0045-6535.

45. Rizzo, L. (2011). Bioassays as a tool for evaluating advanced oxidation processes in water and wastewater treatment. Water Research, Vol. 45, No. 15 (1 Oct 2011), pp. 4311-43401, ISSN 0043-1354.

46. Sá, J., Agüera, C.A., Gross, S., & Anderson, J.A. (2009). Photocatalytic nitrate reduction over metal modified TiO . Applied Catalysis B: Environmental, Vol. 85, No. 34- (Jan 2009), pp. 192-200, ISSN 0926-3373.

47. Silva, F.M., Lucas Junior, J.,Benicasa, M., & Oliveira, E. (2005). Performance of a water heating system using biogas (in Portuguese).. Engenharia Agrícola, Vol. 25, No. 3, pp. 608-614, ISSN 1808-4389.

48. Sroka, A., Kaminski, W., & Bohdziewicz, J. (2004). Biological treatment of meat industry wastewater. Desalination, Vol. 162, No. 10, pp. 85-91, ISSN 0011-9164.

49. Stanmore, B.R. (2004). The formation of dioxins in combustion systems - a review. Combustion and Flame, Vol. 136, No. 3 (Feb 2004), pp. 398-427, ISSN 0010-2180.

50. Schievano, A., D'Imporzano, G., & Adani, F. (2009). Substituting energy crops with organic wastes and agro-industrial residues for biogas production. Journal of Environmental Management, Vol. 90, No. 8 (Jun 2009), pp. 2537-2541, ISSN 0301-4797.

51. Tambosi, J.L., De Sena, R.F., Gebhardt, W., Moreira, R.F.P.M., José, H.J., & Schröder, H.Fr. (2009). Physicochemical and advanced oxidation processes – A comparison of elimination results of antibiotic compounds following an MBR treatment. Ozone: Science & Engineering, Vol. 31, No. 6 (2009), pp. 428–435, ISSN 0191-9512.

52. Tritt, W.P., & Schuchardt, F. (1992). Materials flow and possibilities of treating liquid and solid wastes from slaughterhouses in Germany. A review. Bioresource Technology, Vol. 41, No. 3 (1992), pp. 235-245, ISSN 0960-8524.

53. US EIA – United States Energy Information Administration. (2011). Annual Energy Outlook 2011 with projections to 2035. Date of access: 12 Jun 2011, Available from: <http://www.eia.gov/forecasts/aeo/>.

54. US EPA - United States Environmental Protection Agency. (2004). Technical development document for the final effluent limitations guidelines and standards for the meat and poultry products point source category. EPA-821-R-04-011, Washington, DC.

55. US EPA - United States Environmental Protection Agency. (Dec 2000). Code of Federal Regulation. Title 40, Part 60. In: US Environmental Protection Agency. Date of access: 16 Jul 2009, Available from:<http://www.epa.gov/epacfr40/chapt-I.info>.

56. van den Berg, M., Birnbaum, L.S., Denison, M., De Vito, M., Farland, W., Feeley, M., Fiedler, H., Hakansson, H., Hanberg, A., Haws, L., Rose, M., Safe, S., Schrenk, D., Tohyama, C., Trischer, A., Tuomisto, J., Tysklind, M, Walker, N., & Peterson, R.E. (2006). The 2005 World Health Organization reevaluation of human and mammalian Toxic Equivalent Factors for dioxins and dioxin-like compounds. Toxicological Sciences, Vol. 93, No. 2 (Oct 2006), pp. 223-241, ISSN 1096-6080.

57. Van Paasen, S.V.B, Cieplik, M.K., & Phokawat, N.P. (2006). Gasification of non-woody biomass - economic and technical perspectives of chlorine and sulphur removal from product gas (non-confidential version), In: ECN-C-06-032, 2006. Date of access: 15 Apr 2010, Available from: <http://www.ecn.nl/docs/library/report/2006/ e06032.pdf>.

58. Virmond, E. (2007). Aproveitamento do lodo de tratamento primário de efluentes de um frigorífico como fonte alternativa de energia. Master's Thesis, Federal University of Santa Catarina, Florianópolis, Brazil. Date of access: 01 Aug 2012, Available from: <http://www2.enq.ufsc.br/teses/m176.pdf > (in Portuguese).

59. Virmond, E., Floriani, S.L., Moreira, R.F.P.M., José, H.J., 2008. Co-combustion of wood chips with the biomass obtained from the physico-chemical treatment of slaughterhouse wastewater for steam generation - A case study. Proceedings of the 23rd International Conference on Solid Waste Technology and Management, ISSN 1091-8043, Philadelphia, Pennsylvania, USA, March 30- April 2, 2008, pp. 1177-1188.

60. Virmond, E., Schacker, R.L., Albrecht, W., Althoff, C.A., Souza, M., Moreira, R.F.P.M., & José, H.J. (2010). Combustion of apple juice wastes in a cyclone combustor for thermal energy generation (ES2009-90152). Journal of Energy Resources Technology, Vol. 132, No. 4 (Dec 2010), pp. 041401-041409, ISSN 0195-0738.

61. Virmond, E., Schacker, R.L., Albrecht, W., Althoff, C.A., Souza, M., Moreira, R.F.P.M., & José, H.J. (2011). Organic solid waste originating from the meat processing industry as an alternative energy source. Energy, Vol. 36, No. 6 (Jun 2011), pp. 3897-3906, ISSN 0360-5442.

62. Virmond, E., et al. (2012a). Characterisation of agroindustrial solid residues as biofuels and potential application in thermochemical processes. Waste Management (2012), http://dx.doi.org/10.1016/j.wasman.2012.05.014>, article in press.

63. Virmond, E., et al. (2012b). Valorization of agroindustrial solid residues and residues from biofuel production chains by thermochemical conversion: a review, citing Brazil as a case study. Brazilian Journal of Chemical Engineering, Accepted for publication, ISSN 0104- 6632.

64. Watanabe, N., Yamamoto, O., Sakai, M., &Fukuyama, J. (2004). Combustible and incombustible speciation of Cl and S in various components of municipal solid waste. Waste Management, Vol. 24, No. 6 (2004), pp. 623-632, ISSN 0956-053X.

65. Wehbe, N., Jaafar, M., Guillard, C., Herrmann, J.-M., Miachon, S., Puzenat, E., & Guilhaume, N. (2009). Comparative study of photocatalytic and nonphotocatalytic reduction of nitrates in water. Applied Catalysis A: General, Vol.368, No. 1-2 (31 Oct 2009), pp. 1-8, ISSN 0926-860X.

66. Werther, J., Saenger, M., Hartge, E-U., Ogada, T., & Siagi, Z. (2000). Combustion of agricultural residues. Progress in Energy and Combustion Science , Vol. 26, No. 1 (Feb 2000), pp. 1-27, ISSN 0360-1285.

67. Werther, J. (2007). Gaseous emissions from waste combustion. Journal of Hazardous Materials, Vol. 144, No. 3 (18 Jun 2007), pp. 604-613, ISSN 0304-3894.

68. WHO - World Health Organization. (2005). Project for the re-evaluation of human and mammalian toxic equivalency factors (TEFs) of dioxins and dioxin-like compounds. In: International Program on Chemical Safety (IPCS), WHO Headquarters. Geneva, Switzerland.

69. Zhang, F., Jin, R., Chen, J., Shao, C., Gao, W., Li, L., & Guan, N. (2005). High photocatalytic activity and selectivity for nitrogen in nitrate reduction on Ag/TiO catalyst with fine silver clusters. Journal of Catalysis, Vol. 232, No. 2 (10 Jun 2005), pp. 424-431, ISSN 0021-9517.

70. Zhang, G., Li, J., Cheng, H.R., Li, X.D., Xu, W.H., & Jones, K.C. (2007). Distribution of organochlorine pesticides in the northern South China Sea: implications for land outflow and air–sea exchange. Environmental Science and Technology, Vol. 41, No 11 (4 May 2007), pp. 3884–3890, ISSN 0013-936X.

71. Zichari, S.M. (2003). Removal of hydrogen sulfide from biogas using cow-manure compost. Master's Thesis, Faculty of the Graduate School of Cornell University, USA. Available from: http://www.green-trust.org/Al%20Rutan/MS-Thesis-Steve-Zicari.pdf

31 Chapter MARINE ALGAE FOR CULINARY AND INDUSTRIAL USES

1. INTRODUCTION: AN OVERVIEW

Marine macroalgae, sometimes referred to as seaweeds, are plant-like creatures that typically adhere to rocks or other solid substrates in coastal regions. The categorization into divisions is predicated on several attributes, including pigmentation, the chemical composition of photosynthetic storage products, the arrangement of photosynthetic membranes, and additional morphological characteristics. Traditionally, they are categorized into four distinct groups, empirically identified since the mid-nineteenth century based on color: blue-green algae (phylum: Cyanophyta, approximately 1500 species), red algae (phylum: Rhodophyta, around 6000 species), brown algae (phylum: Ochrophyta, class: Phaeophyceae, roughly 1750 species), and green algae (phylum: Chlorophyta, classes: Bryopsidophyceae, Chlorophyceae, Dasycladophyceae, Prasinophyceae, and Ulvophyceae, approximately 1200 species). Nevertheless, each of these categories contains microscopic, if not unicellular, representatives. All seaweeds are unicellular at certain stages of their life cycles, existing as spores or zygotes, and may exhibit temporary planktonic characteristics. Blue-green algae are prevalent on temperate rocky and sandy coastlines and have been infrequently recognized in seaweed floras. Seaweeds are distributed globally across oceans and seas, with none identified as toxic (Bold and Wyne, 1985; Guiry, 2009; Lobban and Harrison, 2000). What is the significance of seaweed? Many individuals are unaware of the ecological and commercial significance of marine macroalgae. Indeed, seaweeds are essential primary producers in marine aquatic food webs. They are abundant in minerals and important trace elements, as well as raw materials for the pharmaceutical and cosmetics industries (Chapman, 1970). Seaweed is a highly flexible foodstuff extensively utilized for direct human intake. It is taxonomically classed as algae, representing a food group that is not typically consumed in its uncooked form to any significant degree in Western countries. Humankind is no strangers to the use of algae as a food source. Even if seaweeds have been used as a human food since ancient times, particularly in the region bounded by China, the Korean Peninsula and Japan. But the commercial exploitation of this resource is only a few decades old, after World War II, when the focus was set on a possible insufficient protein supply due to the rapid increase of the world population. Today those countries are the largest consumers of marine algae as food. However, as nationals from these countries have migrated to other parts of the Earth, the demand for seaweedfor food has followed them, for example, in some parts of the North and South America. Nowadays, seaweeds are major coastal resources which are valuable to human consumption and environment in many countries. Edible seaweeds were widely consumed, especially in Asian countries (e.g., Japan, China, Korea, Taiwan, Singapore, Thailand, Brunei, Cambodia, and Vietnam, but also in South Africa, Indonesia, Malaysia, Belize, Peru, Chile, the Canadian Maritimes, Scandinavia, South West England, Ireland, Wales, California, Philippines, and Scotland) as fresh, dried, or ingredients in prepared foods. Their photosynthetic mechanism is similar to that of land-based plants. They are generally more efficient in converting solar energy into biomass, mainly because of their simple cellular structure and being submerged in an aqueous environment with access to water, CO2, and other nutrients. Same time, macroalgae are considered as the food supplement for 21st century, because they contain proteins, lipids, polysaccharides, minerals, vitamins, and enzymes. In common, seaweeds are rich in vitamins A, E, C, and Niacin with similar content in red algae (Rhodophyta), brown algae (Ochrophyta) and green algae (Chlorophyta). The concentration of vitamins B12, B1, panthothetic acid, folic, and folinic acids are generally higher in greens and reds than in browns. The brown algae possess organic iodine in greater amounts. Marine algae are similar to oats in protein and carbohydrate values. The green and red algae appear higher in crude protein far tested about 2 to 4 %. All algae contain high content of cabohydrates (sugar and starches) in polysaccharide biochemical structure which is a natural nontoxic colloidal substance that has been used as mucilaginous material referred to as gel. The nutrients composition of seaweed vary and is affected by species, geographic area, season and temperature of water. These sea-vegetables are of nutritional interest as they are low calorie food, but rich in vitamins, minerals and dietary fibres. Seaweeds, which have traditionally been used by the Western food industry for their polysaccharide extractives 'alginate, carrageenan and agar' also contain compounds with potential nutritional benefits. Seaweeds have recently been approved in France for human consumption (as vegetables and condiments), thus opening new opportunities for the food industry. These seaweed ingredients must meet industrial and technical specifications and consumer safety regulations. It is also an ingredient for the global food and cosmetics industries and is used as fertilizer and as an animal feed additive. Total annual value of production is estimated at almost US$ 6 billion of which food products for human consumption represent US $ 5 billion. Total annual use by the global seaweed industry is about 8 million tonnes of wet seaweed. Seaweed can be collected from the wild but is now increasingly cultivated. It falls into three broad groups based on pigmentation; brown, red and green seaweed. Use of seaweed as food has strong roots in Asian countries such as China, Japan and the Republic of Korea, but demand for seaweed as food has now also spread to North America, South America and Europe. China is by far the largest seaweed producer followed by the Republic of Korea and Japan but seaweeds are today produced in all continents. Red and brown seaweeds are utilized to manufacture hydrocolloids such as alginate, agar, and carrageenan, which serve as thickening and gelling agents. Currently, almost 1 million tonnes of wet seaweed are harvested and processed to yield roughly 55,000 tonnes of hydrocolloids, valued at nearly US $600 million. (McHug, 2003).

2. HISTORICAL BACKGROUND ON THE USE OF SEAWEEDS

The use of seaweed as food has been traced back to the fourth century in Japan and the sixth century in China. In 1750's, an English physician successfully used ash from kelp (Phaeophyceae) which is rich in iodine to treat goiter. Kelp was also used to treath obesity in 19 th. century, and agar was used as a laxative. Seaweeds were used as a source of iodine. And their crude extracts were used for clarification in brewing. Another hydrocolloid, carrageen, found initially in the red seaweed Chondrus crispus was known in Ireland since 1810. Alginic acid, a hydrocolloid found in all brown seaweeds, was discovery first by Charles Stanford in the 1880s. Development of a large scale alginate industry began California and in Scotland in the late 1920s and early 1930s, respectively. Laminaria japonica was cultivated in China from the 1950s. The hydrocolloids have found increasing industrial and food applications in those years. Today China, Japan and the Republic of Korea are the largest consumers of seaweed as food. However, as nationals from these countries have migrated to other parts of the world, the demand for seaweed for food has followed them, as, for example, in some parts of the United States of America and South America. Increasing demand over the last fifty years outstripped the ability to supply requirements from natural (wild) stocks. Research into the life cycles of these seaweeds has led to the development of cultivation industries that now produce more than 90 percent of the market's demand. In Ireland, Iceland and Nova Scotia (Canada), a different type of seaweed has traditionally been eaten, and this market is being developed. Some government and commercial organizations in France have been promoting seaweeds for restaurant and domestic use, with some success. An informal market exists among coastal dwellers in some developing countries where there has been a tradition of using fresh seaweeds as vegetables and in salads. Various red and brown seaweeds are used to produce three hydrocolloids: agar, alginate and carrageenan. A hydrocolloid is a non-crystalline substance with very large molecules and which dissolves in water to give a thickened (viscous) solution. Alginate, agar and carrageenan are water-soluble carbohydrates that are used to thicken (increase the viscosity of) aqueous solutions, to form gels (jellies) of varying degrees of firmness, to form water-soluble films, and to stabilize some products, such as ice cream (they inhibit the formation of large ice crystals so that the ice cream can retain a smooth texture). Seaweeds as a source of these hydrocolloids dates back to 1658, when the gelling properties of agar, extracted with hot water from a red seaweed, were first discovered in Japan. Extracts of Irish Moss, another red seaweed, contain carrageenan and were popular as thickening agents in the nineteenth century. It was not until the 1930s that extracts of brown seaweeds, containing alginate, were produced commercially and sold as thickening and gelling agents. Industrial uses of seaweed extracts expanded rapidly after the Second World War, but were sometimes limited by the availability of raw materials. In the 1950's, it was found that Gracilaria spp. treated with alkali produced higher strengthgels. After several years, Gracilaria can be cultivated successfully on a commercial scale, it is now used more widely. Once again, research into life cycles has led to the development of cultivation industries that now supply a high proportion of the raw material for some hydrocolloids. Today, approximately 1 million tonnes of wet seaweed are harvested and extracted to produce the above three hydrocolloids. Total hydrocolloid production is about 55 000 tonnes, with a value of US $ 585 million. Alginate production (US $ 213 million) is by extraction from brown seaweeds, all of which are harvested from the wild; cultivation of brown seaweeds is too expensive to provide raw material for industrial uses. Agar production (US $ 132 million) is principally from two types of red seaweed, one of which has been cultivated since the 1960-70s, but on a much larger scale since 1990, and this has allowed the expansion of the agar industry. Carrageenan production (US $ 240 million) was originally dependent on wild seaweeds, especially Irish Moss, a small seaweed growing in cold waters, with a limited resource base. However, since the early 1970s the industry has expanded rapidly because of the availability of other carrageenan-containing seaweeds that have been successfully cultivated in warm-water countries with low labour costs. Today, most of the seaweed used for carrageenan production comes from cultivation, although there is still some demand for Irish Moss and some other wild species from South America. Seaweed meal, used an additive to animal feed, has been produced in Norway, where its production was pioneered in the 1960s. It is made from brown seaweeds that are collected, dried and milled. Drying is usually by oil-fired furnaces, so costs are affected by crude oil prices. Approximately 50 000 tonnes of wet seaweed are harvested annually to yield 10 000 tonnes of seaweed meal, which is sold for US $ 5 million. Fertilizer uses of seaweed date back at least to the nineteenth century. Early usage was by coastal dwellers, who collected storm- cast seaweed, usually large brown seaweeds, and dug it into local soils. The growth area in seaweed fertilizers is in the production of liquid seaweed extracts. These can be produced in concentrated form for dilution by the user. Several can be applied directly onto plants or they can watered in, around the root areas. There have been several scientific studies that prove these products can be effective. In

1991, it was estimated that about 10 000 tonnes of wet seaweed were used to make 1 000 tonnes of seaweed extracts with a value of US $ 5 million. However, the market has probably doubled in the last decade because of the wider recognition of the usefulness of the products and the increasing popularity of organic farming, where they are especially effective in the growing of vegetables and some fruits. Cosmetic products, such as creams and lotions, sometimes show on their labels that the contents include "marine extract", "extract of alga", "seaweed extract" or similar. Usually this means that one of the hydrocolloids extracted from seaweed has been added. Alginate or carrageenan could improve the skin moisture retention properties of the product. Pastes of seaweed, made by cold grinding or freeze crushing, are used in thalassotherapy, where they are applied to the person's body and then warmed under infrared radiation. This treatment, in conjunction with seawater hydrotherapy, is said to provide relief for rheumatism and osteoporosis (De Roeck-Holtzhauer, 1991).

3. SOURCES OF SEAWEED

A seaweed may belong to one of several groups of multicellular algae: the red algae, brown algae and green algae,. As these three groups are not thought to have a common multicellular ancestor, the seaweeds are a polyphyletic group. In addition, some tuft-forming blue green algae (Cyanobacteria) are sometimes considered as seaweeds — "seaweed" is a colloquial term and lacks a formal definition. Two specific environmental requirements dominate seaweed ecology. These are the presence of seawater (or at least brackish water) and the presence of light sufficient to drive photosynthesis. Another common requirement is a firm attachment point. As a result, seaweeds most commonly inhabit the littoral zone and within that zone more frequently on rocky shores than on sand or shingle. Seaweeds occupy a wide range of ecological niches. The highest elevation is only wetted by the tops of sea spray, the lowest is several meters deep. In some areas, littoral seaweeds can extend several miles out to sea. The limiting factor in such cases is sunlight availability. The deepest living seaweeds are some species of red algae. A number of species such as Sargassum have adapted to a fully planktonic niche and are free-floating, depending on gas-filled sacs to maintain an acceptable depth. Others have adapted to live in tidal rock pools. In this habitat seaweeds must withstand rapidly changing temperature and salinity and even occasional drying

3.1 Red seaweeds

Red seaweeds have had a more diverse evolution than the green and the brown. Many species cannot stand desiccation and dominate the inter-tidal rock pools. Others tolerate desiccation, such as the purple laver which can often be seen stretched out like a dry black film over mussle beds on rocky beaches. Red seaweeds such as Polysiphonia lanosa are epiphytes, these are plants that grow on other plants for physical support. In this case the epiphyte benefits from the host's buoyancy lifting it closer to the sunlight. The red colour of the seaweeds is due to the larger amount of red phycoblin pigments overriding the green pigment chlorophyll. The main biomass of red algae worldwide is provided by the Corallinaceae and Gigartinaceae. The red algae Gelidium, Gracilaria, Pterocladis and other many red algae are used in the manufacture of the important agar, used widely as a growth medium for microorganism and other biotechnological and food applications. Another important red seaweed alga is Eucheuma used in the production of Carrageenan, an important product used in cosmetics, food processing and industrial uses, as well as a food source. Some of the most significant caargeenan species include Betaphycus gelatinae, Eucheuma denticulatum, and several species of the genus Kappaphycus including K. alvarezii (Lobban and Harrison, 1994).

3.2 Brown seaweeds

Laminaria sp. 'kombu', Undaria sp. 'wakame', Hizikia fusiforme 'hiziki' is edible and an important resource Asia countries especiaqlly China and Japan. They are consumed raw, boiled or dried material with sweetened green beans, jelly, crushed ice, and coconut milk in Southern Vietnam (Tsutsui et. al., 2005). Laminaria sp. was in plentiful supply in Japan, mainly from the northern island of Hokkaido, where several naturally growing species were available.

Undaria sp. has been harvested from natural resources for many years in the China, Japan and Korean region. Another algae Cladosiphon okamuranus 'mozuku' as salad in Okinawa-Japan (Thoma, 1997; Zhang et.al., 2007; Zhu et.al., 2009). Sargassum sp. is known as horsetail and it is eaten as soup or dressed with soybean sauce, or after being processed in Korea (Madlener, 1997) and in Hawaii (Novaczek, 2001). In the Pasific region, Rosenvingea sp. or slippery cushion, Turbinaria or spiny leaf are eaten as soup or omelet Colpomenia sp. or paperly sea bubble as chop soup, stew or salad. Hydroclatharus sp. or sea colonder, Dictyota sp. or brown, Padina or sea fan ribbon weeds as a food dressing, soup or stew (Novaczek, 2001).

3.3 Green seaweeds

Green seaweeds are found on both sandy and rocky beaches. Many can tolerate low salinity and will colonise areas where rivers meet the sea. The green colour of the seaweed is due to the green pigment chlorophyll required for the photosynthesis of light. Using only chlorophyll means that green seaweeds require good levels of light and therefore will not thrive in shadowed areas or too any depth. It does give them an advantage, the ability to live higher up shore without competition from the red or brown seaweeds. The green saeaweeds Ulva sp., Enteromorpha sp., Monostroma sp., Caulerpa sp., Codium sp., are commonly known as source of food. In Asia countries especially Japan, dried fronds of edible Monostroma sp. and

Enteromorpha sp are being known as 'aonori-green laver-ele ele-lulua-lumi boso'. These algae are eaten by humans as edible raw, dried, or cooked. They used in preperation of 'nori-jam' soup (Lobban and Harrison, 1994; Novaczek, 2001).

4. NUTRITONAL COMPOSITION OF EDIBLE SEAWEEDS

Proximate composition (moisture, ash, protein and oil content), total dietary fibre content and physicochemical properties of three brown and two red edible Spanish seaweeds, namely: Himanthalia elongata (sea spaghetti), Bifurcaria bifurcata, Laminaria saccharina (sweet kombu), Mastocarpus stellatus and Gigartina pistillata were studied. Ashes (24.9–36.4%) were high in all samples. Protein content ranged from 10.9 to 25.7%, being much higher for Laminaria (25.7%) followed by the red seaweeds (15.5–21.3%). Minor components were lipids (0.3–0.9%) in all samples except for Bifurcaria (5.6%).In conclusion, these seaweeds can be estimated as a good source of food fibre, protein and minerals for human consumption (Gómez-Ordóñez et al., 2010). Mineral content was determined in several brown (Fucus vesiculosus, Laminaria digitata, Undaria pinnatifida) and red (Chondrus crispus, Porphyra tenera)edible marine sea vegetables. Seaweeds contained high proportions of ash (21.1–39.3%) and sulphate (1.3–5.9%). In brown algae, ash content (30.1–39.3%) was higher than in red algae (20.6–21.1%). Edible brown and redseaweeds could be used as a food supplement to help meet the recommended daily intake of some essential minerals and trace elements (Rupérez, 2002). Sea spaghetti (Himanthalia elongata), Wakame (Undaria pinnatifida), and Nori (Porphyra umbilicalis), on fatty acid composition, amino acid profile, protein score, mineral content and antioxidant capacity in low-salt meat emulsion model systems. The addition of seaweeds caused an increase in ω-3 polyunsaturated fatty acids (PUFA) and a decrease in the ω-6/ ω-3 PUFA ratio. In general, addition of seaweeds to products increased the concentrations of K, Ca, Mg and Mn. The presence of Nori caused an increase in levels of serine, glycine, alanine, valine, tyrosine, phenylalanine and arginine, whereas Wakame and Sea Spaghetti produced no significant changes in amino acid profiles in the model systems. López-López et.al., 2009). The nutritional compositions of 34 edible seaweed products of the Laminaria sp.,Undaria pinnatifida, Hizikia fusiforme and Porphyra sp. varieties were analyzed.The marine macroalgae varieties tested demonstrated low lipid contents with 2.3 ± 1.6 g/100 g semi-dry sample weight(s.w.) and proved to be a rich source of dietary fibre (46.2 ± 8.0 g/100 g s.w). The pure protein content of seaweed products varied widely (26.6 ± 6.3 g/100 g s.w. in red algae varieties and 12.9 ± 6.2 g/100 g s.w. in brown algae varieties). All essential amino acids were detected in the seaweed species tested and red algae species featured uniquely high concentrations of taurine when compared to brown algae varieties (Dawczynski et al., 2007). The total lipid, protein, ash and individual fatty acid contents of edible seaweeds that had been canned (Saccorhiza polyschides and Himanthalia elongata) or dried (H. elongata, Laminaria ochroleuca, Undaria pinnatifida, Palmaria sp. and Porphyra sp.) Total lipid content ranged from 0.70±0.09 to 1.80±0.14 g/(100 g dry weight). The four most abundant fatty acids were C16:0, C18:1ω9, C20:46 and C20:5ω3. Unsaturated fatty acids predominated in all the Brown seaweeds studied, and saturated fatty acids in the red seaweeds, but both groups are balanced sources of ω3 and ω6 acids. Ash content ranged from 19.07±0.61 to 34.00±0.11 g/(100 g dry weight), and protein content from 5.46±0.16 to 24.11±1.03 g/(100 g dry weight) (Sanchez-Machado, et. al.,2004).

5. EDIBLE SEAWEED IN FOODS

Red macro-algae (Gracilaria spp.) are used as a fresh food in Hawaii. Species commonly marketed include G. coronopifolia, G. parvispora, G. salicornia and G. tikvahiae, however, these seaweeds have a short postharvest life of about 4 days (Paul and Chen, 2008). Seaweeds are a rich source of phytochemicals having anti-oxidant and antimicrobial properties. Presence of fibres and minerals helps in improving the mineral content reduce the salt content. The adding of seaweeds or their extracts to food products will help in reducing the utilization of chemical preservatives (Gupta and Abu-Ghannam, 2011). Edible seaweeds contain various bioactive compounds with potential health benefits and their use as functional ingredients opens up new prospects for food processing, meat product formulations included. Seaweeds basically contain high proportions of polysaccharides along with various other potentially beneficial compounds such as good-quality protein and essential fatty acids, particularly long-chain n-3 polyunsaturated fatty acids (PUFAs). Alginates are the most abundant ionicpolysaccharides present in brown seaweeds (Fernández- Martín et al., 2009). Some seaweed polysaccharides are used by food industry as texture modifiers because of their high viscosity and gelling properties. In Asia seaweeds have been used for centuries in salads, soups and as low calorie dietetic foods. The diatery fibre which constitutes 25-75 % of the dry weight of marine algae and represents their major component, is primarily souble fibre.(Jiménez-Escrig and Sánchez-Muniz, 2000). In particular, miyeok (Undaria pinnatifida) is often served in soup,salad, and sidedishes.

Gamma irradiation at 10 kGy is sufficient to sterilizefreeze-dried miyeokguk without significant deterioration in the sensory quality,and thus,the freeze-dried and irradiated miyeokguk at 10 kGy fulfills themicrobiological requirements as spacefood (Song et al., 2012). The sausages were produced with two types of carrageenan (i- and j-) in four levels (0%, 1%, 2% and 3%). Carrageenan had a better effect on such characteristics as pH, weight loss and lipid oxidation of the sausages, as well as, on sensory attributes. The carrageenan level of 3% negatively affected the firmness of the sausages.Carrageenan added at levels up to 2% had a positive effect on the physicochemical and microbiological characteristics of the lowfat fermented sausages. (Koutsopoulos et al., 2008). Cultivated Ulva rigida was utilized by using marination technology. Fresh and boiled (at 100°C for 2 min.) Ulva rigida were marinated with two different formulations by using 2 % lemon salt and 2 % vinegar. The marination of Ulva rigida were made at room temperature for 20 days. Marinated fresh and boiled ulva rigida by using lemon salt and vinegar can be an alternative for human foods (Kılınç et al., 2011). Breads were made by using Lemna minor (Tekogul et al., 2011) and Ulva rigida (Turan et al., 2011). The shelf-life of breads by using Ulva rigida were determined as unacceptable on day 5 at room temperature whereas on day 10 at 4°C. When compared with control groups, the shelf-life of breads containing Ulva rigida were determined longer shelf-life. Breads prepared with Ulva rigida extended the shelf-life of breads for 2 days in two different storage period. Lemna minor extended the shelf-life of breads. The shelf-life of breads with Lemna minor were extended the acceptable limit on day 8 at room temperature whereas on day 12 at 4 °C. But control group extended this acceptable limit on day 3 at room temperature, on day 8 at 4 °C (Tekogul et al., 2011).

5.1 Fermented seaweed

Brown edible seaweeds as a sole source of nutrition for the growth of lactic acid bacteria. Growth kinetics of lactic acid bacteria (LAB; Lactobacillus plantarum) was studied using three species of edible Irish brown seaweeds Himanthalia elongata, Laminaria digitata and Laminaria saccharina. The results of this study present an indication of the potential of fermentation of seaweeds using LAB with a possibility towards the development of a range of functional foods (Gupta et al., 2011). Low molecular weight polysaccharides from seaweed as prebiotics. Gelidium seaweed showed significant increase in bifidobacterial populations. Agar and alginate bearing seaweeds indicate prebiotic potential (Ramnani et al., 2012). Brown macroalgae contain high concentration of mannitol and laminarian. Clostrium acetobutylicum ferments these seaweed extract substrates to butanol. Seaweed fermentation exhibited triauxic growth: glucose-mannitol- laminarin.Butanol yields in seaweed and pure glucose fermentations were comparable (Huesemann et al., 2012).

5.2 Seaweeds used as fertilizer and biogas production

Seaweed are used as a fertilizer which is suitable for use in organic agriculture (López- Mosquera et al., 2011). Energy-rich methane can be harnessed from seaweed deposits by anaerobic digestion. However, the high heavy metal content in the seaweed and its digestates limits their use as fertilisers. The efficient utilisation of seaweed for biogas production, and the partial heavy metals mobilisation to enable the metal removal for improved fertiliser quality (Nkemka and Murto, 2012). The red alga Chondracanthus squarrulosus was cultured under semi-controlled conditions to valuate growth (biomass production) with agricultural fertilizers (ammonium nitrate, ammonium sulphate and urea) vs. analytical grade inorganic salts; sodium nitrate (analytical grade) and seawater were used as controls (Pacheco-Ruiz et al., 2004).

6. CONCLUSION AND OUTLOOK

Seaweeds are being studied on the use of many industrial applications such as food, cosmetics, chemistry, paint, medicine, etc., at nowadays. In Western countries has traditionally concentrated on the extraction of compounds used by pharmaceutical, cosmetics, and food industries (Wijesinghe and Jeon, 2012a). Biologically active compounds of seaweeds (phlorotannins, carotenoids, alginic acid, fucoidan, peptides) have been demonstrated to play a significant role in prevention of certain degenerative diseases such as cancer, inflammation, arthritis, diabetes and hypertension. Therefore, seaweed derived active components, whose immense biochemical diversity looks like to become a rich source of novel chemical entities for the use as functional ingredients in many industrial applications such as functional foods, pharmaceuticals and cosmeceuticals (Wijesinghe and Jeon, 2012b). Commercially available varieties of marine macroalgae are commonly used to as 'seaweeds' Conventional, macroalgae can be classified as brown algae (Phaeophyta), red algae (Rhodophyta), and green algae (Chlorophyta), depending on their nutrient and chemical composition. Red and brown algae are mainly used as human food sources. The protein content of seaweed species varies greatly and demonstrates a dependence on such factors as season and environmental growth conditions. For example, the protein content of brown algae species, e.g., Laminaria japonica, Hizikia fusiforme or Undaria pinnatifida, is relatively low with 7–16 g/100 g dry weight (d.w.) (Jurković et al., 1995; Kolb et al., 1999; Rupérez and Saura-Calixto, 2001). On the other hand, red algae, e.g., Palmaria palmata (Dulse) and Porphyra tenera contain 21–47 g protein/100 g d.w. (Fleurence, 1999; Rupérez and Saura-Calixto, 2001).The protein in algae contains all essential amino acids (EAA) and all EAA are available throughout the year although seasonal variations in their concentrations are known to occur (Galland-Irmouli et al., 1999). For example, the proportion of EAA is 45–49% in Hizikia sp. and Eisenia bicyclis (Arame). In both these brown algae varieties, Ecological factory is the first limiting EAA, followed by Lys (Kolb et al., 1999). The EAA contents of some species (e.g., Porphyra sp.) can be compared with those of soy and egg protein (Fleurence, 1999; Galland-Irmouli et al., 1999). In addition to, high concentrations of Arg, Asp and Glu peptides are found in many macroalgae species (Fleurence, 1999). The fat content of marine macroalgae accounts for 1–6 g/100 g d.w. (Fleurence et al., 1994; Jurković et al.,

1995; Herbreteau et al., 1997). In some brown algae varieties, such as Hizikia sp. and Eisenia bicyclis (Arame), only 0.7–0.9 g/100 g d.w. of fat content were found (Kolb et al., 1999). Brown seaweeds are rich in fucose rich sulfated polysaccharides fucoidans (Wijesinghe and Jeon, 2012a). Polysaccharides produced by marine seaweeds form the basis of an economically important and expanding global industry. Key products are agars, agaroses, algins, and carrageenans.

These are used on ingredients in food, pharmaceuticals and diverse consumer products and industrial processes (Renn, 1997). Red algae (e.g., Porphyra sp.) have high concentrations of eicosapentaenoic acid (C20:5, ω -3, EPA),with 48.0–51.0% of total fatty acid methyl esters (FAME), and marginal concentrations of arachidonic acid (C20:4, ω-6, ARA), with 2.1–10.9% of total FAME and, linoleic acid (C18:2, ω -6, LA), with 1.3–2.5% of total FAME (Fleurence et al., 1994; Takagi et al., 1985). In contrast, brown algae (e.g., Laminaria sp., Undaria sp., Hizikia sp.) have high concentrations of oleic acid (C18:1, ω -9, OA) with 4.1–20.9% of total FAME, LA with 4.0–7.3% of total FAME as well as α-linolenic acid (C18:3, ω -3, ALA) with 3.6–13.8% of total FAME but low concentrations of EPA with 5.9–13.6% of total FAME (Fleurence et al., 1994; Takagi et al., 1985). Interestingly, in Porphyra sp., Laminaria sp., and Undaria sp., the concentrations of docosahexaenoic acid (C22:6, ω -3, DHA) and docosapentaenoic acid (C22:5, ω -3, DPA) were below the detection limit (less than 0.1% of total FAME (Fleurence et al., 1994; Takagi et al., 1985). The types and abundance of carbohydrates vary strongly between algae species. Typical carbohydrates in red algae varieties consist of floridean starch (α-1.4-binding glucan), cellulose, xylan, and mannan. The water-soluble fibre fraction is formed by sulfur- containing galactans, e.g., agar and carrageen (Jiménez-Escrig and Sánchez-Muniz, 2000; Van den Hoek et al., 1993). The typical carbohydrates in brown algae varieties consist of fucoidan, laminaran (β-1.3-glucan), cellulose, alginates, and mannitol. Brown algae fibres are mainly cellulose and insoluble alginates. Alginates are Ca, Mg, or Na salts of alginic acid (1.4-linked polymer of β-d-mannuronic acid and α-l-guluronic acid). The amorphous, slimy fraction of brown algae fibres consists mainly of water-soluble alginates and/or fucoidan. Main reserve polysaccharides of Phaeophyta are laminaran (β-1.3-glucan) and mannitol (Kolb et al., 1999; Dawczynski et al., 2007). The typical algae carbohydrates are not digestible by the human gastrointestinal tract and, therefore, they are dietary fibres. The content of total dietary fibre ranges from 33–50 g/100 g d.w. (Jiménez-Escrig and Cambrodon, 1999, Lahaye, 1991). Accordingly, the fibre content of seaweed varieties is higher than those found in most fruits and vegetables. The human consumption of algal fibre has been proven to be health-promoting and it benefits are well documented in the scientific literature. The consumption of this dietary fibre has been related to the following health promoting effects: 1) its consumption promotes the growth and protection of the beneficial intestinal flora

(Fujii et al., 1992, Goni et al., 2001), 2) its consumption, in combination with high glycemic load foods, reduces the overall glycemic response, macroalgae fibre acts as a hypoglycaemic (Goni, Valdivieso, & Garcia-Alonso, 2000), 3) its consumption greatly increases stool volume (Jiménez-Escrig & Sánchez-Muniz, 2000) and 4) its consumption reduces the risk of colon cancer (Guidel-Urbano & Goni, 2002). In addition, seaweeds varieties are rich sources of vitamin C, vitamin B-complex, e.g., folic acid and B12, and vitamin A precursors, such as β-carotene (McDermid and Stuercke, 2003, Takenaka et al., 2001; Watanabe et al., 2002). Because the seaweed species are rich in beneficial nutrients, in countries such as China, Japan, and Korea, they have been commonly utilised in human alimentation (since ancient times) (Lahaye, 1991). For example, Japanese people consume more than 1.6 kg algae (d.w.) per year per capita (Fleurence, 1999). In addition to their importance as traditional the Asian foods, macroalgae species are utilised industrially as a source of hydrocolloids, such as agar, carrageen, and alginate (Jiménez-Escrig & Sánchez- Muniz, 2000). Over the past few decades, the consumption of seaweed products has increased in European countries. Currently, approximately 15–20 edible algae strains are being commonly marketed for consumption in Europe. These seaweed varieties differ greatly in their quality, colour, consistency, and nutrient content. Nowadays, for this reason the studies of algology evaluate and compares the nutrient and chemical contents of many commercially available seaweed products which were locally purchased in European food stores and speciality shops (Herbreteau et al., 1997; Dawczynski et al., 2007).

REFERENCES

1. Bold, H. C, & Wyne, M. J. (1985). Introduction to the algae: Structure and Reproduction, Second Edition, New Jersey, Prentice-Hall, Inc., USA, 720.
2. Chapman, V. J. (1970). Seaweeds and Their Uses, Chapman & Hall. London, 334.
3. Dawczynski, C, Schubert, R, & Jahreis, G. (2007). Amino acids, fatty acids, and dietary fibre in edible seaweed products, J. of Food Chemistry, 103
4. De Roeck-holtzhauer, J. (1991). Uses of Seaweeds in Cosmetics. In: Seaweed Resources in Europe: Uses and Potential. M. G. Guiry andn G. Blunden (editors). John Wiley and Sons, Ltd. England, 83-93.
5. Fujii, T, Kuda, T, Saheki, K, & Okuzumi, M. (1992). Fermentation of water-soluble polysaccharides of brown algae by human intestinal bacteria in vitro, Nippon Suisan Gakk, 58, 149-152.
6. Fleurence, J, Gutbier, G, Mabeau, S, & Leray, C. (1994). Fatty acids from 11 marine macroalgae of the French Brittany coast, Journal of Applied Phycology, 6, 527-532.

7. Fleurence, J. (1999). Seaweed proteins: Biochemical, nutritional aspects and potential uses, Trends in Food Science and Technology, 10, 25-28.

8. Fernández-martín, F, López-lópez, I, & Cofrades, S. Jiménez Colmenero F. ((2009). Influence of adding Sea Spaghetti seaweed and replacing the animal fat with olive oil or a konjac gel on pork meat batter gelation. Potential protein/alginate association. Meat Science, 83, 209-217.

9. Galland-irmouli, A. V, Fleurence, J, Lamghari, R, Luçon, M, Rouxel, C, & Barbaroux, O. (1999). Nutritional value of proteins from edible seaweed Palmaria palmata (Dulse), Journal of Nutritional Biochemistry, 10, 353-359.

10. Guiry, M. (2009). Seaweed site, seaweed web site, National University of Ireland, Galway, URL.: http://www.seaweed.ie/algae/seaweeds.html.

11. Goni, I, Valdivieso, L, & Garcia-alonso, A. (2000). Nori seaweed consumption modifies glycemic response in healthy volunteers, Nutrition Research, 20, 1367-1375.

12. Goni, I, Guidel-urbano, M, Bravo, L, & Saura-calixto, F. (2001). Dietary modulation of bacterial fermentative capacity by edible seaweeds in rats, Journal of Agriculture and Food Chemistry, 49, 2663-2668.

13. Gómez-ordóñez, E, Jiménez-escrig, A, & Rupérez, P. (2010). Dietary fibre and physicochemical properties of several edible seaweeds from the northwestern Spanish coast. Food Research International, 43, 2289-2294.

14. Gupta, S, & Abu-ghannam, N. (2011). Recent developments in the application of seaweeds or seaweed extracts as a means for enhancing the safety and quality attributes of foods. Innovative Food Science and Emerging Technologies, 12, 600-609.

15. Gupta, S, Abu-ghannam, N, Scannell, A, & Growth, G. M. and kinetics of Lactobacillus plantarum in the fermentation of edible Irish brown seaweeds. Food and Bioproducts Processing, 89, 346-355.

16. Herbreteau, F, Coiffard, L. J. M, Derrien, A, & De Roeck-holtzhauer, Y. (1997). The fatty acid composition of five species of macroalgae, Botanica Marina, , 40, 25-27.

17. Huesemann, M. H, Kuo, G, Urquhart, A, Gill, G. A, & Roesijadi, G. (2012). Acetone-butanol fermentation of marine macroalgae. Bioresource Technology, 108, 305-309.

18. Jurkovic, N, Kolb, N, & Colic, I. (1995). Nutritive value of marine algae Laminaria japonica and Undaria pinnatifida, Die Nahrung, 39, 63-66.

19. Jiménez-escrig, A, & Sánchez-muniz, F. J. (2000). Dietary fibre from edible Seaweeds:Chemical structure, physicochemical properties and effects on cholesterol metabolism, Nutrition Research, 20, 585-598.

20. Kolb, N, Vallorani, L, & Stocchi, V. (1999). Chemical composition and evaluation of protein quality by amino acid score method of edible brown marine algae Arame (Eisenia bicyclis) and Hijiki (Hijikia fusiforme), Acta Alimentaria, 28, 213-222.

21. Koutsopoulos, D. A, Koutsimanis, G. E, & Bloukas, J. G. (2008). Effect of carrageenan level and packaging during ripening on processing and quality characteristics of low- fat fermented sausages produced with olive oil. Meat Science, 79, 188-197.

22. Kilinç, B, Turan, G, & Tekogul, H. (2011). Kültürü Yapılan Deniz Yosunu (Ulva Rigida) ile Farklı Formülasyonlarda Marinat Üretimi ve Duyusal Değerlendirmesi", 16. Ulusal Su Ürünleri Sempozyumu, 25-27 Ekim, sf. 272-273, Antalya. Lahaye, M. (1991). Marine algae as sources of fibers: Determination of solubleandinsolubledietary fiber contents in some, sea vegetables, Journal of Science and Food Agriculture, 54, 587-594.

23. Lobban, S. C, & Harrison, J. P. (1994). Seaweed Ecology and Physiology, Cambridge University Press, 386 p.

24. López-lópez, I, Bastida, S, Ruiz-capillas, C, Bravo, L, Larrea, M, Sánchez-muniz, F, Cofrades, S, & Jiménez-colmenero, F. (2009). Composition and antioxidant capacity of low-salt meat emulsion model systems containing edible seaweeds. Meat Science, 83, 492-498.

25. Lobban, S. C. (2000). Seaweed Ecology and Physiology, Cambridge University Press, The Edinburg Bulding, Cambridge CB2 2RU, UK, ISNB-0521403340., 359.

26. López-mosquera, M. E, Fernández-lema, E, Villares, R, Corral, R, Alonso, B, & Blanco, C. (2011). Composting fish waste and seaweed to produce a fertilizer for use in organic agriculture. Procedia Environmental Sciences, , 9, 113-117.

27. Madlener, J. C. (1977). Seavegetable Book, Carlkson N. Potter, Inc., New York, USA, 288 p.

28. Mchug, J. D. (2003). A Guite to the Seaweed Industry, FAO Fisheries Technical Papers Rome-Italy, 9-25104-958-0(441), 105.

29. Mcdermid, K. J, & Stuercke, B. (2003). Nutritional composition of edible Hawaiian seaweeds, Journal of Applied Phycology, 15, 513-524.

30. Novaczek, I. (2001). A guide to the Common and Edible and Medicinal Sea Plants of the Pacific Island, University of the South Pasific, 40pp.

31. Nkemka, V. N, & Murto, M. (2012). Exploring strategies for seaweed hydrolysis: Effect on methane potential and heavy metal mobilisation. Process Biochemistry. In press.

32. Pacheco-ruiz, I. Zertuche-Gonza´lez, J.A., Arroyo-Ortega, E., Valenzuela-Espinoza, E. ((2004). Agricultural fertilizers as alternative culture media for biomass production of Chondracanthus squarrulosus (Rhodophyta, Gigartinales) under semi-controlled conditions. Aquaculture, 240, 201-209.

33. Paull, R. E, & Chen, N. J. Postharvest handling and storage of the edible red seaweed Gracilaria. Postharvest Biology and Technology, , 48, 302-308.

34. Ramnani, P, Chitarrari, R, Tuohy, K, Grant, J, Hotchkiss, S, Philp, K, Campbell, R, Gill, C, & Rowland, I. (2012). In vitro fermentation and prebiotic potential of novel low molecular weight olysaccharides derived from agar and alginate seaweeds. Anaerobe, 18, 1-6.

35. Renn, D. (1997). Biotechnology and the red seaweed polysaccharide industry: status, needs, and prospects. Trends in Biotechnology, , 15, 9-14.

36. Rupérez, P. (2002). Mineral content of edible marine seaweeds. Food Chemistry, 79, 23-26.

37. Rupérez, P, & Saura-calixto, F. (2001). Dietary fibre and physicochemical properties of edible Spanish seaweeds, European Food Research Technology, 212, 349-354.

38. Sa´nchez-MachadoD.I., Lo´ pez-Cervantes, J., Lo´ pez-Herna´ndez, J., Paseiro-Losada, Fatty acids, total lipid, protein and ash contents of processed edible seaweeds. Food Chemistry, 85: 439-444., 2004.

39. Song, B. S, Park, J. G, Kim, J. H, Choi, J. I, Ahn, D. H, Hao, C, & Lee, J. W. (2012). Development of freeze-dried miyeokguk, Korean seaweed soup,as spacefood sterilized by irradiation. Radiation Physics and Chemistry, , 81, 1111-1114.

40. Tekogul, H, Turan, G, & Kilinç, B. (2011). Su Mercimeği (Lemna Minor) Katkılı Bayatlamayan Ekmek Üretimi, 16. Ulusal Su Ürünleri Sempozyumu, Ekim, sf. 274, Antalya., 25-27.

41. Turan, G, Tekogul, H, Kilinç, B, & Cirik, Ş. (2011). Demir Deposu Kültür Deniz Marulu (Ulva Rigida) İle Hazırlanan Yosunlu Ekmeklerin Duyusal Olarak Değerlendirilmesi ve Raf Ömrü Tespiti, 16. Ulusal Su Ürünleri Sempozyumu, Ekim, sf.275, Antalya, 25-27.

42. Thoma, T. (1997). Cultivation of the brown alga, Cladosiphon okamuranus 'okinawa-Mozuku ' in Seaweeds Cultivation and Marine Ranching, Ohno, M. & Cricthely, A. (Eds) JICA, Yokosuka, , 51-56.

43. Takenaka, S, Sugiyama, S, Ebara, S, Miyamoto, E, Abe, K, & Tamura, Y. (2001). Feedingdried purple laver (Nori) to vitamin B12-deficient rats significantly improves vitamin B12-status, British Journal of Nutrition, 85, 699-703.

44. Tsutsui, I, Huybh, Q. N, Nguyen, H. D, Arai, S, & Yoshida, T. (2005). The Common Marine Plants of Sothern Vietnam, Japan Seaweed Association, Kochi, 250 pp.

45. Watanabe, F, Takenaka, S, Kittaka-katsura, H, Ebara, S, & Miyamoto, E. (2002). Characterization and bioavailability of vitamin B12-compounds from edible algae, Journal of Nutritional Science and Vitaminology, 48, 325-331.

46. Wijesinghe, W. A. J. P, & Jeon, Y. J. ((2012a). Biological activities and potential industrial applications of fucose rich sulfated polysaccharides and fucoidans isolated from brown seaweeds: A review. Carbohydrate Polymers, 88, 13-20.

47. Wijesinghe, W. A. J. P, & Jeon, Y. J. extraction (EAE) of bioactive components: A useful approach for recovery of industrially important metabolites from seaweeds:A review. Fitoterapia , 83, 6-12.

48. Zhang, X. C, Zheng, L. H, Liu, S. W, Tang, X. R, & Wang, J. F. (2004). Introduction to new economical seaweed, Cladosiphon okamuranus, Period. Ocean. Univ. China, 34, 807-810.

49. Zhu, Q, & Zhang, X. A., K.K.I.U. ((2009). Ultrastructure Of Sporophyte of Cladosıphon okamuranus Tokida (Ectocarpales, Phaeophyceae), *Bangladesh J. Bot.* 38(2), 177-180.

INDEX

www.ingramcontent.com/pod-product-compliance
Lightning Source LLC
LaVergne TN
LVHW081940200926
848484LV00009B/2479